Chemical Processes for Pollution Prevention and Control

Chemical Processes for Pollution Prevention and Control

By
Paul Mac Berthouex
Linfield C. Brown

CRC Press is an imprint of the
Taylor & Francis Group, an **informa** business

CRC Press
Taylor & Francis Group
6000 Broken Sound Parkway NW, Suite 300
Boca Raton, FL 33487-2742

Printed on acid-free paper

International Standard Book Number-13: 978-1-138-10632-1 (Hardback)

Visit the Taylor & Francis Web site at
http://www.taylorandfrancis.com

and the CRC Press Web site at
http://www.crcpress.com

Contents

Preface

This book is about how chemical reactions and transformations are used for pollution prevention and control. *Prevention* means conducting our industrial, commercial, and personal business in ways that do not create pollutants. *Control* means capturing and treating pollution that cannot be avoided.

Pollution prevention and control systems use two kinds of processes: *transformations* and *separations*. *Separations* rely on differences in physical or chemical properties of materials so one material can be removed from another. Transformations are often used to create property differences that facilitate separation. For example, a precipitation reaction converts soluble metal ions to insoluble particles that are separated from the liquid by filtration.

This book is about how *chemical reactions and transformations* are used to destroy or rearrange materials to make them less toxic, more valuable, or less damaging when they are released to the environment.

We need accurate estimates of the mass and composition of waste streams in order to propose solutions that stand a chance of implementation. Skills in process chemistry and material balances are essential in deducing beneficial changes in composition and estimating the quality of treated waste streams.

Traditional environmental engineering courses include much of the same information that you will find in this book, but it is not integrated over media, technologies, and industries. Lessons on converting surface water to drinking water typically discuss coagulation of colloids and fine particles to enhance settling and filtration. Lessons on phosphorus removal from wastewater discuss solubility and precipitation. Neutralization and the adjustment of pH for process control discuss acid–base chemistry and equilibrium. And almost all of the examples and applications are directed toward municipal practice. Little or nothing is said about industrial problems.

A goal of this book is to integrate the synthesis and analysis chemical processes for municipal and industrial problems, and to see how the same principles apply in all cases. Engineers at municipal wastewater treatment plants administer industrial pretreatment regulations, and they will benefit from knowing about the production of wastes in industry and how wastes can be treated, reduced, or eliminated. Chemical and mechanical engineers who work on manufacturing need the same knowledge. That is pollution prevention at the source, and pollution control at the treatment site.

We like to give problems and solutions in context whenever possible. Instead of simply saying "reduce lead from 1 mg/L to 0.05 mg/L," we prefer to create a sense of application to realistic situations. Examples and case studies are used throughout the book to do this.

Chapter 1 introduces the basic concepts of chemical process design as used for pollution prevention and control. It explains why chemical processes are important and useful. Students who intend to study the complete book will find it is a good orientation. Readers who want an easy introduction to the subject should enjoy this chapter. They can use it to discover which additional subjects would make interesting reading.

Chapter 2 describes the pollutants that are the focus of most treatment systems: particulates in air, solids in water, BOD, COD, and heavy metals. Students may have this knowledge from prior studies, but a quick review will be beneficial.

Chapter 3 is an introduction to organic chemicals (but not organic chemistry). We describe some classes of problem organic chemicals that have a special importance in environmental protection work.

Chapter 4 explains how pollutants are quantified to calculate the mass and mass flow rate of the materials that move through a system. A variety of units for concentrations and mass flows are used, depending on whether the material is a gas, liquid, or solid. Almost all the calculations in this book are done with SI units, but there are a few exceptions, because the USCS system is still used in the United States.

Chapter 5 is about *stoichiometry*, or the science of using a *balanced chemical reaction* to keep track of the atoms, molecules, and compounds that enter and leave a reactor. It shows the composition of the reactants and the products, and it provides the means to calculate the mass of all the relevant chemical species. Stoichiometry provides the material balance of a chemical reaction.

Chapter 6 explains *empirical stoichiometry*, which is our way learning about the chemistry when we cannot calculate the required information from fundamental theory. This is necessary when the precise chemical composition of a waste stream is unknown, or when the pollutant is measured with a qualitative property, like turbidity or color. Empirical stoichiometric data typically come from experiments. Statistical experimental design is a strategy for extracting maximum information from a minimum of experiments. This is a great help when there are several variables to be investigated.

Chapters 7 through 9 are about three important kinds of chemistry: *acid–base reactions* to control pH, *precipitation reactions* to make soluble compounds insoluble (or occasionally to dissolve compounds), and *oxidation–reduction reactions* to destroy toxic compounds. Many solved examples help to explain the fundamental concepts. Those students who are interested more in how the chemistry is used to solve problems, and less in the calculations, will find many instructive examples and case studies.

Chapter 10 is about *green chemistry,* or the art and science of devising manufacturing processes that are less polluting. An important strategy is to find a reaction sequence that converts nearly all of the reactants into the desired product. Other strategies are to eliminate toxic feedstock, reagents, and solvents.

A Note to Non-Engineering Students: We hope that non-engineers will use this book. Pollution control engineers bring logic and order and solid quantitative information to the discussion about how public and private funds will be used to solve problems so better decisions will be made. Implementing the proposed solution goes beyond engineering design into public policy and business management, so the overall result will be better if the people in these related areas understand some basic tools of the engineer.

We want this book to be used by students who want to understand *how* chemical processes are used, but who do not want to do chemical calculations. Students who are less interested in calculations should not be shy about reading around the calculations. Look for the broader and general ideas by finding examples and case studies that match your interests.

Special thanks go to Dale Rudd (UW-Madison, Department of Chemical Engineering) for his support and ideas over many years. Dr. Rodolfo Perez helped in many ways, both technical and editorial. Rachel Craven, UW-Madison biomedical engineering student, helped with editing and checking.

Paul Mac Berthouex
Emeritus Professor, Department of Civil and Environmental Engineering
The University of Wisconsin–Madison

Linfield C. Brown
Emeritus Professor, Department of Civil and Environmental Engineering
Tufts University

About the Authors

Paul Mac Berthouex is emeritus professor of civil and environmental engineering at the University of Wisconsin–Madison. He has taught a wide range of environmental engineering courses, and was twice awarded the Rudolph Hering Medal from the American Society of Civil Engineers for the most valuable contribution to the environmental branch of the engineering profession, in addition to the Harrison Prescott Eddy Medal of the Water Pollution Control Federation. Berthouex was project manager for two Asian Development Bank projects in Indonesia and one in Korea. He also has design experience in Germany, Nigeria, India, and Samoa, and has been a visiting professor in England, Denmark, New Zealand, and Taiwan. He is co-author of *Statistics for Environmental Engineers*, several books on pollution control, and numerous articles in refereed journals.

Linfield C. Brown is emeritus professor and former chairman of civil and environmental engineering at Tufts University. He taught courses on water quality modeling, water and wastewater chemistry, industrial waste treatment, and engineering statistics, and was the recipient of the prestigious Lillian Liebner and Seymour Simches awards for excellence in teaching and advising. He has served as consultant to the U.S. Environmental Protection Agency and the National Council for Air and Stream Improvement. He is the author or co-author of over 60 technical papers and reports, has offered over two dozen workshops in the United States, Spain, Poland, England, and Hungary on water quality modeling with QUAL2E, and is the coauthor of three books on statistics and pollution prevention and control.

1 The Chemical Process Design Problem

1.1 INTRODUCTION

Think of a modern city as a giant chemical process or a giant organism. It needs to be provided with food and energy, and it must rid itself of wastes. Figure 1.1 shows the transformation of water, food, and fuel into sewage, air pollutants, and solid wastes, and also into useful products. The inputs are broken down by various means and reorganized into different components, many of which are wastes. Coal becomes carbon dioxide, water, ash, sulfur dioxide, and other pollutants. Food becomes waterborne waste and refuse. Unwanted newspapers, cans of tomatoes, computers, and automobiles are discarded as refuse, which is often processed to recover reusable material.

Albert Einstein said that the environment was "everything that is not me." For *me* to be healthy and happy, everything that is *not me* must be healthy and happy. The challenge is to maintain a healthy balance between the demands of 7 billion people and the natural cycles of essential nutrients and to include protection from hazardous substances.

1.2 CHEMICAL PROCESSES

Interesting and important chemical reactions occur in nature. Photochemical smog and acid rain are examples of chemistry in the atmosphere. Pollution problems in lakes, rivers, sediments, soil, and groundwater are caused by a failure to control the discharge of pollutants, either intentional or accidental. Contamination of soil and groundwater is caused by careless disposal and spills of chemicals.

The focus of pollution prevention and control is to minimize the flow of pollutants to the natural environment—the atmosphere, sediments, soil, streams, lakes, and oceans. This book is about using chemical processes to control pollutants before they can cause problems in the natural environment.

Table 1.1 classifies treatment processes according to whether they are best suited to handling inorganic or organic substances in soluble, particulate, or gaseous form. The italicized names are chemical processes. The rest are separations.

This book is mostly about soluble inorganic chemicals—the top left portion of Table 1.1. Selected topics about soluble and gaseous organic chemicals are also included.

Inorganic atoms and molecules can be rearranged, dissolved, and made to precipitate or agglomerate, and they can be diluted and concentrated. They are not biodegradable or combustible.

Organic compounds are biodegradable (at least most of them are). The biological reactions are oxidations and reductions, but usually we do not know specifically which chemicals are being reacted because wastewater is a mixture of many substances. Biological treatment processes must operate within fairly narrow limits of pH and temperature. These factors make them different than the most widely used inorganic chemical treatment processes. Biological transformations are not included in this book.

Organic chemicals vaporize and condense, and they adsorb and desorb. This suggests a class of treatment methods that are more separations than chemical reactions.

Air pollution control is mostly about physical separations, and to a lesser extent about chemical reactions and transformations. Some air pollution control processes capture the pollutants in a liquid stream. The absorption of sulfur dioxide and acids into alkaline solution is an example. Thus, an

FIGURE 1.1 The daily metabolism of a city of 1,000,000 people (Wolman, 1965). Not all inputs and outputs are listed. (Photo credit: Pixabay.)

TABLE 1.1
Classification of Treatment Processes according to Their Suitability for Handling Different Substances

Form of Pollutant	Inorganic	Organic
Soluble	*Neutralization*	*Biological oxidation (aerobic)*
	Coagulation/flocculation	*Biological fermentation (anaerobic)*
	Precipitation	*Chemical oxidation*
	Oxidation/reduction	*Thermal combustion*
	Adsorption	*Catalytic combustion*
	Absorption	Absorption
	Ion exchange	Adsorption
	Reverse osmosis	Distillation
		Condensation
		Ultrafiltration
Particulate	Screening	Screening
	Settling	Settling
	Flotation, gravity	Flotation, gravity
	Flotation, dissolved air	Flotation, dissolved air
	Filtration, granular media	Filtration, granular media
	Filtration, membrane	Filtration, membrane
	Filtration, cloth media	Filtration, cloth media
	Centrifugation	Centrifugation
	Cyclone and hydrocyclone	Cyclone and hydrocyclone
	Wet scrubbing	
Gaseous	Absorption	Absorption
	Adsorption	Adsorption
		Chemical oxidation
		Thermal combustion
		Catalytic combustion
		Condensation

air pollution problem becomes a water pollution problem. Dust handling may involve some chemical treatment for stabilization or, in some cases, to recover valuable metals.

An important part of many soil and groundwater remediation projects is separating the pollutants from the natural media so they can be treated using the processes that are common for wastewater and gaseous emissions.

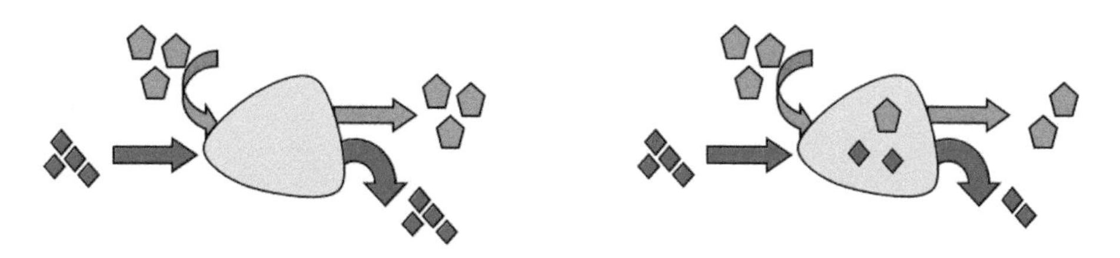

FIGURE 1.2 The law of conservation of mass says that all material entering a system must leave the system or accumulate within the system. The material balance (or mass balance) must account for all material flow.

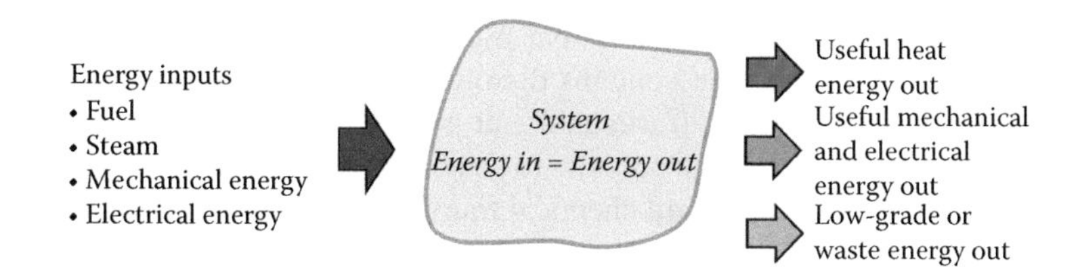

FIGURE 1.3 The law of conservation of energy says that all energy entering a system must accumulate in or leave the system. The energy balance is the engineer's tool for making an accounting of all energy flow.

1.3 PROCESS ANALYSIS

Pollution prevention and control engineering involves the organization of artificial things to increase our level of health and happiness. As systems are created, in the engineer's mind and on the drawing board, they must be evaluated. The quantities of all substances that flow through the system must be determined. The energy requirements must be calculated. These are the activities of *process analysis*. By analysis, we mean doing the process design calculation, not the laboratory process of identifying and quantifying chemical substances. The tools for doing the analysis are the material balance (Figure 1.2) and the energy balance (Figure 1.3). For details and example calculations, see Berthouex and Brown (2014).

1.4 PROCESS SYNTHESIS

The inventive part of design is *process synthesis*, through which we seek to satisfy the metabolism of our society with the least amount of damage to human health and environmental quality. Synthesis is the combining of diverse elements into a coherent whole that will produce the materials we consume without being wasteful of raw materials and with manufacturing methods that produce the smallest possible amounts of waste. And it means being clever about capturing and processing the waste products that are produced.

The synthesis of pollution prevention and control systems involves linking *reactors* and *separation processes* in support of each other.

Chemical and biochemical transformations are promoted and controlled in *reactors* in order to (1) upgrade an otherwise useless material to a material that is too valuable to waste, (2) destroy or inactivate a dangerous or harmful substance, and (3) facilitate the separation of one material from a bulk flow of other materials.

Chemical transformations are almost always integrated with separation processes. For example, a dissolved pollutant is converted to a solid particle by precipitation, and the solid is then separated from the liquid. The removed solid material may be a useful product, or it may be an undesirable substance, the removal of which raises the quality of the water so it can be safely discharged or reused.

Separation processes selectively remove one species of material from another (solids from a liquid, for example). This provides the means to concentrate the input to a reactor or to remove substances that would interfere with the reaction or damage the equipment. Separations are also used to remove from the reactor's output any substances that have a high value, that are detrimental to downstream processes, or that cannot be safely discharged to the environment.

Figure 1.4 shows a simple reactor–separator system. Substance A is fed into a reactor and, after some reaction time, it is converted into a mixture of B and C. That the mixture may be saleable or useful, but more often value is added by separating the two materials. The desired product, B, must be purified by removing C. Material C, a by-product of the desired reaction, may have some value once it has been separated, or it may be an innocuous waste product or a waste that has a high cost of disposal.

Figure 1.5 shows a real treatment process that uses three chemical transformations and two separation processes to treat water that contains dissolved lead and sulfuric acid. Hydrated lime, $Ca(OH)_2$, is added to neutralize the sulfuric acid and precipitate the dissolved lead as $Pb(OH)_2$. A second chemical transformation uses a polymer to coagulate the $Pb(OH)_2$ particles so they can be removed efficiently in the settler. The third chemical transformation is the addition of hydrochloric acid (HCl) to neutralize excess $Ca(OH)_2$ before the effluent is discharged. The two separation processes are settling to remove the precipitated solids and centrifugation to thicken the sludge that is removed from the settling tank.

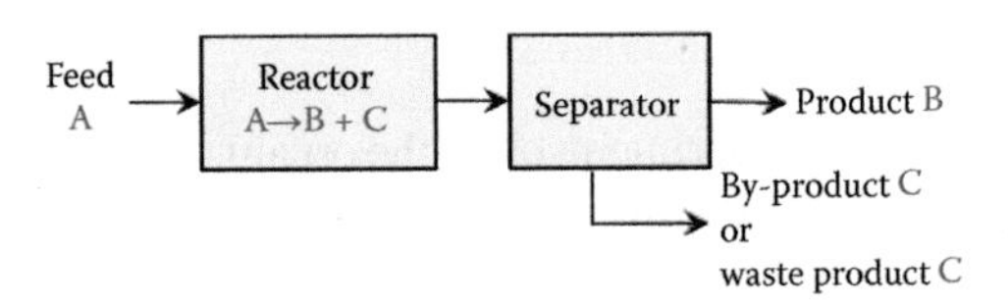

FIGURE 1.4 A simple reactor–separator process. A chemical transformation converts substance A into B and C. B may be a useful material.

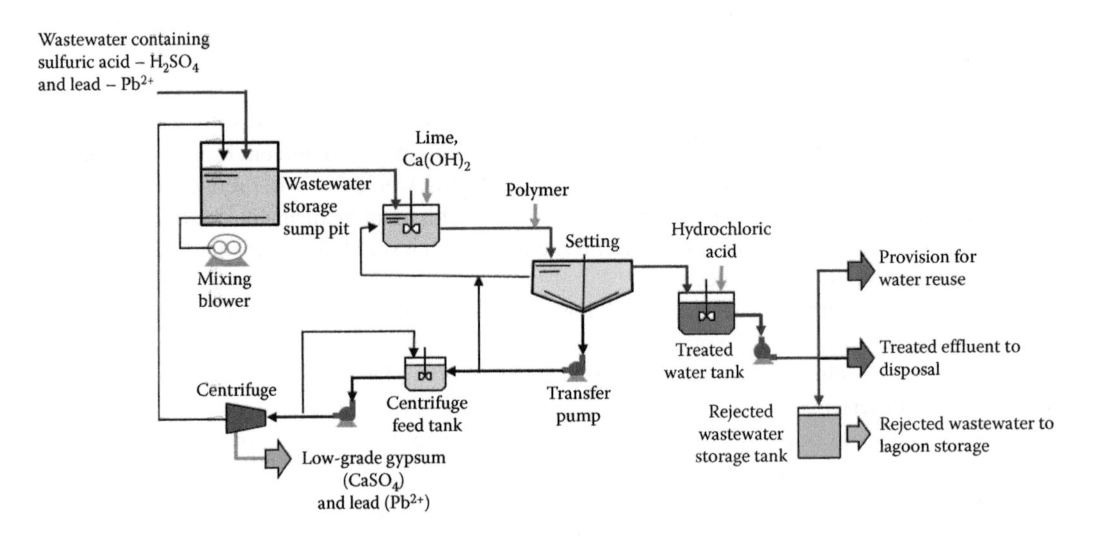

FIGURE 1.5 A wastewater treatment process that uses three chemical transformations (precipitation, coagulation, and neutralization) and two separation processes (settling and centrifugation).

1.5 STOICHIOMETRY

Chemistry, whether inorganic, organic, or biochemical, is concerned with *stoichiometry, equilibrium theory,* and *kinetics.*

Stoichiometry tells us what reacts and what is produced, and in what proportions. It answers the following questions:

- How much lime (CaO) must be added to wastewater to remove soluble phosphate (PO_4^{3-}) as calcium phosphate precipitate?
- How much CaO is required to neutralize sulfur dioxide (SO_2) in flue gas?
- How much sulfite (SO_3^{2-}) is required to reduce hexavalent chromium (Cr^{6+}) to trivalent chromium (Cr^{3+}) so it can be removed by precipitation?

The chemical sentence

$$1CH_4 + 2O_2 \rightarrow 1CO_2 + 2H_2O$$

defines a balanced chemical reaction that says the combustion of methane (CH_4) with oxygen (O_2) produces carbon dioxide (CO_2) and water (H_2O). The balanced equation also gives the proportions of reactants and products. One molecule of CH_4 reacts with two molecules of oxygen (O_2) to produce one molecule of CO_2 and two molecules of water (H_2O). Or 1,000,000 molecules of CH_4 react with 2,000,000 molecules of O_2, and so on.

The balanced equation defines the material balance of the reaction. The number of atoms of carbon (C) on the left-hand side (the reactants) equals (balances) the number on the right-hand side (the products)—one atom of C on each side in this case. The same is true for hydrogen (H) and oxygen (O). If this is true, the mass of H is the same on both sides, and the same is true for C and O. Also, the mass of the products must equal the mass of the reacting ingredients.

The stoichiometric coefficients indicate moles of material involved in the reaction. One mole of CH_4 will react with 2 moles of O_2 to form 1 mole of CO_2 and 2 moles of H_2O. It is not correct to say that 1 kg of CH_4 will yield 1 kg of CO_2, or 1 kg of CH_4 reacts with 2 kg of O_2. The mass of a mole of CH_4 is 16 g and the mass of a mole of CO_2 is 44 g. The masses are calculated using the molar masses of the reactants and products. This is the subject of Chapter 5.

By convention, balanced equations show the reactants on the left and the products on the right, but they do not show nonreactive substances. If the oxygen in our simple example is supplied in the form of air, nitrogen is also present, but it is nonreactive and is therefore not shown in the reaction. This means that the balanced stoichiometric equation is almost always a simplification of what happens in a real process.

The two sides of the balanced stoichiometric equation are merely the endpoints. There may be intermediate steps. Stoichiometry is like a road map for a trip that shows only the starting point of Madison, Wisconsin, and the destination of Randolph, Vermont, without showing the sharp curves, steep grades, construction zones, truck traffic, road accidents, and slippery spots in-between. The map serves a purpose, but there is much to be learned by actually making the trip.

The reaction may need to be pushed along by increasing pressure or temperature, or by adding a catalyst, or by adding an excess of one or more reactants.

The stoichiometry is correct for the molecules that do react, but typically less than 100% of what enters the reactor is converted into product. The yield of product may be, and typically is, much less than 100% of the stoichiometric prediction.

The ideal reaction is

$$A + B \rightarrow C$$

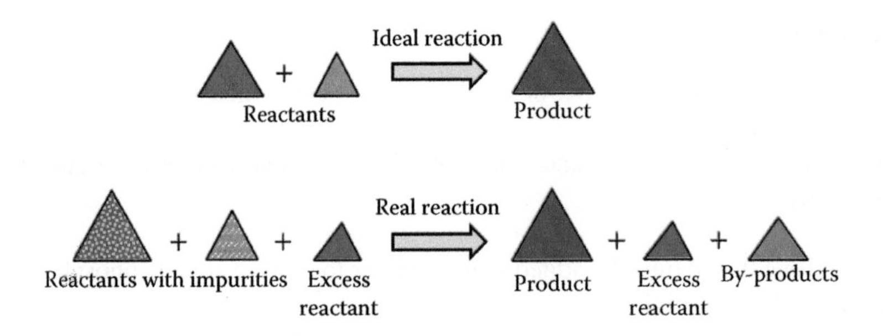

FIGURE 1.6 An ideal reaction converts all reactants into the desired product. Real reactions do not yield 100% product.

In practice, it is more likely to resemble the expanded equation in Figure 1.6. Impurities in the feedstock leave as contaminants in the product or as waste. By-products may be waste or they may have some values:

$$A + B + \text{excess } A + \text{impurities} \rightarrow C + \text{excess } A + \text{by-products} + \text{waste}$$

Clean chemistry or *green chemistry* considers the ultimate effect of the chemical product and manufacturing by-products on human health. The chemical process is selected or modified to eliminate or reduce the generation of unwanted by-products and waste during manufacturing. It is better to prevent waste than to create it and then clean it up.

Many waste treatment problems are solved using *empirical stoichiometry* because we lack the detailed information or knowledge needed to calculate the outcome from basic chemical principles. We have to "make the trip"—do the experiment and see what happens. Wastewater can contain hundreds or thousands of chemical compounds, most of which are not individually identified and quantified. There may be solid particles of various kinds in suspension. A reagent added to remove a targeted substance might react simultaneously with other similar substances. Therefore, instead of calculating, we experiment, observe, and measure.

1.6 EQUILIBRIUM

A solid substance put into water may dissolve totally or partially. The soluble ions may then interact with the water and other soluble species. If the solution is undisturbed, the reactions will find a balance where in which there is no net change in the concentrations of the reactants and products. This condition is called *equilibrium.*

Equilibrium theory deals with stable concentrations of reactants and products under specific conditions. It answers the following questions:

- Ammonia (NH_3) in water will exist as unionized NH_3 and ionized NH_4^+. The NH_3 form is toxic to fish, but NH_4^+ is not. At a given temperature and pH, how much of the ammonia will exist in the toxic form?
- What is the lowest possible concentration of total copper in water that can be achieved by precipitating copper hydroxide, $Cu(OH)_2$, assuming all that all solid precipitate can be removed from the water?
- How much calcium can be removed from water if lime (CaO) is added or if the temperature is increased?
- Will calcium carbonate ($CaCO_3$) in water be soluble or will it tend to precipitate in pipes and water heaters and cause maintenance problems?

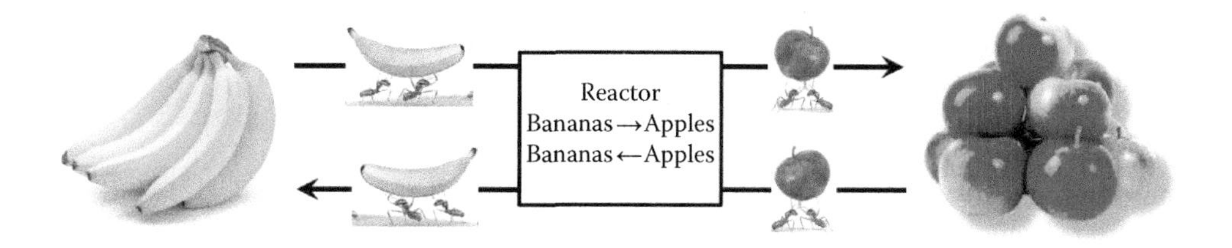

FIGURE 1.7 Cartoon explanation of equilibrium.

Equilibrium is a concept that applies to reversible reactions. Figure 1.7 shows a cartoon explanation of equilibrium. A reversible reaction converts apples to bananas and bananas to apples. Ants deliver apples and bananas to the reactor at the same rate, so there is equilibrium. The inventory of bananas on the left and the inventory of apples on the right do not change, even though there is material moving in both directions.

The reaction

$$HCl + NaOH \rightarrow Na^{+} + Cl^{-} + H_2O$$

is not reversible. It goes in one direction—from left to right—as indicated by the arrow. (Reactions are conventionally written as going from left to right.) The products do not spontaneously recombine and go back to HCl and NaOH. At equilibrium, everything is converted to the right-hand side species. These kinds of problems are solved using stoichiometry. No knowledge of equilibrium theory or kinetics is needed.

In contrast, these are reversible reactions:

$$NH_3 + H^{+} \leftrightarrows NH_4^{+},$$

$$HCO_3^{-} \leftrightarrows H^{+} + CO_3^{2-}$$

$$Ca^{2+} + CO_3^{2-} \leftrightarrows CaCO_3\ \text{(solid)}.$$

Equilibrium chemistry can be used to estimate the balance between ionized and unionized chemical species. In acid–base reactions, equilibrium is reached almost instantaneously. Whether equilibrium is reached in an instant or over a longer time, equilibrium determines the theoretical limit on the final concentrations of all the involved chemical species.

Equilibrium theory applies to all reactions in gases, liquids, or solids. Many equilibrium problems involve a solid phase, a chemical precipitate. Whatever the conditions, if the system is undisturbed, the soluble and insoluble species will find equilibrium.

When the reactions are reversible, we can disturb the equilibrium and shift the balance to make more or less of the solid form. Increasing the concentration of a reactant will push the reaction toward the right—toward making more solid product. Removing some of the product will allow more to form. Reducing one of the reactants will shift the reaction toward dissolving the solid.

For example, in the calcium carbonate solubility reaction

$$Ca^{2+} + CO_3^{2-} \leftrightarrows CaO_3\ \text{(solid)},$$

we can make more or less solid $CaCO_3$ by changing the concentration of Ca^{2+} or CO_3^{2-}. Increasing either one will shift the reaction to the right and more $CaCO_3$ will form. The system can be pushed back to the left by changing the conditions slightly so $CaCO_3$ will dissolve. Without going into detail, the most convenient way to shift the reaction is to change the pH of the solution. There is a second equilibrium reaction that is not shown between CO_3^{2-} (carbonate) and HCO_3^- (bicarbonate). Increasing the pH will create more CO_3^{2-} from HCO_3^-; lowering the pH will reduce CO_3^{2-} and less $CaCO_3$ will precipitate.

Equilibrium chemistry also controls the balance between solid and dissolved species. If one of the dissolved chemicals on the left were a pollutant, driving the reaction to the right would be an effective way to remove it as a solid that is easily removed from solution by settling or filtration.

Equilibrium theory predicts the scientifically possible lower limit on soluble material. This limit is not achieved in a real process. For example, a treatment process to remove copper from water by precipitation will not reach this theoretical lower level because (1) the solution usually does not reach equilibrium and (2) not all particulate copper solids can be removed from the water. The final effluent concentration is therefore determined by the interaction of chemical and separation processes.

1.7 KINETICS AND REACTOR DESIGN

Kinetics refers to the rate at which reactants are converted to products and how rapidly the reaction approaches its equilibrium condition. It answers the following questions:

- How long will it take for sodium metabisulfite to reduce toxic hexavalent chromium to benign trivalent chromium?
- What reaction time is required for a suspension of microorganisms to remove 95% of the biodegradable organic matter in a municipal wastewater prior to discharge to a river?

In preliminary analysis of most problems, it is sufficient to know whether the reaction occurs in seconds, minutes, or hours. Figure 1.8 shows the approximate reaction times for some chemical and biological reactions. As a rough guide, acid–base reactions are fast (seconds if the reactants are soluble), oxidation–reduction reactions are fast (seconds to a few minutes), and precipitations are slow (5–30 min). This book is mostly about reactions of this kind.

Reactions in nature are often slow because they occur at ambient temperature and at low concentrations. Many wastewater treatment processes also operate at ambient conditions because it costs too much to heat and cool water.

Reactor design is how kinetics and stoichiometry are put into practice. Some equipment is needed to hold the gas, liquid, or solids that are active in the reaction. The reactor contents must be maintained at particular conditions that favor a high yield of product with a minimal use of reagents and energy. If the required reaction time is 5 min, the required reactor volume is one-tenth the size for a reaction that requires 50 min, assuming that the reaction conditions are the same. The required reactor volume is calculated from models that predict how temperature, pressure, and other factors affect the required detention time and conversion efficiency.

These are important topics, but they are not discussed in this book.

1.8 GREEN CHEMISTRY

Green chemistry (*clean manufacturing*) is about changing the chemistry of manufacturing to reduce or eliminate the creation of pollutants. Several measures of chemical efficiency have been proposed (see Chapter 10). One of these is the yield of a reaction. Another is the *E-Factor*—the mass of waste generated per unit mass of useful product.

Figure 1.9 shows the M&M model of the chemical process industry. Assume that the green M&Ms are an edible product and all other colors are waste. The E-Factors are very large for the

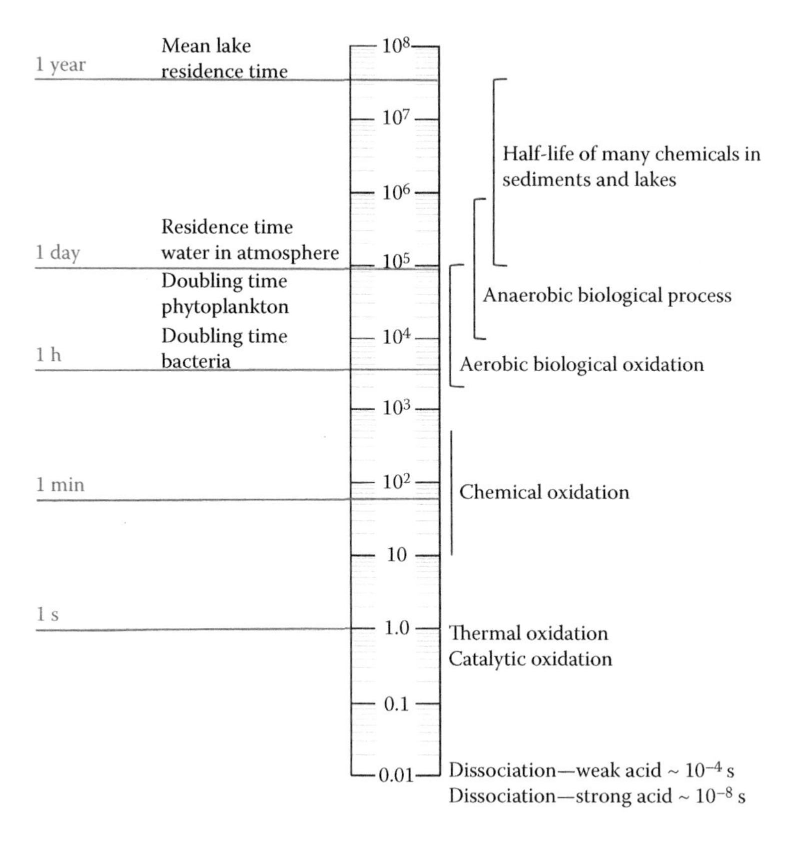

FIGURE 1.8 Time (in seconds) required for some chemical and biological processes. The numbers in the center of the scale are powers of 10. ($10^0 = 1$ s, 1 min = $10^{1.778} = 60$ s, and so on.)

E-Factor = $\frac{\text{Waste (kg)}}{\text{Product (kg)}}$	M&M model	Industry
0.1		Petrochemicals
1		Bulk chemicals, polymers, and plastics
10		Bulk chemicals, polymers, and plastics
100	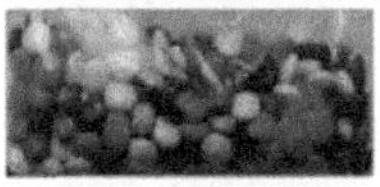	Fine chemicals
250		Pharmaceuticals

FIGURE 1.9 The M&M model of the chemical process industries. Green M&Ms are useful products; the other colors are waste.

fine-chemical and pharmaceutical industries. This is where a chemist will find opportunities for doing green chemistry.

Yield is the amount of product obtained in a chemical reaction. The idealized balance that stoichiometric reaction predicts is the *theoretical yield.* The *actual yield* measures what is obtained under realistic reaction conditions when using actual chemical feedstock.

The absolute yield can be given as the mass of product in grams or in moles. The percentage yield (or fractional yield or relative yield), which serves to measure the effectiveness of a synthetic procedure, is calculated by dividing the actual yield of the desired product obtained by the theoretical yield (the units for both must be the same):

$$\text{Percent yield} = \frac{\text{Actual yield}}{\text{Theoretical yield}} \times 100\%$$

It is easy to misjudge an industry as inefficient, wasteful, or even environmentally irresponsible when in fact a process is operated as efficiently as current technology allows. A reaction yield of 90% of the theoretical possible would be considered excellent. A yield of 80% would be very good. Even a yield of 50% may be considered adequate.

This three-step reaction series has a 90% yield of product for each step: 90% of A is converted to B, 90% of B is converted to C, etc.,

$$\mathrm{A} \xrightarrow{90\%} \mathrm{B} \xrightarrow{90\%} \mathrm{C} \xrightarrow{90\%} \mathrm{D}$$

The overall percentage yield for the synthesis is 100(0.9)(0.9)(0.9) = 73%.

The synthesis of some pharmaceuticals involves 10–20 steps. If the yield of each step is 80% (probably unreasonably high) for a 10-step sequence, the overall fractional yield = $(0.8)^{10} = 0.107$, or 10.7%. For a 20-step sequence at 80% per step, it is 1.2%.

The goal is always to increase yield, and one way to do this is to increase the yield of one or more steps. The three-step reaction with a 50% yield at each step

$$\mathrm{A} \xrightarrow{50\%} \mathrm{B} \xrightarrow{50\%} \mathrm{C} \xrightarrow{50\%} \mathrm{D}$$

has a fractional yield = 100(0.5)(0.5)(0.5) = 12.5%.

Suppose the yield of one of the steps (any step) is increased from 50% to 90%, such as

$$\mathrm{A} \xrightarrow{50\%} \mathrm{B} \xrightarrow{50\%} \mathrm{C} \xrightarrow{90\%} \mathrm{D}$$

This would increase the overall fractional yield to 100(0.5)(0.5)(0.9) = 22.5%.

This is a worthy improvement. It will save raw materials, reduce the load on separation processes, and decrease waste output (perhaps including a variety of toxic substances).

Another way to increase yield is to find new synthesis routes that involve fewer steps. This can be more productive than increasing the yield of any one step. A new synthesis route that eliminates one step to give

$$\mathrm{A} \xrightarrow{50\%} \mathrm{B}^* \xrightarrow{50\%} \mathrm{D}$$

will increase the fractional yield to 100(0.5)(0.5) = 25%. This is slightly better than increasing the yield of one step to 90%. By eliminating one step *and* increasing the efficiency of the remaining step from 50% to 90%, the new yield would be 100(0.5)(0.9) = 45%.

In general, if a multistep reaction has a 50% yield at each step, eliminating one step will double the overall fractional efficiency. If each step has a different efficiency of conversion, eliminating the most inefficient step will pay huge gains.

This is one of the principles of green chemistry.

1.9 ABOUT THIS BOOK

Chapter 2 defines and describes the traditional pollutants. Chapter 3 explains the naming conventions for organic chemicals and gives a short discussion of some compounds that have caused problems in recent years. Chapter 4 explains the units that are used to quantify pollutants, including concentration, volume flow rate, and mass flow rate. These chapters will be useful to the new student of pollution prevention and control. Those who have our previous books or those who have a good prerequisite knowledge of the subject may be comfortable going directly to Chapter 5.

Chapters 5 through 9 discuss stoichiometry, equilibrium in acid–base systems and precipitation, and oxidation–reduction reactions. These chapters are quantitative and many example calculations have been provided. If you have no interest in the calculations, you can omit most of those details and still understand the important ideas about how the processes are used and how they work.

Statistical experimental design is a strategy for extracting maximum information from a minimum of experimental runs. This is a great help when dealing with empirical stoichiometry and when there are several variables to be investigated. The basic ideas are introduced in case studies of Chapter 6.

Chapter 10 discusses changing the chemistry of manufacturing to reduce or eliminate the creation of pollutants. This is sometimes called *clean manufacturing* or *green chemistry*. This chapter is mostly descriptive. A few of the examples are about organic chemicals. The information in Chapter 3 should be sufficient for those who have little knowledge of this subject.

The many examples and case studies are intended to make your study more interesting and enjoyable. We hope you find this to be true.

2 Pollution and Pollutants

2.1 POLLUTANTS

Pollutants are substances that, when released into the environment, will degrade air, water, or soil and harm or kill plants and animals. This degradation may happen quickly or over time. Some pollutants, such as arsenic, lead, and cyanide, are outright poisons. Some damage the brain, liver, or kidneys. Some cause cancer, mutations, and birth defects. Some cause weakness and dizziness, and skin irritations. These substances are defined in the regulations as *toxic* or *hazardous*.

A broader definition of pollutants could include substances that can damage machinery or interfere with the performance of a manufacturing or a waste treatment process. By this definition, a "pollutant" is any substance that must be removed or controlled for any reason.

Inorganic salts can cause problems, even when they are not toxic. High total dissolved solids (TDS) can inhibit the microorganisms that are essential in biological wastewater treatment processes. High TDS will cause problems in boilers and cooling towers, two of the largest industrial uses for water. Semiconductor and pharmaceutical industries need extremely pure water. Food industries have special needs. Some industries need process water with up to eight levels of purity, ranging from deionized water to city water (or well water) to river water (for cooling).

Pollutants come in many physical and chemical forms, and almost always as a mixture. There will be dissolved chemicals in water, particles in liquids and gases, solids mixed with other solids, and gases mixed with air. They will be organic and inorganic. This is true at the source, after passing through treatment processes, and in the natural environment.

Specific pollutants must be identified and quantified as the first step in defining the problem. This chapter defines collective, or lumped, characteristics and nutrients. Chapter 3 discusses toxic metals and organic chemicals.

2.2 ELEMENTS OF LIFE

The six most important atoms for life are carbon (C), hydrogen (H), oxygen (O), nitrogen (N), phosphorus (P), and sulfur (S). Figure 2.1 shows that C, H, O, N, P, and S are organized into increasingly more complex structures in living organisms. Any chemical or substance that disrupts this organization is a toxicant. Any substance that disrupts the ability of organisms to grow and reproduce in a natural and healthy way is a pollutant.

Many other elements are essential, but are required only in trace amounts. Some are listed in Table 2.1.

Three kinds of *macromolecules* that are essential for life are carbohydrates, proteins, and fats. The other essential macromolecule is DNA.

Carbohydrates are composed of carbon, hydrogen, and oxygen. Their empirical formula is $C_m(H_2O)_n$, where m could be different from n. Another name for carbohydrate is *saccharide*. The smaller (lower molecular mass) carbohydrates are referred to as sugars. Their names end with the suffix *-ose*. For example, grape sugar is glucose, cane sugar is sucrose, and milk sugar is lactose. *Ribose*, a five-carbon sugar, is the major building block in the double helix backbone of DNA.

Proteins are constructed from amino acids and contain C, H, O, and N in the form of amine ($-NH_2$). Some proteins also contain sulfur and phosphorus. Proteins perform a vast array of functions within living organisms, including catalyzing metabolic reactions, replicating DNA, responding to stimuli, and transporting molecules from one location to another. Animals must consume protein because they cannot synthesize the amino acids that are essential to building new cells, DNA and RNA.

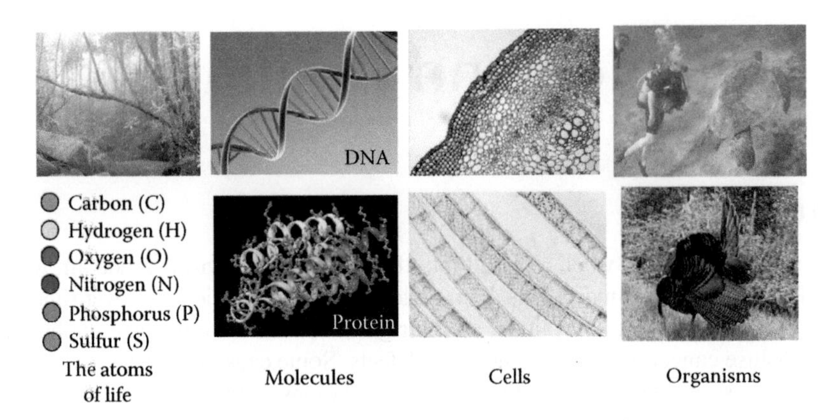

FIGURE 2.1 The six atoms of life, C, H, O, N, P, and S, are organized into increasingly more complex structures in living organisms.

TABLE 2.1
Elements That Are Essential for Life

		Major Components of Biological Cells
Carbon	C	Required for organic compounds
Hydrogen	H	Required for water and organic compounds
Oxygen	O	Required for water and organic compounds; necessary for aerobic organisms
Nitrogen	N	Required for many organic compounds, especially amino acids, and proteins
Phosphorus	P	Essential for biochemical synthesis and energy transfer
Sulfur	S	Required for some proteins and other biological compounds
		Other Essential Elements
Iron	Fe	Essential for hemoglobin and many enzymes
Manganese	Mn	Required for activity of several enzymes
Fluoride	F	Growth factor in rats; constituent of teeth and bones
Potassium	K	Principal cellular cation
Sodium	Na	Principal extracellular cation
Calcium	Ca	Major component of bone; required for some enzymes
Magnesium	Mg	Required for activity of many enzymes; in chlorophyll
Chlorine	Cl	Principal cellular and extracellular anion
Iodine	I	Essential constituent in thyroid hormones
Silicon	Si	Structural element in diatoms
Boron	B	Essential in some plants
Chromium	Cr	Essential in higher animals; related to action of insulin
Cobalt	Co	Required for activity of several enzymes; in vitamin B12
Copper	Cu	Essential in oxidative and other enzymes and hemocyanin
Selenium	Se	Essential for liver functions
Molybdenum	Mo	Required for activity of several enzymes
Vanadium	V	Essential in lower plants, certain marine animals, and rats
Zinc	Zn	Required for activity of many enzymes

Fats are the third important nutrient. The words oil, fat, and lipid are often used interchangeably. *Lipid* is the general term. Oils are liquids at normal room temperature; fats usually are solids at normal room temperature.

DNA, the molecular code for life and reproduction, contains C, H, O, P, and N arranged into two strands to form a double helix. The two DNA strands are composed of simpler units called *nucleotides*. The nucleotides are joined to one another in a chain between the sugar of one nucleotide and the phosphate of the next, resulting in an alternating sugar–phosphate backbone. Each nucleotide is a molecule of *guanine* (G), *adenine* (A), *thymine* (T), or *cytosine* (C), combined with a monosaccharide sugar called *deoxyribose* and a phosphate group. The nitrogenous bases of the two separate polynucleotide strands bond to make double-stranded DNA. Bonding happens according to the base-pairing rules: C links with G and A links with T.

2.3 AGGREGATE OR LUMPED MEASUREMENTS

The aggregate, or lumped, measurements characterize groups of substances without distinguishing individual species. *Turbidity* and *color* are unusual because they are measured in calibrated arbitrary units and not as mass concentrations. They are related to other quantitative measures, but are not always strongly correlated with them.

Lumped measurements are used to measure solids and organics in wastewater, including particulate matter (PM), suspended solids (SS), and *total dissolved solids* (TDS). SS are defined as all solids that can be captured on a filter of a specified pore size. These solids may be organic or inorganic, putrescible or biologically inert, or large enough to settle under the force of gravity or buoyant enough to float. PM in air is measured as the total concentration or the total mass, and this measure does not differentiate the kinds of solids, such as metal fumes, fly ash, dust, or pollen.

The measures of organic carbon play a dominant role in defining the performance of biological pollution control systems, and the response of rivers and lakes to wastewater inputs. Such measurements are defined by the analytical procedures. Chemical oxygen demand (COD) measures all the organic compounds that are oxidized in a boiling acid solution. Biochemical oxygen demand (BOD) measures the compounds that are oxidized in ambient aerobic conditions that are favorable for the growth of bacteria. The tests do not differentiate between carbohydrates, proteins, and fats. Synthetic organic chemicals, including toxics, will be measured in the COD test. They may be measured in the BOD test, unless they are toxic to bacteria, in which case they would invalidate the test.

2.4 TURBIDITY

Turbidity is a measure of water clarity. It measures how much the passage of light through the water is attenuated by material suspended in water. It is not measured in mass units.

Turbidity is caused by colloids and other finely divided particles that are suspended in water. Colloids are microscopically dispersed particles. The size range of colloids is 0.001–1 μm (somewhere between a molecule and a bacterium). They are too small to be seen by the naked eye. (The eye can see particles to about 40 μm.) The micrometer, μm, also commonly known as the micron, is 1×10^{-6} m or 1×10^{-3} mm. The symbol μ denotes the standard SI prefix "micro" (= 10^{-6}).

Other suspended materials that cause turbidity include soil particles (clay, silt, and sand), algae, plankton, microbes, and other substances. These materials are typically in the size range of 0.004 mm (clay) to 1.0 mm (sand).

It has been estimated that 50%–70% of the organic matter in domestic wastewater is colloidal matter. In water treatment, color, turbidity, viruses, bacteria, algae, and organic matter are primarily in the colloidal form or behave like colloids.

Particles in a colloidal suspension do not settle. A 1 μm colloidal particle would settle, by gravity in perfectly quiescent conditions, 1 m in about 3.5 years. The particle surfaces carry an electrical charge, usually negative, that prevents the particles from coming together (coagulating) to form larger particles. Clays and finely divided metal oxides and sulfides have this characteristic. Coagulation is promoted by adding positively charged ions, usually aluminum or iron, to reduce the repulsive ionic forces. (Unlike charges attract, but like charges repel.) Chapter 6 will discuss

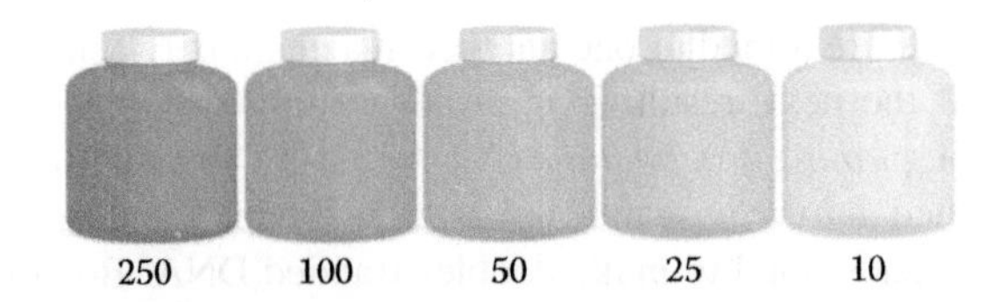

FIGURE 2.2 Rough scale of turbidity, measured in NTU, in river water or wastewater effluent.

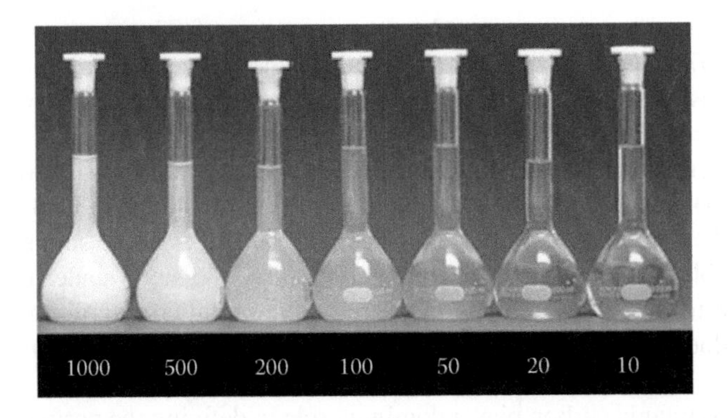

FIGURE 2.3 Standardized turbid suspensions for NTU measurements. (Courtesy of Hach Chemical.)

empirical stoichiometry and the use of jar tests to understand chemical coagulation and the removal of turbidity.

Colloids pass through the filters that are used to measure SS and will be measured as dissolved (filterable) solids even though they are not dissolved. In fact, water can have a high turbidity and zero SS. Turbidity is not the same as dissolved solids either, because colloids are not dissolved (they are particles). Dissolved salts do not cause turbidity, but they can cause color.

Colloidal size is on the order of the wavelength of visible light, so colloids will scatter incident visible light. The best way to quantify colloidal concentration is by *nephelometry*, or the measurement of light scattering. Turbidity is measured by shining a standard light beam through a specified depth of water and measuring the amount of light by scattering (Tyndall effect) that is due to the turbidity-causing particles. The unit of measurement is the Nephelometric Turbidity Unit (NTU). Figure 2.2 shows turbid water samples, measured in NTU, as they might appear if collected from a river. Figure 2.3 shows an array of turbidity standard solutions that are used to calibrate nephelometers.

2.5 COLOR

Color is another parameter that cannot be measured in mass units. It has hue (blue, green, etc.) and intensity. A red solution can be pinkish or deep red. We may or may not be able to see through a colored sample.

The color of a filtered sample can be expressed as follows:

- *Hue,* designated by the *dominant wavelength*
- *Degree of brightness*, designated by *luminance*
- *Saturation* (pale, pastel, etc.), designated by *purity*

The relationship between dominant wavelengths and hue is shown in Table 2.2.

TABLE 2.2
Relationship between Dominant Wavelengths of Visible Light and Hue

Hue	Violet	Blue	Blue–Green	Green	Greenish–Yellow
Wavelength range (nm)	400–465	465–482	482–497	497–530	530–575
Hue	**Yellow**	**Yellow–Orange**	**Orange**	**Orange–Red**	**Red**
Wavelength range (nm)	530–575	580–587	580–587	580–587	580–587

FIGURE 2.4 Natural waters colored by decomposing vegetation. (From left to right: maple leaves, oak leaves, fern, hemlock, and distilled water.)

Color can be caused by dissolved substances or by suspended matter. In the jargon of water quality measurement, *true color* is caused by dissolved substances. *Apparent color* includes color caused by both dissolved and suspended materials. It will change if the suspended substances are filtered out.

The true color is measured in the *filtered sample* after the turbidity from suspended or colloidal matter has been removed through filtration or other liquid–solid separation processes. Figure 2.4 shows samples of natural water that have true color arising from decomposing vegetation.

Spectrophotometry is the analytical method of choice for domestic and industrial wastewaters with complex and varied color components. A spectrophotometer generates visible light at different wavelengths, so the filtered samples can be compared on a specific color scale.

Color is also measured using the platinum-cobalt method. The unit of measurement is the *platinum-cobalt unit* (PCU). This method is used to assess the color of potable water and of water in which color is due to naturally occurring materials such as organic acids from leaves, bark, roots, humus, and peat materials. It is not useful for highly colored industrial wastewaters.

2.6 ODOR

Many people believe, "If it smells, it must be bad." There are colorless and odorless gases that are dangerous; carbon monoxide is one such example. There are bad-smelling compounds that are safe, such as amines that cause a *fishy* odor. Some are smelly *and* toxic; the most important is probably hydrogen sulfide (H_2S). Some people are offended by odors that are noticeable but not disgusting, especially if the odor is frequent or intense.

The odor *detection threshold* is the lowest concentration of odorant that will elicit a response in a given percentage (usually taken as 50%) of a population without reference to odor quality. A different threshold is the minimum concentration that is recognized as having a characteristic odor

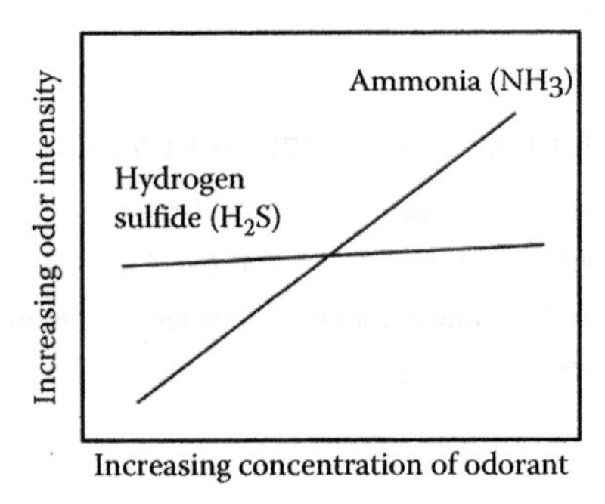

FIGURE 2.5 Odor intensity is proportional to the concentration of the odorant in the air. Hydrogen sulfide has a high intensity even at very small concentrations.

quality. Odor character is how the substance smells; descriptors included fishy, nutty, creosote, turpentine, rancid, sewer, and ammonia. Hedonic tone describes how pleasant or unpleasant the odor is.

Intensity refers to the perceived strength of the odor. It increases with concentration. Figure 2.5 shows that hydrogen sulfide has a high intensity, even at very small concentrations. The intensity of ammonia is more strongly a function of concentration. If H_2S and NH_3 coexist, H_2S will dominate at low concentrations, whereas NH_3 may be sensed as having the stronger odor at high concentrations.

2.7 PARTICULATES AND DISSOLVED SOLIDS IN WATER

Solids in water and wastewater are measured as mass per volume, usually mg/L. Most applications use lumped or aggregate measures, such as total solids, total volatile solids (organic solids), and total fixed solids (inert solids). These measures give no information about the composition of the individual solid particles.

Total solids are the residue left in an evaporating dish after the sample has been dried to a constant weight at 103°C–105°C. They include particles and dissolved materials. *Volatile solids* are the material that will burn off at 550°C. *Fixed solids* are the ash that will not burn at a temperature of 550°C.

Volatile solids are the solid organic fraction of the total solids. All carbohydrates, fats, proteins, and synthetic chemicals will be measured as volatile solids. When a biological treatment process reduces the mass of total solids, the lost mass is volatile solids that have been converted to gas (usually methane or carbon dioxide). If the volatile solids decrease and fixed solids increase, it is because organic compounds have been mineralized.

Suspended solids are particles that can be captured on a filter of 2.0 μm (or smaller) pore size. *Dissolved solids* pass through the filter. More correctly, these are called *filterable solids* because this fraction includes very small particles (*colloids*) as well as truly dissolved chemicals. Dissolved solids can be measured by using filters with a finer pore size or by using an ultracentrifuge to remove the particles. The distinction between filterable solids and dissolved solids is unimportant in most applications. We will use the term *dissolved solids.*

The relations among the various solids measurements are as follows:

$$\text{Total solids}\,(\text{TS}) = \text{Volatile solids}\,(\text{VS}) + \text{Fixed solids}\,(\text{FS})$$

$$\text{Total solids}\,(\text{TS}) = \text{Total suspended solids}\,(\text{TSS}) + \text{Total dissolved solids}\,(\text{TDS})$$

$$\text{Total suspended solids}\,(\text{TSS}) = \text{Volatile suspended solids}\,(\text{VSS}) + \text{Fixed suspended solids}\,(\text{FSS})$$

$$\text{Total dissolved solids}\,(\text{TDS}) = \text{Volatile dissolved solids}\,(\text{VDS}) + \text{Fixed dissolved solids}\,(\text{FDS})$$

$$\text{Total fixed solids} = \text{Fixed dissolved solids}\,(\text{FDS}) + \text{Fixed suspended solids}\,(\text{FSS})$$

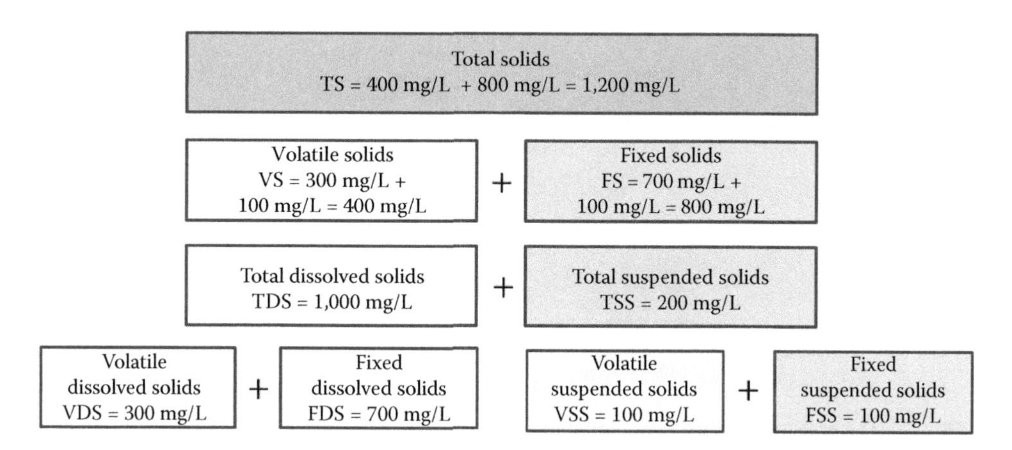

FIGURE 2.6 The relations of the different classes of solids that are commonly used to characterize wastewater.

Figure 2.6 shows a possible classification of the solids that are commonly measured to characterize wastewater.

In most applications where solids are to be separated from fluids (water or air), there are not a few discrete sizes that can be neatly shown as an ordered property list. Rather, there is a continuous distribution of particle sizes that describes how particle size changes continuously over a range of diameters.

Stormwater, municipal wastewater, and river water will carry particles that range in size from colloidal (0.5 μm) to 1 mm and larger. The particles also vary in specific gravity from 1.00 to 2.65. The large, dense particles are easily removed by screening and gravity settling. Other particles, including colloids, can be removed by coagulation-assisted settling and filtration. Membrane filters can remove bacteria and large organic molecules.

2.8 PARTICULATES IN AIR

An aerosol is a collection of liquid or solid particles suspended in air. The particles may be dry or moist, acidic or basic, inert or reactive, fibrous or granular or elastic. Particles do not have to be solids; they can be liquids or fumes or smoke. Some air particulates coalesce or condense from nonsolid materials. The U.S. Environmental Protection Agency (EPA) says, "Particles are a mixture of mixtures."

The most important particle characteristic is size, which can range from 1 nm to 100 μm. A sample of ambient air will contain a range of particle sizes. Figure 2.7 helps us understand the physical scale of the PM_{10} and $PM_{2.5}$ classifications that are used in air pollution regulations. In the future, scientists will measure PM_1 and perhaps even smaller particles.

Particles larger than 10 μm are generally captured in the nose and throat. Particles that are smaller than 10 μm are known as PM_{10}. They can be inhaled into the deepest part of the lungs. Even smaller particles (≤2.5 μm), known as $PM_{2.5}$, can pass from the lungs into the blood. Accumulating research and experience indicate that these superfine particles are causing many serious health problems and millions of premature deaths.

The 10-μm size refers to the aerodynamic diameter. That means, whatever the particle's shape may be, it behaves like a spherical particle with a diameter of 10 μm. Particles captured in a certain sampling device are, by definition, PM_{10} or $PM_{2.5}$. PM_{10} is smaller than the limit of human vision (about 40 μm). It is about the size of a small pollen or red blood cell. A large bacillus is 5 μm.

Particulate emissions control is a solid–gas separation problem. Baghouse filters and wet scrubbers are the most used separation technologies. A baghouse filter and a wet scrubber can capture

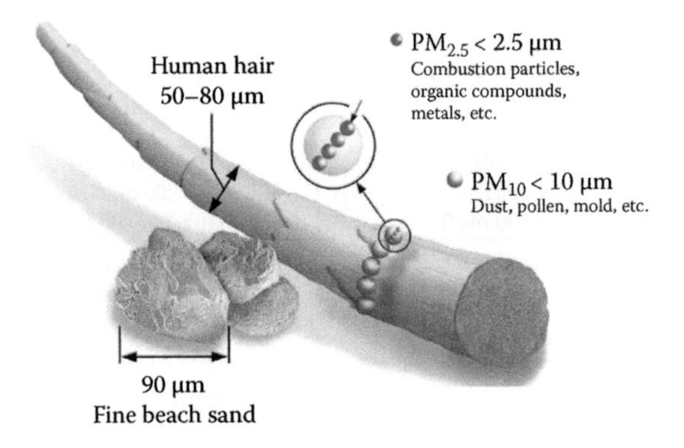

FIGURE 2.7 The scale of particulate air pollutants. (Courtesy of the U.S. EPA.)

98%–99% of particles larger than 5 μm. Electrostatic precipitators are slightly less efficient. ("Precipitation" in this context means "falling," as in rain falling. The particles have a surface charge than can be used to cause them to "fall" toward an electrically charged surface where they are collected.)

The chemical composition and chemical stability of the particles are relevant to toxicology and lung disease and to the disposal of dust and fly ash. Chemical processing is relevant to the recovery of metals, for example from foundry dust.

2.9 DISSOLVED ORGANIC MATTER AND NATURAL ORGANIC MATTER

Dissolved organic matter (DOM) is a broad classification for organics of varied origin and composition in water. It includes *natural organic matter* (*NOM*), which originates mostly from decomposition processes. *Dissolved* is an operational classification for organics that pass through a filter of 0.45-μm pore size (0.22-μm is sometimes used).

Ultraviolet absorbance (UVA) is an effective measure of the quantity of dissolved organics in water due to the strong UV absorption properties of organic molecules at the 254-nm wavelength. It can be used for many applications, such as UV disinfection, coagulation, granular activated carbon (GAC) filtration, water and wastewater treatment plant efficiency and effectiveness, chlorination, and other water and wastewater treatment processes.

Chlorine has long been used as a primary disinfection method for water and wastewater treatment. Harmful *disinfection by-products* can form when chlorine in drinking water reacts with NOM. The most important are the four compounds known as *trihalomethanes* (THMs). These are chloroform, bromodichloromethane, dibromochloromethane, and bromoform. The U.S. EPA standard for total THMs in drinking water is not to exceed 80 parts per billion (ppb).

TOC, COD, and BOD are other parameters used to characterize organics in water and wastewater. These are defined and described in the next sections, 2.10–2.12. These tests are complicated, time-consuming, and expensive, and making precise measurements at very low concentrations is difficult. In certain applications, UVA can be used as a supplement or surrogate.

2.10 TOTAL ORGANIC CARBON

Total organic carbon (TOC) measures the amount of carbon bound in organic compounds and is often used as a nonspecific indicator of water quality or wastewater strength. A typical analysis for TOC measures both the total carbon (TC) and the so-called "inorganic carbon" (IC); the latter represents the content of dissolved carbon dioxide and carbonic acid salts. Subtracting the IC from the TC yields TOC. That is, TOC = TC − IC.

TABLE 2.3
Degradability of Medium-Strength Municipal Wastewater

Parameter	Biodegradable	Inert	Total
COD, total	570	180	750
COD, soluble	270	30	300
COD, particulate	300	150	450
BOD	350	0	350
Total-N	43	2	15
Organic-N	12	2	15
Ammonia-N	75	45	20
Total-P	14.7	0.3	15

Source: Henze, M. et al., 2008.
Note: Concentrations in mg/L (g/m^3).

2.11 CHEMICAL OXYGEN DEMAND

Chemical oxygen demand is another lumped or aggregate measure of organics. COD measures all compounds that can be oxidized to carbon dioxide with a strong chemical oxidizing agent (dichromate) under acidic conditions and high temperature (~150°C). This includes all carbohydrates, fats, proteins, and most synthetic organic compounds. It also includes reduced inorganic substances, such as sulfides, sulfites, and ferrous iron (Fe^{2+}). NH_3-N will not be oxidized in the COD test.

Table 2.3 shows some typical values for medium-strength municipal wastewater (Henze, 2008).

The COD test has two advantages over the BOD test, which is described in the next section. It can be completed in a few hours, compared to 5 days for the BOD_5 test. It can be measured on wastewaters that are toxic to the bacteria on which the BOD test depends. This makes it useful for testing certain kinds of industrial wastewaters. COD can be measured on the whole wastewater or on the "soluble" (filterable) fraction of wastewater.

2.12 BIOCHEMICAL OXYGEN DEMAND

There are no specific regulatory limits for carbohydrates, fats, and proteins or for the individual kinds of carbohydrates (glucose, starch, etc.) or for proteins and their amino–acid building blocks (glycine, tryptophan, etc.). There are regulations for the *aggregate* of these kinds of biodegradable organic chemicals. They all can be decomposed and metabolized by bacteria and other microbes. Almost all wastewater treatment plants include processes to accomplish this biodegradation under controlled conditions.

The BOD test measures the consumption of oxygen by bacteria as they degrade the organic compounds. The mass of oxygen consumed is proportional to the mass of organic compounds that are metabolized by the organisms as they consume the oxygen. This gives the BOD test two interpretations: (1) a direct measure of oxygen-consuming potential and (2) an indirect measure of the concentration of biodegradable organic compounds in the water or wastewater.

BOD is commonly used to quantify the strengths of the influent and effluent to a wastewater treatment plant or treatment process. Percentage BOD removal is used to measure the efficiency of a treatment process. This is common because almost every municipal wastewater treatment plant and many industrial plants are based on biological processes that are designed specifically to remove the kinds of organic compounds that are measured in the BOD test.

Every wastewater treatment plant has an effluent limit for BOD, which is typically between 5 and 30 mg/L. Whatever BOD is discharged from a treatment plant represents an oxygen demand that can reduce the dissolved oxygen (DO) in a lake or stream. Low DO concentrations will not support a healthy aquatic population.

The 5-day BOD test is a bioassay that is done in sealed BOD bottles (Figure 2.8) that contain *diluted wastewater.* The dilution water–wastewater mixture is aerated to give an initial DO concentration close to 9 mg/L. (The saturation concentration of oxygen in clean water is 9.18 mg/L at 20°C). The samples are incubated at a standard temperature of 20°C for 5 days to measure the *5-day BOD*, or BOD_5. The DO concentration is measured after 5 days of incubation at a standard temperature of 20°C.

The DO depletion (ΔDO_{Bottle}) from the initial to the final concentration is proportional to the amount of biodegradable organic matter in the sample. The BOD of the *bottle contents* is

$$BOD_{Bottle} = \Delta DO_{Bottle} = [\text{initial DO}] - [\text{final DO}]$$

The final DO must be 2 mg/L or greater for the measurement to be valid. Therefore, the DO depletion in the bottle must be less than 6–7 mg/L; conversely, the BOD of the diluted mixture cannot exceed 6–7 mg/L. These limits are the reason for incubating a diluted mixture.

The *dilution factor* (*DF*) is the ratio of the BOD bottle volume to the volume of wastewater in the bottle. DF = $(V_{Bottle})/(V_{Sample})$. DF = 30 means a 10-mL sample in a 300-mL bottle; DF = 300/10. DF = 2 means that the incubated mixture is 50% effluent and 50% dilution water: DF = 300/150.

The BOD of the undiluted wastewater is

$$BOD_{Wastewater} = (\text{DO depletion})\left(\frac{V_{Bottle}}{V_{Sample}}\right) = (\Delta DO_{Bottle})(DF)$$

The effluent from a well-operated activated sludge process should be <10 mg/L BOD_5. A DF of 2 should be suitable. A BOD_5 concentration of 200 mg/L to 300 mg/L is typical for untreated municipal wastewater. A DF of 25–50 is needed so the mixture has $BOD_{Bottle} \leq 6$–7 mg/L.

A longer incubation time of 20–30 days is used to measure the *ultimate* BOD. The ultimate BOD is proportional to the quantity of biodegradable organic compounds that were present at the start of the incubation period. The 5-day BOD is approximately two-thirds of the ultimate BOD for municipal wastewater.

The BOD test is problematic in many ways. One is that it takes 5 days to get a result. Another concern, because the test is a bioassay, is the toxicity of some industrial wastewaters. COD can be used as a surrogate for biodegradable organics in wastewater even though it will measure some compounds that are not biodegradable.

Example 2.1 Measuring BOD

Four 300 mL BOD bottles are used to measure the 5-day BOD of a municipal wastewater. The test conditions are given in Table 2.4. Three of the four dilutions (bottles) give valid test results. These are averaged to estimate the 5-day BOD of 297 mg/L, which is rounded to 300 mg/L.

TABLE 2.4
Five-Day BOD Test Data and BOD Calculation

Bottle ID No.	Volume of Wastewater V (mL)	Dilution Factor DF = 300/V	Initial DO (mg/L)	Final DO (Day 5) (mg/L)	DO Depletion ΔDO (mg/L)	5-day BOD = DF × ΔDO (mg/L)
8	2	150	9.1	7.3	1.8	270
15	4	75	9.1	5.1	4	300
18	6	50	8.9	2.5	6.4	320
21	10	30	9	1.8 ≤ 2.0	Invalid	Invalid

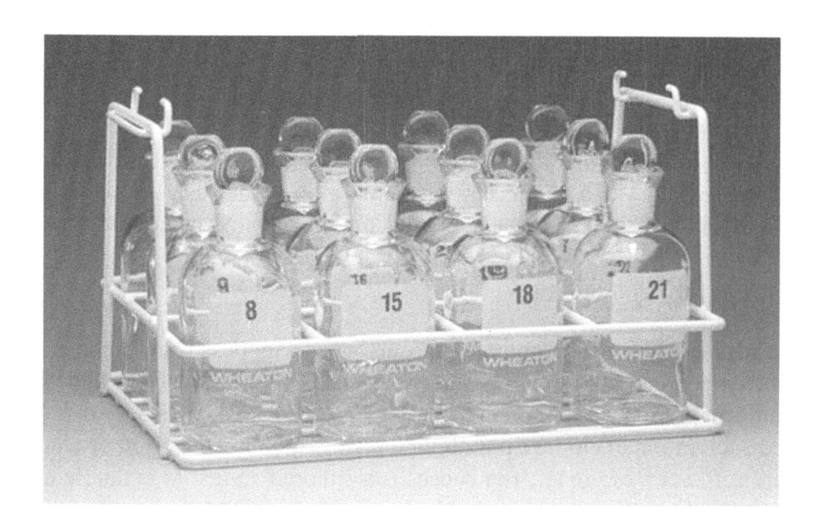

FIGURE 2.8 A rack of bottles used for measuring the BOD. (Courtesy of Wheaton.)

Example 2.2 Inhibition of the BOD Test

A high pH (pH 10) and highly saline industrial wastewater contained propylene glycol and ethylene glycol. The company hoped to develop a biological process to treat this wastewater.

Preliminary BOD tests, shown in Figure 2.9, were done on the pure compounds—not on the wastewater mixture. The pH in the BOD test bottles was 7.0 and the salinity was near zero. Nutrients and a bacterial inoculum were added. Dissolved oxygen was provided. Incubation was at 20°C. In other words, everything possible was done to create a healthy environment for the biological tests.

The tests showed that both compounds were biodegradable. The theoretical ultimate BODs (BOD_{ult}) are 129 mg/L for the 100 mg/L ethylene glycol solution and 168 mg/L for the propylene glycol. The measured values approached these theoretical limits. This does not prove that a biological treatment process can be developed, but it suggests that more work in that direction could be worthwhile.

In Figure 2.9, the dotted line represents the typical BOD curve. Interpret BOD as the *amount of oxygen consumed by the biological reaction*. The slope of this curve is the *rate of oxygen consumption*. The rate is greatest at the beginning and it steadily decreases until the biodegradable organics have been consumed. The curves for the two test compounds do not have this typical shape.

The curve for ethylene glycol shows severe inhibition. The reaction needs about 6 days to show significant biological activity. This is because the bacteria seeded into the BOD bottle are

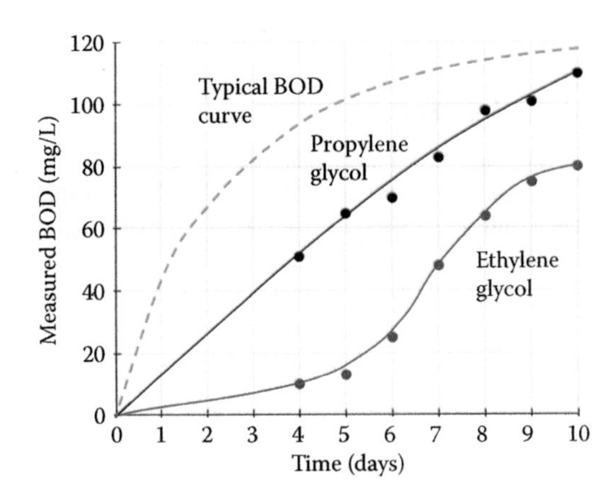

FIGURE 2.9 Ten-day BOD tests on propylene glycol and ethylene glycol.

not acclimated to using ethylene glycol as a food source, and that is the only food source they have. After a few days, they become acclimated, and after one week they are active.

Propylene glycol shows some inhibition, but much less than ethylene glycol.

2.13 IRON AND MANGANESE

Iron (Fe) and manganese (Mn) are common in groundwater supplies. They can also be introduced by corrosion of iron pipes and fixtures. They are important because of the aesthetic problems they cause and the amount of money spent to remove them from water supplies.

They cause discoloration of clothes, plumbing fixtures, and manufactured products. The water shown in Figure 2.10 is an example. Iron causes rust-colored stains. Manganese causes brownish stains in laundered clothing, and leaves black particles on fixtures. Soaps and detergents do not remove these stains, and the use of chlorine bleach and alkaline builders (such as sodium

FIGURE 2.10 Red water caused by presence of iron that has oxidized when exposed to air.

carbonate) can actually intensify the stains. Both Fe and Mn also affect the taste of beverages, including coffee and tea.

The U.S. EPA recommends a maximum of 0.3 mg/L total iron and 0.05 mg/L manganese. These are not enforceable because iron and manganese cause no health problems.

Iron is found in natural water in the form of ferrous ions (Fe^{2+}). The dissolved form of manganese is Mn^{2+}. The oxidized forms, Fe^{3+} and Mn^{4+}, form insoluble hydroxides, $Fe(OH)_3$ and MnO_2. One means of removing Fe and Mn is to oxidize them using aeration, which provides oxygen and strips out carbon dioxide and raises the pH. Other means of oxidation are chlorine (Cl_2), sodium hypochlorite (NaOCl), potassium permanganate ($KMnO_4$), and ozone (O_3). The oxidized and insoluble Fe and Mn particles are removed by filtration.

2.14 HARDNESS, CALCIUM, AND MAGNESIUM

Hardness is caused by compounds of calcium (Ca) and magnesium (Mg), and by a variety of other metals. Hard water is not a health risk, but economically it is one of the most important water quality parameters. Removing hardness from water creates residues (sludge or brine) that are managed as wastes.

Hard water requires more soap and synthetic detergents for home laundry and washing than soft water does. It causes calcium and magnesium deposits in water heaters, boilers, cooling towers, and other industrial equipment. Pipes can become clogged so badly that water flow is severely reduced and pipes must be replaced. The deposits also waste energy. Figure 2.11 illustrates the problem.

The United States Geological Survey uses this classification for hard and soft water:

Soft water	<60 mg/L as $CaCO_3$
Moderately hard	61–120 mg/L as $CaCO_3$
Hard	121–180 mg/L as $CaCO_3$
Very hard	more than 181 mg/L as $CaCO_3$

About half the locations tested in the U.S. have hardness of 120 mg/L $CaCO_3$ or more. More than 85% of U.S. homes have hard water.

Hardness is reported as the equivalent amount of calcium carbonate ($CaCO_3$). The equivalents for calcium and magnesium are as follows:

$$40 \text{ mg/L } Ca^{2+} = 100 \text{ mg/L hardness as } CaCO_3$$

$$24.3 \text{ mg/L } Mg^{2+} = 100 \text{ mg/L hardness as } CaCO_3$$

Total hardness measures the sum of calcium hardness and magnesium hardness.

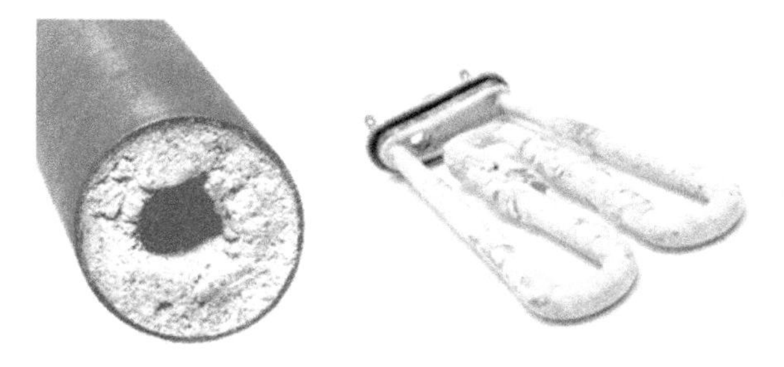

FIGURE 2.11 Scale in a water pipe and on a heating element; it is due to deposition of calcium and magnesium minerals deriving from hard water.

$$\text{Total hardness}\left(\text{mg/L CaCO}_3\right) = 2.5\left(\text{mg/L Ca}^{2+}\right) + 4.1\left(\text{mg/L Mg}^{2+}\right)$$

This classification can be converted to a molar basis using 1 mmol $CaCO_3$ = 100.1 mg/L $CaCO_3$.

Hardness that is caused by dissolved calcium bicarbonate, $Ca(HCO_3)_2$, and magnesium bicarbonate, $Mg(HCO_3)_2$, can be reduced by adding lime, $Ca(OH)_2$, to form insoluble compounds $CaCO_3$ and $Mg(OH)_2$. It can also be reduced by heating or boiling because these minerals, unlike most, become less soluble at higher temperatures. This makes hardness a serious problem in industrial boilers, heat exchangers, distillation towers, and hot water systems, including household hot water heaters and coffee makers.

2.15 pH

Of all measures of water quality, pH is likely known to the greatest number of people, at least in a general or intuitive way. It will be explained in detail in Chapter 7, which is about acid–base reactions. Those details are not needed here, but this basic information is helpful:

- pH is an intensity measurement of acidity and alkalinity.
- It is reported in pH units, on a scale from 0 to 14.
- pH 7 is neutral.
- Low pH indicates acidic conditions.
- High pH indicates alkaline conditions.

The definition of pH is

$$\text{pH} = -\log_{10}[\text{H}^+]$$

where $[H^+]$ is the molar concentration of hydrogen ion in solution.

Natural waters have a pH that is close to neutral, and so do most municipal wastewaters. Water that has been affected by acid rain will have a pH lower than 7. Industrial wastewaters can have a pH from 2 to 12, which can change very quickly.

2.16 ALKALINITY

Alkalinity is the measure of the capacity of water to neutralize acids or to *buffer* the effect of acid addition. *Buffer* means to resist shock or change. A *well-buffered system* has a stable chemistry.

Pure water (mineral-free) has no buffering capacity. The pH drops immediately when acid is added.

Most alkalinity in natural water is caused by bicarbonates (HCO_3^-) and carbonates (CO_3^{2-}). Total alkalinity is the sum of HCO_3^- and CO_3^{2-}, expressed as $CaCO_3$. Many other compounds will neutralize acid, but they are usually found at concentrations that are low enough to make specific identification unnecessary.

Alkalinity is measured by adding a strong acid of known strength to a sample of water or wastewater. The procedure for doing this is called titration and the results may be plotted as a *titration curve*. A typical titration curve for water that contains *both* carbonate and bicarbonate alkalinity is shown in Figure 2.12. There are two *equivalence points*, or breakpoints, where the pH changes rapidly with small additions of acid.

When the water contains both forms of alkalinity, there will be a break (change in slope) in the titration curve at pH 8.4 and another at pH 4.5. At pH = 4.5 all alkalinity has been destroyed, and

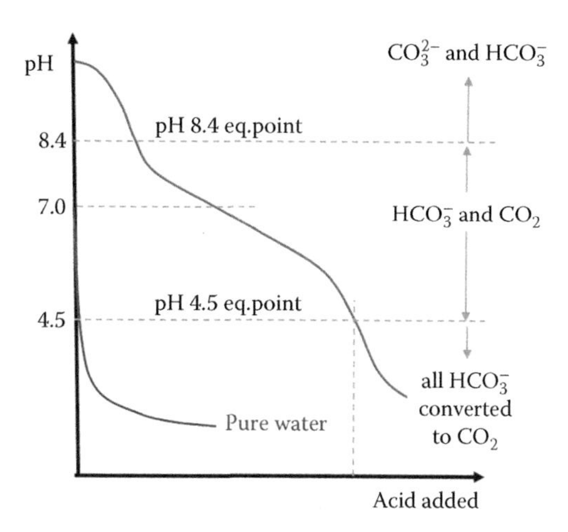

FIGURE 2.12 A reduction in pH, as acid is added to measure the alkalinity of water. All carbonate and bicarbonate have been neutralized at about pH 4.5.

this is the *endpoint* of the measurement. The amount of acid added to reach the endpoint is proportional to the alkalinity of the sample.

Alkalinity is reported as mg/L of equivalent calcium carbonate ($CaCO_3$). This is the same reporting convention used for hardness; it is convenient when dealing with hardness removal, because carbonate (CO_3^{2-}) and bicarbonate (HCO_3^-) are key to the water softening process. Hard waters typically have high alkalinity. This is not surprising, because the hardness (Ca and Mg) comes from water that has dissolved calcite ($CaCO_3$) and dolomitic limestone ($CaMg(CO_3)_2$).

Alkalinity and pH are related, but not directly. As mentioned, all the alkalinity has been neutralized at pH 4.5. Therefore, pH below the neutral value of 7.0 does not mean zero alkalinity. Water can have a high alkalinity, 500–1000 mg/L $CaCO_3$, and a pH at or near 7.0.

These are some typical alkalinity ranges:

Rainwater	<10 mg/L $CaCO_3$
Typical surface water (rivers and lakes)	20–200 mg/L $CaCO_3$
Surface water with alkaline soils	100–500 mg/L $CaCO_3$
Groundwater	50–1000 mg/L $CaCO_3$
Seawater	100–500 mg/L $CaCO_3$

2.17 NITROGEN

Nitrogen compounds can be air or water pollutants. In the context of water quality, they are nutrients. In air quality, they are irritants and important actors in the chemistry of ozone formation.

In common usage, "nutrients" means all the things found in food, including carbohydrates, proteins, fats, sodium, iron and other minerals, and vitamins. *Nutrients* in the context of pollution control refers to compounds of *nitrogen* and *phosphorus*.

Nitrogen is essential for plants and animals to synthesize protein. Nitrogen occurs in dissolved, particulate, and gaseous forms.

Nitrogen gas (N_2) comprises 78% of the atmosphere, but it cannot be used directly by plants or animals. Nitrogen gas is generally considered to be inert. It is not combustible. It will dissolve in water but it does not react with water. Certain plants (legumes) can fix nitrogen from the atmosphere and convert it to ammonium (NH_4^+).

Nitrogen in raw (untreated) sewage is mostly in the form of organic nitrogen and ammonia. The decomposition of organic compounds can release ammonia. After biological treatment, it mainly occurs in an oxidized form, nitrate NO_3^-. All forms of nitrogen are reported as mg/L N.

Organic nitrogen, or *Total Kjeldahl Nitrogen* (TKN) is a lumped measurement of all compounds with the empirical chemical formula C_xH_y–NH_2. TKN represents the sum of organic nitrogen compounds and ammonium nitrogen:

$$\text{TKN} = \text{org-N} + \text{NH}_4\text{-N}$$

Total ammonia is the sum of *ammonia* (NH_3) and *ammonium* ion (NH_4^+). The form of ammonia changes between NH_3 and NH_4^+ as the pH of the water changes. The TKN test is done at acid conditions so all ammonia is in the form of ammonium (NH_4^+).

As the pH increases, the fraction of total ammonia that is in the NH_3 form increases and the NH_4^+ fraction decreases. This is important because the un-ionized NH_3 is toxic to aquatic life and NH_4^+ is not. At pH 7.5 about 2% of the total ammonia is NH_3; at pH 8.5 the percentage is about 12%. These percentages are at 20°C; they increase at higher temperatures.

Nitrite (NO_2^-) and *nitrate* (NO_3^-) are formed from ammonia when it is oxidized. The biological oxidation reaction is called *nitrification*

$$NH_4^+ + 2.5O_2 \rightarrow NO_3^- + 2H_2O$$

Nitrate is the form of nitrogen most used by plants for growth. It is completely soluble and is easily lost from soil into groundwater. It can be converted in a process called *denitrification* to nitrogen gas. This provides a means to remove nitrate from wastewater. Common sources of excess nitrate reaching lakes and streams include septic systems, animal feed lots, agricultural fertilizers, manure, industrial wastewaters, sanitary landfills, and garbage dumps. Nitrite is an intermediate form of oxidized nitrogen, and is readily converted to nitrate in the presence of oxygen.

Drinking water standards regulate nitrate, usually at a level of 10 parts per million (ppm), measured as nitrate-nitrogen. That means 10 mg/L of N in the form of nitrate compounds. Higher levels present a health risk to infants.

Example 2.3 Measuring Total Nitrogen

The laboratory report gives nitrogen concentrations as:

$$\text{Ammonia} = 20\ \text{mg/L NH}_3\text{-N}$$

$$\text{Nitrate} = 20\ \text{mg/L NO}_3\text{-N}$$

$$\text{Total Kjeldahl Nitrogen (TKN)} = 60\ \text{mg/L as N}$$

Having all results reported as Nitrogen (N) is convenient because they can be added and subtracted. It is also inconvenient because a conversion is needed to determine the concentration of the actual compound.

Organic Nitrogen = TKN − Ammonia = 60 − 20 = 40 mg/L as N
Total Nitrogen = Organic Nitrogen + Ammonia + Nitrate = 40 + 20 + 20 = 80 mg/L as N.

The molecular masses are used to calculate the concentrations of ammonia and nitrate:

$$\text{MM (g/mol)} \quad H=1 \quad N=14 \quad O=16 \quad N=14$$

$$NH_3 = 14 + 3(1) = 17$$

$$NO_3^- = 14 + 3(16) = 62$$

$$(20 \text{ mg/L } NH_3\text{-N})(17/14) = 24.3 \text{ mg/L } NH_3$$

$$(20 \text{ mg/L } NO_3\text{-N})(62/14) = 88.6 \text{ mg/L } NO_3\text{-N}$$

This conversion cannot be done for organic nitrogen because that is a collective measurement. It represents a collection of unidentified organic nitrogen compounds.

Example 2.4 Nitrogen Fertilizer

A liquid fertilizer contains 18% NH_3, 18% NH_4NO_3, and 6% CH_4N_2O (urea), by mass. Calculate the percentage of nitrogen (N) of the fertilizer solution.

Basis = 100 kg aqueous solution

Masses	0.18 (100 kg) = 18 kg NH_3
	0.18 (100 kg) = 18 kg NH_4NO_3
	0.06 (100 kg) = 6 kg $CO(NH_2)_2$

Atomic masses of N = 14, hydrogen = 1, oxygen = 16, carbon = 12

Molecular masses	NH_3 = 14 + 3(1) = 17
	NH_4NO_3 = 2(14) + 4(1) + 3(16) = 80
	$CO(NH_2)_2$ = 1(12) + 2(14) + 4(1) + 1(16) = 60
Mass fraction of N	in NH_3 = 14/17 = 0.824
	in NH_4NO_3 = 28/80 = 0.350
	in $CO(NH_2)_2$ = 28/60 = 0.467
Mass N	in NH_3 = 0.824 (18 kg) = 14.83 kg N
	in NH_4NO_3 = 0.350 (18 kg) = 6.3 kg N
	in $CO(NH_2)_2$ = 0.467(6) = 3.40 kg N

Total N = 23.53 kg/100 kg liquid nitrogen solution = 23.53%

Note: The specific gravity of this concentrated solution will be 1.3–1.4, for a specific weight of 1.3–1.4 kg/L (10.8–11.7 lb/gal).

The important nitrogen compound in air pollution is nitrogen oxide gas (NO_X). This is a mixture of nitric oxide (NO) and nitrogen dioxide (NO_2) that forms when fossil fuels are burned at high temperatures. The major source in urban areas is vehicle exhaust. Nitrogen dioxide, the most common form, is a reddish brown gas that has an unpleasant smell and causes respiratory problems.

Nitrous oxide (N_2O), a third form of gaseous nitrogen, is not found in automobile exhaust. It is used as anesthetic (laughing gas). It supports combustion and is sometimes used to enhance the power output of automobile engines in racing cars.

NO_X can be removed from vehicle exhaust and stack gases by catalytic conversion. The end product is nitrogen gas (N_2). Catalytic converters on internal combustion engines also convert CO to CO_2.

Shale oil, from fracking, contains 2% nitrogen. This must be removed prior to burning or NO_X will be released to the atmosphere.

Ammonia (NH_3) along with NO_X and SO_2, can form acidic secondary pollutants that can cause problems a large distance from the primary source. The primary source of ammonia released to the atmosphere is agricultural (manure handling and fertilizer).

2.18 PHOSPHORUS

Phosphorus (P) is a vital nutrient for converting sunlight into usable energy, and it is essential to cellular growth and reproduction. P contributed to lakes by human activity is a major cause of excessive algae growth and degraded lake water quality. P occurs in dissolved organic and inorganic forms or is attached to sediment particles.

Phosphorus builds up in the sediments of a lake. When it remains in the sediments it is generally not available for use by algae; however, various chemical and anaerobic biological processes can allow sediment phosphorus to be released back into the water.

All P has the same oxidation state—it does not participate in oxidation–reduction reactions. There are no naturally occurring gaseous forms of phosphorus. All non-phosphate P is human-made; pesticides are an example.

Almost all P is found in the inorganic form of phosphate (PO_4^{3-}). This form is preferred for plant growth, but other forms can be used when phosphates are unavailable.

There are three forms of phosphates: *orthophosphate* (PO_4^{3-}), *condensed phosphates* (polymers of PO_4^{3-}), and *organic phosphates* (nucleotides, sugar phosphates, etc.). *Total phosphorus* is the sum of these three forms.

In the U.S., the per capita contribution from human and food waste is about 0.5 kg P/year. The typical concentration of P in raw sewage is 5 mg P/L, of which 4 mg P/L is orthophosphate and 1 mg P/L is organic phosphate. Modern wastewater treatment plants can produce an effluent with <1.0 mg P/L, and some can go as low as 0.1 mg P/L.

2.19 SULFUR

Sulfur (S) is also important in air and water pollution prevention and control. Two forms of sulfur are common in water and wastewater: sulfate (SO_4^{2-}) and hydrogen sulfide (H_2S). When oxygen is present, sulfur tends to be oxidized to form sulfate. In anaerobic conditions (lack of oxygen), sulfate can be reduced, and H_2S will form.

Both are a nuisance in drinking water, but they do not pose a health risk. Sulfates in high concentrations can cause a laxative effect and a bitter taste. The U.S. EPA-recommended maximum concentration is 250 mg/L. This is a secondary standard and is not enforceable. Hydrogen sulfide at 1–2 mg/L produces an offensive "rotten egg" odor, which can be especially noticeable in a shower. An odor can be detected by most persons at 0.5 mg/L. H_2S can be eliminated by oxidation with chlorine or aeration.

H_2S has caused serious odor problems for Kraft pulp and paper mills, in oil fields and oil refineries, and in tanneries and coking facilities. Air pollution control systems have minimized this problem. It can also cause odors in wastewater treatment plants, as it is formed by the anaerobic decomposition of organic matter.

H_2S is a component of biogas produced in anaerobic sludge digesters. It must be removed before the gas can be burned in boilers or internal combustion engines because H_2S combines with water to form sulfuric acid. This reaction also can cause corrosion of sewers and other facilities.

H_2S is toxic. A particular danger arises because at 100–150 parts per million by volume (ppmv; in air) the olfactory nerve is paralyzed, with the result that a person may be warned by the foul odor but then quickly lose the ability to smell the gas and be fooled into thinking the danger has abated. Concentrations over 1000 ppmv cause immediate collapse with loss of breathing, even after a single breath. Between 100 and 1000 ppmv, a variety of serious health problem can occur very quickly.

Deaths due to hydrogen sulfide usually happen to workers who enter a sewer or septic tank without taking proper precautions.

Sulfur dioxide (SO_2) is formed during the burning of sulfur-containing fuels, such as coal and oil. About 80% of the SO_2 put into the atmosphere in the U.S. comes from power plants, amounting to 26.5 million tons of SO_2 per year. Oil from the North Sea contains 1% sulfur by weight; Middle Eastern oils contain 4–5% sulfur. This is in the form of organic sulfur, mostly thiophene (C_4H_4S), which is combustible. Some of this can be removed in the refining process. The oil processed by an average refinery (about 100,000 barrels per day) yields roughly 325 million kilograms of SO_2 per year.

Most metals in ore are in the form of sulfides; SO_2 is produced from refining aluminum, copper, zinc, and iron ores, among others. It is also released from hot springs and volcanoes in large quantities. SO_2 is colorless, but it does have an odor. It can cause breathing difficulties and it is toxic to plants. When it combines with moisture in the atmosphere, sulfuric acids will form. This is a cause of acid rain, a serious problem in the twentieth century that has been largely solved by using low-sulfur fuels and installing SO_2 capture systems on power plants and at sites in selected other industries.

Acids that form in the atmosphere from SO_2 and NO_X have caused severe damage to crops and forests by increasing the acidity (lowering the pH) of streams, lakes, and soil.

Atmospheric SO_2 reacts with hydroxide radical ($OH^{\bullet}$) to form sulfur trioxide (SO_3), which, in the presence of water vapor, is quickly converted to sulfuric acid (H_2SO_4) (Masters and Ela, 2007). In a similar manner, nitrogen dioxide (NO_2) reacts with hydroxide radical to form nitric acid (HNO_3).

Atmospheric carbon dioxide reacts with water to form weakly acidic carbonic acid (H_2CO_3). Clean water saturated with CO_2 will have pH 5.6. In the 1970s, it was observed that some lakes had pH values as low as 4.

2.20 TOXIC METALS

Toxic chemicals are a challenge, but not because we lack the technology to remove or destroy them in liquid effluents and gaseous emissions. The technology exists. The challenge comes from the rich menu of toxic compounds that come in many forms and cause diverse harmful effects.

The dose makes the poison. A toxic chemical will cause harm only when an organism is exposed at certain levels and durations of exposure. Chemicals that are poisons in high doses may be essential nutrients in small amounts for normal healthy life. Many potentially poisonous substances can be excreted or metabolized if the dose does not exceed some critical level.

The list of toxic and hazardous organic chemicals in environmental regulations is vastly longer than the list of toxic inorganics. This is in part because there are a great many toxic organic chemicals. The organic toxic chemicals need to be identified and regulated individually because small changes in their molecular structures can change their physical properties and their toxicity.

There are also a great many inorganic compounds that contain toxic metals, but it is the soluble ionized metal and not the compounds that are dangerous. This is because the compounds, at least to some extent, dissolve and ionize in water and in the blood. The metal ions do the damage, so we need only identify the metals and not all the compounds.

Table 2.5 is an abbreviated list of toxic chemicals that are regulated under the U.S. Safe Drinking Water Act. The *maximum contaminant level* (MCL) is the legal threshold limit on the amount of a substance that is allowed in public drinking water systems. It is the level that the U.S. EPA believes can be consumed over a lifetime with no adverse health effects. The inorganic chemicals are listed as elements, rather than compounds, because we assume they will exist as dissolved ions in drinking water. In rivers and lakes, they may be bound to organic matter, and this will reduce their bioavailability as toxins (Berthouex and Brown, 2013).

TABLE 2.5
A Short List of Toxic Chemicals That Are Regulated by the U.S. EPA

Inorganic Chemicals	MCL (mg/L)	Health Significance
Antimony	0.006	Alters cholesterol and glucose levels
Arsenic	0.01	Dermal and nervous system effects
Barium	2.0	Circulatory system effects, high blood pressure
Beryllium	2.0	Cancer risk and damage to bones and lungs
Cadmium	0.004	Concentrates in liver, kidney, pancreas, thyroid
Chromium	0.005	Skin sensitization, liver, and kidney effects
Copper	0.1	Nervous system damage and kidney effects
Cyanide (as free cyanide)	1.3	Spleen, liver, and brain effects
Fluoride	0.2	Skeletal damage
Lead	4.0	Nervous system damage and kidney effects
Mercury (inorganic)	0.015	Nervous system damage and kidney effects
Nitrate (as N)	0.002	Nervous system and skin sensitization
Nitrite (as N)	10.0	Methemoglobinemia
Selenium	1.0	Hair or fingernail loss, circulatory problems
Thallium	0.05	Gastrointestinal effects
Organic Chemicals	**MCL (mg/L)**	**Health Significance**
Benzene	0.005	Anemia, risk of cancer
Benzo(a)pyrene (PAHs)	0.075	Reproductive difficulties, risk of cancer
Carbon tetrachloride	0.6	Liver problems, risk of cancer
Chlorobenzene	0.007	Liver or kidney problems
Ethylbenzene	0.1	Liver or kidney problems
Polychlorinated biphenyls (PCBs)	0.1	Skin changes, cancer risk
Pentachlorophenol	0.1	Liver and kidney problems, cancer risk
Tetrachloroethylene	0.005	Liver problems, cancer risk
Toluene	1.0	Nervous system, kidney, or liver problems
1,2,4-Trichlorobenzene	0.2	Changes in adrenal glands
1,1,1-Trichloroethane	0.005	Liver problems, cancer risk
Trichloroethylene	0.005	Liver problems, cancer risk
Vinyl chloride	0.002	Increased risk of cancer
Xylenes (total)	10.0	Nervous system damage

MCL = maximum contaminant level.

The so-called *heavy metals* can cause a variety of toxic responses. Some of the human health problems that can arise from acute and chronic exposure to five common toxic metals are described in Table 2.6.

Arsenic (As) poisoning is usually related to ground water that naturally contains high concentrations of oxides of arsenic. A 2007 study found that over 137 million people in more than 70 countries are probably affected by arsenic poisoning from drinking water.

Cadmium (Cd) is an extremely toxic metal found in industrial workplaces. Overexposure may occur even in situations where only trace quantities of cadmium are found. Cadmium is used extensively in electroplating, although the nature of the operation does not generally lead to overexposure. Cadmium is also present in the manufacturing of some types of batteries. It has been used in some industrial paints and could be a hazard when sprayed. Removal of cadmium paints by scraping or blasting may pose a significant hazard. Exposures to cadmium are addressed in specific standards for industry.

TABLE 2.6
Acute or Chronic Exposure to Cadmium, Mercury, Lead, Chromium, and Arsenic Can Cause a Variety of Serious Health Problems

Heavy Metal	Acute Exposure	Chronic Exposure
Cadmium	Pneumonitis (lung inflammation)	Lung cancer
		Osteomalacia (softening of bones)
		Proteinuria (excess protein in urine; possible kidney damage)
Mercury	Diarrhea	Stomatitis (inflammation of gums and mouth)
	Fever	Nausea
	Vomiting	Nephrotic syndrome (nonspecific kidney disorder)
		Neurasthenia (neurotic disorder)
		Parageusia (metallic taste)
		Pink Disease (pain and pink discoloration of hands and feet)
		Tremor
Lead	Encephalopathy (brain dysfunction)	Anemia
	Nausea	Encephalopathy
	Vomiting	Foot drop/wrist drop (palsy)
		Nephropathy (kidney disease)
Chromium	Gastrointestinal hemorrhage (bleeding)	Pulmonary fibrosis (lung scarring)
	Hemolysis (red blood cell destruction)	Lung cancer
	Acute renal failure	
Arsenic	Nausea	Diabetes
	Vomiting	Hypopigmentation/hyperkeratosis
	Diarrhea	Cancer
	Encephalopathy	
	Multi-organ effects	
	Arrhythmia	
	Painful neuropathy	

Mercury (Hg), in its elemental state, exists as a vapor or liquid. In its mercuric state (Hg^{2+}), it can form either inorganic salts or *organomercury* compounds. The organic forms are the most dangerous. Important organomercury compounds are the *methylmercury* cation (CH_3Hg^+), *ethylmercury* cation ($C_2H_5Hg^+$), *dimethylmercury* [$(CH_3)_2Hg$], *diethylmercury* [$(CH_3CH_2)_2Hg$], and merbromin ("Mercurochrome").

Outbreaks of methylmercury poisoning occurred in several places in Japan during the 1950s due to industrial discharges of mercury into rivers and coastal waters. Symptoms typically include sensory impairment (vision, hearing, speech), disturbed sensation, and a lack of coordination. The best-known instance was in Minamata, where more than 600 people died from what became known as Minamata disease.

Lead (Pb) poisoning is a medical condition in humans and other vertebrates. Julius Caesar's engineer, Vitruvius, reported, "[W]ater is much more wholesome from earthenware pipes than from lead pipes. For it seems to be made injurious by lead, because white lead is produced by it, and this is said to be harmful to the human body." In 2013, the World Health Organization estimated that lead poisoning resulted in 143,000 deaths annually and "contributed to 600,000 new cases of children with intellectual disabilities."

Lead interferes with a variety of body processes and is toxic to many organs and tissues including the heart, bones, intestines, kidneys, and reproductive and nervous systems. It interferes with the development of the nervous system and is therefore particularly toxic to children, causing potentially permanent learning and behavior disorders. Symptoms include abdominal pain, confusion, headache, anemia, irritability, and in severe cases, seizures, coma, and death.

There is an effective therapy, called *chelation*, for Pb poisoning that has been administered to thousands of children. An organic chemical called a *chelate* binds metal ions and converts them to an inert form that is passed out of the body in urine. When calcium EDTA, Na_2[Ca-EDTA], is used to chelate lead, Pb from the bloodstream replaces the Ca in Na_2[Ca-EDTA] to give Na_2[Pb-EDTA]. Chelation is also used as treatment for mercury and arsenic poisoning.

2.21 CONCLUSION

Many pollutants are measured as collective or lumped parameters. This is because it is impractical or unnecessary to identify the individual compounds or to sort out the specific materials that are in the wastewater.

Turbidity is caused by colloids and finely divided particles. It is measured by passing a light beam through a water sample and measuring the light attenuation. Turbidity reduction is a useful measure of process performance even though we do not know what kinds of particles are being removed. Color is a similar measurement.

The mass concentration can be measured for total solids, TDS, SS, settleable solids, and volatile solids. Knowing the mass concentration of the solids gives no information about the composition, shape, or density of the particles. Volatile solids are combustible organics, but knowing the mass concentration says nothing about the specific compounds or classes of compounds.

TOC, COD, and BOD are lumped measures of organic compounds. The BOD test measures biodegradable compounds. The COD test measures chemically oxidizable compounds, including oxidizable inorganics like iron. TOC measures combustible carbon compounds, including carbonates.

Nitrogen and phosphorus occur in several forms, both organic and inorganic. It is convenient to measure and report the concentrations of all forms as mg/L of N or mg/L of P. For example, 17 mg/L NH_3 contains 14 mg/L N and would be reported as 14 mg/L NH_3-N.

The toxic heavy metals cause a variety of health problems. It is not necessary for environmental regulations to list all the possible compounds of these metals, because it is the ionized form of the metals that is toxic.

3 Organic Pollutants

3.1 A BRIEF INTRODUCTION TO NAMING ORGANIC CHEMICALS

Environmental regulations list hundreds of chemical pollutants. Almost all are organic and toxic. The U.S. Environmental Protection Agency (EPA) Clean Water Act lists 148 pollutants. The Resource Conservation & Recovery Act lists 502; the Clean Air Act 189; the Occupational Safety & Health Act 450; and the Emergency Planning & Community Right-to-Know Act 599.

Many have names that seem overwhelmingly complicated, such as benzene, toluene, xylenes, carbon tetrachloride, 1,2-dichlorobenzene, 1,2-dichloroethane, dichloromethane, pentachlorophenol, trichloroethene, 1,1,1-trichloroethane, and polychlorinated biphenyls (PCBs).

Some rudimentary knowledge of how these chemicals are named helps to remove the intimidation factor. Fortunately, the naming is systematic and not hard to learn. As Yogi Berra said, "You can learn a lot just by looking." So, start by looking at the names. They suggest what would be helpful to know. (Yogi Berra was a New York Yankee baseball player.)

Three names end with *-methane* or *-ethane*. More generally speaking, they end with *-ane*. One compound ends with *-ethene,* or more generally, with *-ene*. It will be a good start to learn the chemicals that are similar to methane, ethane, and ethene; that is, chemicals that end with *-ane* and *-ene*. These chemicals are all built from carbon chains.

We see *benzene,* and two names include *benzene*. And we see two other simple names—*toluene* and *xylene*(s). All of these are made from benzene and they all have a six-carbon ring. We need to learn about benzene.

Several names include *chloro*, *chlorinated*, or *chloride*. All three terms refer to chlorine (Cl). *Di* means 2, *tri* means 3, *tetra* means 4, and *penta* means 5. So, there are molecules with 2, 3, 4, and 5 chlorines added. *Carbon tetrachloride* is a carbon molecule with 4 chlorines. 1,2-Dichlorobenzene is *benzene* with 2 chlorines added. The 1 and 2 convey precise information about the location of the chlorines in the molecule, which is explained later.

The most intriguing name is *PCBs*. The *bi* in *biphenyl* means two connected phenyl groups. Chlorine is added to make them *chlorinated*. *Poly* means many and *polychlorinated* means many chlorines. PCBs is plural because many different molecules—209 to be exact—can be created by attaching 1 to 10 chlorine atoms to the 12 carbon atoms of the biphenyl base structure. More about this later.

3.2 HYDROCARBONS: THE ALKANES

One broad class of organic compounds is the *aliphatics*. Their basic structure is a carbon chain. The other broad class is the *aromatics*. They have ring structures, mostly of six carbon atoms.

Most of the atoms in organic chemicals are carbon (C) and hydrogen (H). Some chemicals are pure hydrocarbons—every atom is C or H.

Many chemicals that are important in environmental work are modified hydrocarbons. One or more hydrogen atoms are replaced by oxygen (O), nitrogen (N), sulfur (S), or chlorine (Cl). Alcohols and carbohydrates contain O. Proteins and amino acids contain N, and some contain S. Many of the offensively odorous chemicals, such as mercaptans (thiols), certain organic acids, and amines, contain S or N. Many solvents and toxic chemicals contain chlorine or fluorine.

The three kinds of aliphatic hydrocarbons are *-ane* compounds, the *-ene* compounds, and the -OH compounds. These are the *alkanes, alkenes,* and *alcohols*.

TABLE 3.1
The Homologous Series of *Saturated Hydrocarbons*, Also Known as *Alkanes*

Name	Molecular Formula	Melting Point (°C)	Boiling Point (°C)	State at 25°C
Methane	CH_4	−182.5	−164	Gas
Ethane	C_2H_6	−183.3	−88.6	Gas
Propane	C_3H_8	−189.7	−42.1	Gas
Butane	C_4H_{10}	−138.4	−0.5	Gas
Pentane	C_5H_{12}	−129.7	36.1	Liquid
Hexane	C_6H_{14}	−95	68.9	Liquid
Heptane	C_7H_{16}	−90.6	98.4	Liquid
Octane	C_8H_{18}	−56.8	124.7	Liquid
Nonane	C_9H_{20}	−51	150.8	Liquid
Decane	$C_{10}H_{22}$	−29.7	174.1	Liquid
Undecane	$C_{11}H_{24}$	−24.6	195.9	Liquid
Dodecane	$C_{12}H_{26}$	−9.6	216.3	Liquid

The most basic organic compounds are known as *saturated hydrocarbons* or *alkanes*. Their names end with *-ane*. Everyone knows the names *methane* (CH_4), *propane* (C_3H_8), *butane* (C_4H_{10}), and *octane* (C_8H_{18}). A series of such compounds exists, with each adding one carbon atom and two hydrogen atoms. Table 3.1 lists the first 12 of the *homologous series* of alkanes. The general formula is C_nH_{2n+2}.

Notice how the melting points and boiling points systematically increase as the size of the molecules increases. The lightest four molecules are gases at ambient conditions, but they can be compressed to form liquids; liquified propane gas is an example. Pentane and the larger molecules, up to 12 carbons, are liquids at normal temperatures, but the boiling points are low and they readily volatilize. Beyond 12 C atoms, solids are formed.

The structure of the alkanes is very simple. Each carbon atom has four bonding sites. Carbons bond to carbons, and hydrogen atoms fill up the remaining bonds. Methane has one carbon atom, and all four bonding sites are filled with hydrogen atoms.

Figure 3.1 shows the structure of carbon chains with hydrogen attached. The connecting bars represent bonds between carbon and hydrogen atoms. This figure also helps explain the term *saturated*. The molecule is saturated with hydrogen atoms, meaning it contains the maximum possible number of hydrogen atoms.

The stick diagrams under the C–H formulas are a popular simplification of the structural diagrams. Carbon atoms are located at the ends of the sticks and at the vertices of two sticks. Invisible hydrogen atoms are attached to the invisible carbon atoms. Propane has three carbons, indicated by the two stick ends and one vertex. Knowing the general structure allows us to easily picture the carbons and the hydrogens.

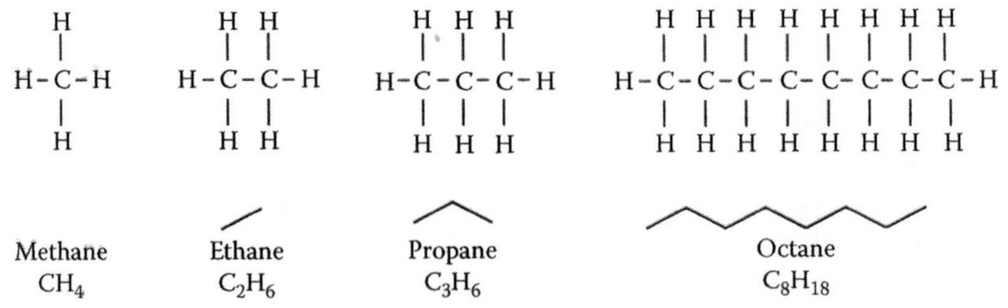

```
  H          H H          H H H          H H H H H H H H
  |          | |          | | |          | | | | | | | |
H-C-H      H-C-C-H      H-C-C-C-H      H-C-C-C-C-C-C-C-C-H
  |          | |          | | |          | | | | | | | |
  H          H H          H H H          H H H H H H H H
```

Methane CH_4 | Ethane C_2H_6 | Propane C_3H_6 | Octane C_8H_{18}

FIGURE 3.1 The structure of saturated hydrocarbons (*alkanes*).

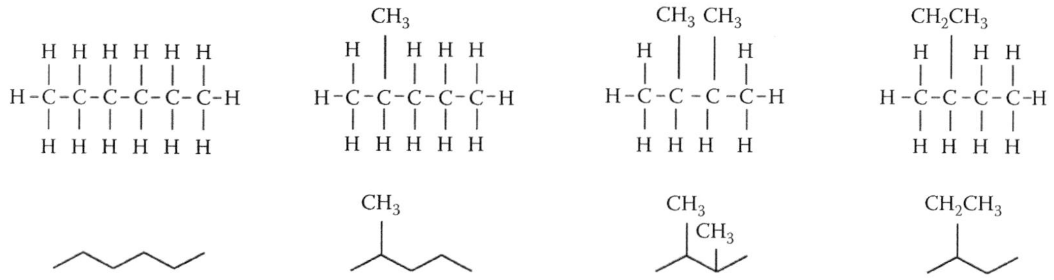

Hexane C_6H_{14} | Methylpentane C_6H_{14} | Dimethylbutane C_6H_{14} | Ethylbutane C_6H_{14}

FIGURE 3.2 Isomers of hexane.

Removing one hydrogen atom from an alkane creates a functional group. A *methyl group*, $-CH_3$, is methane with one H atom removed. A methyl group can be attached to another molecule anywhere an H atom can be removed. Attaching a methyl group to a 4-C compound, like butane, creates a 5-C compound that is a pentane. Attach it at one of the internal carbon atoms and it forms a *branched compound* that will have a name like 2-methyl pentane or 3-methyl pentane. The number indicates which internal carbon has the attached methyl group.

The molecules with three or more carbon atoms can exist in several different structural forms that have the same chemical formula. These different forms are called *isomers*.

Hexane is a straight-chain six-carbon compound. The carbons can be arranged to create branched chains. Figure 3.2 shows hexane and three isomers named for the branching functional group (–methyl or –ethyl) and the length of the longest chain.

Hexane (six carbon atoms) with a methyl group (CH_3) attached to an internal carbon atom is a branched heptane (seven carbons atoms) with the name methylhexane. Hexane with an ethyl group attached to an internal carbon atom is a branched octane (eight carbons), named ethylhexane. Another possibility is hexane with two internal methyl groups (dimethylhexane).

3.3 HYDROCARBONS: THE ALKENES

The next most simple structure is the *alkenes*. Alkenes are alkanes with at least one carbon–carbon double bond. The double bonds are created by removing hydrogen atoms. Like the alkanes, they contain only C and H atoms. The names are formed by removing the *-ane* from the alkane and replacing it with *-ene*. Propane becomes propene, for example. (There is no methene, because a double carbon bond is impossible with only one carbon.)

The alkenes are *unsaturated* because the carbons that lost H^+ atoms form a double bond, as shown by the double line in Figure 3.3. At an end position that has a double bond (double line) only two hydrogens will attach. The double bond can be anywhere in a large molecule. Number prefixes

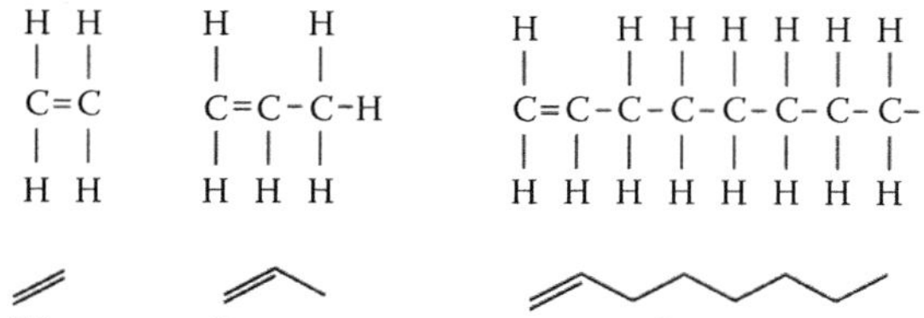

Ethene C_2H_4 | Propene C_3H_6 | Octene C_8H_{16}

FIGURE 3.3 Structure of the *alkenes*, which have a double-bonded carbon.

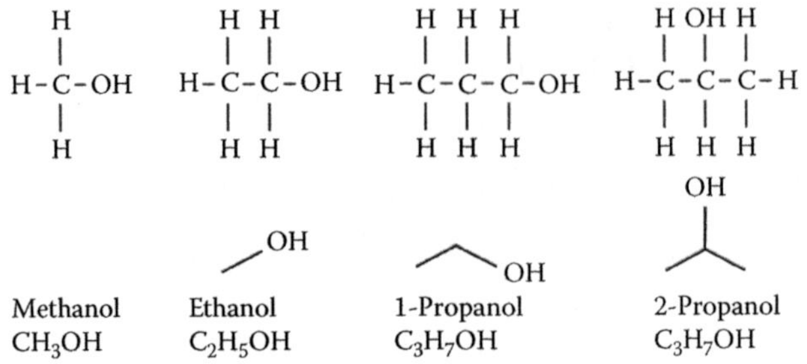

```
    H          H H          H H H          H OH H
    |          | |          | | |          | |  |
  H-C-OH     H-C-C-OH     H-C-C-C-OH     H-C-C--C-H
    |          | |          | | |          | |  |
    H          H H          H H H          H H  H
```

Methanol CH_3OH | Ethanol C_2H_5OH | 1-Propanol C_3H_7OH | 2-Propanol C_3H_7OH

FIGURE 3.4 The structure of simple alcohols.

TABLE 3.2
The Three Best Known Alcohols and Three Well-Known Derivatives

Chemical Formula	IUPAC Name	Common Name
	Monohydric Alcohols	
CH_3OH	Methanol	Wood alcohol
C_2H_5OH	Ethanol	Alcohol
C_3H_7OH	Propanol (Isopropyl alcohol)	Rubbing alcohol
	Polyhydric Alcohols	
$C_2H_4(OH)_2$	Ethane-1,2-diol	Ethylene glycol
$C_3H_6(OH)_2$	Propane-1,2-diol	Propylene glycol
$C_3H_5(OH)_3$	Propane-1,2,3-triol	Glycerol

are used to indicate the position of the double bond. The large molecule shown in Figure 3.3 is 1-octene; 2-octene has the double bond between the second and third carbons in the chain.

3.4 HYDROCARBONS: THE ALCOHOLS

Alcohols contain a hydroxyl group (–OH). The compound names end with *-ol*. The best known are methanol, ethanol, and propanol (isopropyl alcohol). There are two forms of propanol because there two possible positions for the –OH to attach, as shown in Figure 3.4. The well-known polyhydric alcohols, such as ethylene glycol, contain two OH^- groups and are named *poly*-(Table 3.2).

3.5 ORGANIC ACIDS AND ALDEHYDES

The most common organic acids contain a carboxyl group (–COOH). The simple 2, 3, and 4-C organic acids listed in Table 3.3 are formed by the anaerobic decomposition of carbohydrates. These are weak acids. They are soluble in water, but they do not dissociate completely. The larger organic acids are insoluble in water, but are very soluble in organic solvents. Lactic acid, citric acid, oxalic acid, and uric acid are also common.

An aldehyde contains a formyl group with the structure –CHO. The carbon atom shares a double bond with an oxygen atom. Aldehydes are derived from alcohols by dehydrogenation (removing a hydrogen atom). Many aldehydes have pleasant odors. The low-molecular-weight aldehydes are important for the synthesis of many industrial chemicals.

TABLE 3.3
The Low-Molecular-Weight Organic Acids and Aldehydes

Compound	Common Name	IUPAC Name
	Organic Acids	
HCOOH	Formic acid	
CH_3COOH	Acetic acid	Ethanoic acid
CH_3CH_2COOH	Propionic acid	Propanoic acid
$CH_3CH_2CH_2COOH$	Butyric acid	Butanoic acid
C_6H_5COOH	Benzoic acid	
	Aldehydes	
H_2CO	Formaldehyde	Methanal
CH_3CHO	Acetaldehyde	Ethanal
CH_3CH_2CHO	Propionaldehyde	Propanal

TABLE 3.4
The General Structures for Some Homologous Series of Organic Compounds

Name of Series	General Chemical Formula	Characteristic
Alkanes	C_nH_{2n+2}	Single carbon bonds
Alkenes	C_nH_{2n}	At least one double bond
Alkynes	C_nH_{2n-2}	At least one triple bond
Alcohols	$C_nH_{2n+1}OH$	Contains an –OH group
Aldehydes	$C_nH_{2n}O$	Contains a C=O group
Carboxylic acids	$C_nH_{2n}O_2$	Contains a C(O)OH group
Acid chlorides	$C_nH_{2n-1}OCl$	Contains Cl
Amines	$C_nH_{2n+1}NH_2$	Contains nitrogen (N)
Amides	$C_nH_{2n-1}ONH_2$	Contains nitrogen (N)
Nitriles	$C_nH_{2n-3}N$	Contains nitrogen (N)

3.6 OTHER CLASSES OF ORGANIC MOLECULES

Table 3.4 lists other homologous series of organic molecules. These are all *aliphatics,* which means they are built upon carbon chains.

3.7 WHAT HAS BEEN LEARNED SO FAR?

The following sections are about specific organic chemicals that are important in environmental protection. Before moving to that, we will review what has been learned so far.

- Organic molecules are mostly carbon and hydrogen atoms, but they may contain oxygen, sulfur, nitrogen, chlorine, and other atoms.
- Molecules can be diagramed by showing all C, H, and other atoms, or by simplified stick diagrams where the Cs are the stick nodes and the Hs are implied.
- There are *homologous series* of compounds with an increasing number of carbon atoms.
- The simplest series is the alkanes, or saturated hydrocarbons. These are carbon chains or branched carbon chains.

- *Alkenes* are unsaturated hydrocarbons, so named because they contain at least one double bond between two carbon atoms. The double bond is created by removing a hydrogen atom from each of two adjacent carbon atoms.
- Compounds with three or more carbons can exists in different structural forms, called *isomers*, having the same chemical formula and same molecular mass.
- There is a systematic naming convention for all molecules. Our few examples illustrate a few basic naming conventions. The names of all alkanes end in *-ane*, all alkenes end in *-ene*, alcohols end in *-ol*, and so on for *amines* and other homologous series.
- Removing a hydrogen atom from a molecule creates a *functional group*. A *methyl group* ($-CH_3$) is created from methane, an *ethyl group* ($-CH_2CH_3$) is created from ethane. Functional groups are reactive and can be attached to other molecules.
- Number prefixes indicate which carbons in the chain have functional groups (non-hydrogen atoms) attached. For example, 2-methylpentane has a methyl functional group attached to the second carbon in the pentane molecule. This is a six-carbon molecule, a hexane, but calling it hexane reveals nothing about the structure. In contrast, 2-methylpentane reveals a precise structure, a 5-C chain with a 1-C branch at the second carbon position.

3.8 AROMATIC HYDROCARBONS: BENZENE AND THE BTEX CHEMICALS

Aromatic organic compounds are so named because many of them have distinct aromas or fragrances. They are hydrocarbons that contain at least one benzene ring. Benzene is a toxic chemical, and there are other aromatics on the list of restricted and regulated chemicals. But many are beneficial. In fact, aromatic compounds play key roles in the biochemistry of life. All five nucleotides in DNA and RNA are aromatics, and there are four aromatic amino acids.

The *benzene* molecule has six carbon atoms joined in a ring, with one hydrogen atom attached to each carbon. The chemical formula is C_6H_6. Because it contains only hydrogen and carbon, it is a *hydrocarbon*. The right-hand end of Figure 3.5 shows the structure. Two other depictions are also shown. The one in the center is the most popular.

Benzene has an impressive history as a commercial chemical, but its use has diminished because it is a carcinogen. The annual U.S. production of benzene is almost 7 billion metric tons, and the world production is seven times this amount. Today about 80% of benzene is consumed in the production of ethylbenzene, cumene, and cyclohexane. Ethylbenzene is a precursor to styrene, which is used to make polymers and plastics. Cumene is converted to phenol for resins and adhesives. Cyclohexane is used in the manufacture of nylon. Smaller amounts of benzene are used to make some types of rubbers, lubricants, dyes, detergents, drugs, explosives, and pesticides. Benzene is a natural constituent of crude oil and an important component of gasoline. Gasoline in the U.S. cannot contain more than 1.3% benzene by volume.

The first letters of benzene, toluene, ethylbenzene, and xylenes (3 isomers) give these four chemicals the collective name of BTEX. The collective name is convenient because they are often

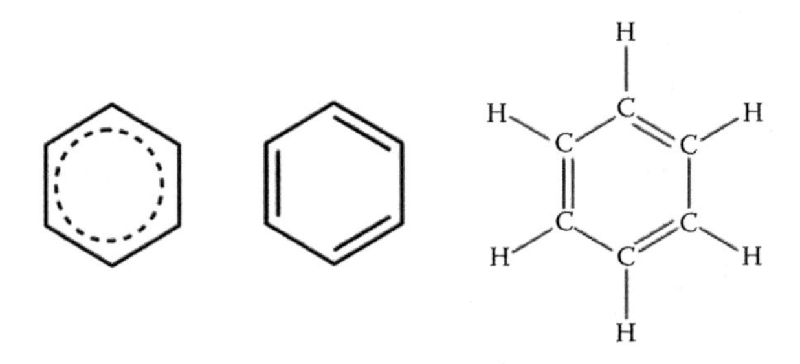

FIGURE 3.5 Three depictions of a benzene ring.

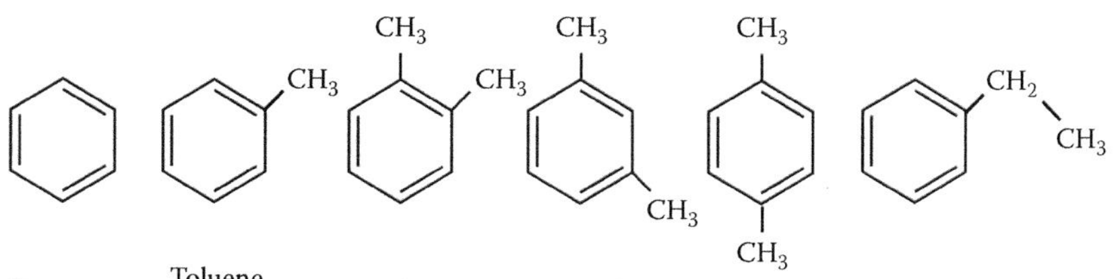

Benzene, Toluene (methyl benzene), *o*-Xylene, *m*-Xylene, *p*-Xylene, Ethylbenzene

FIGURE 3.6 The BTEX chemicals.

found together at sites with contaminated soil and groundwater. They are regulated as carcinogens. Figure 3.6 shows the six BTEX chemicals.

Table 3.5 lists some physicochemical properties. They are colorless liquids and sparingly soluble in water. They have characteristic strong odors and are highly flammable. The major source is petroleum products, such as gasoline. Most contamination problems with BTEX have been near petroleum and natural gas production sites and in association with leaking storage tanks that contained gasoline and similar products.

Toluene (C_7H_8) is benzene with an attached methyl group ($-CH_3$). Another name is *methylbenzene.* Because of symmetry it does not matter which carbon carries the attached group.

Xylene (C_8H_{10}) is a benzene ring with two $-CH_3$ groups. There are three distinct ways to attach the two groups. The prefixes *o, m,* and *p* stand for *ortho, meta, and para,* respectively, which describe the positioning. Other names are 1,2-*dimethylbenzene*, 1,3-*dimethylbenzene*, and 1,4-*dimethylbenzene.*

TABLE 3.5
Physicochemical Properties of BTEX

Property	Benzene, C_6H_6	Toluene, C_7H_8	*m*-Xylene, C_8H_{10}	*o*-Xylene, C_8H_{10}	*p*-Xylene, C_8H_{10}	Ethyl Benzene, C_8H_{10}
Molecular weight (g/mol)	78	92	106	106	106	106
Water solubility (mg/L)	1700	515	—	175	198	152
Vapor pressure (20°C) (mmHg)	95.2	28.4	—	6.6	—	9.5
Specific density (20°C)	0.8787	0.8669	0.8642	0.8802	0.8610	0.8670
Octanol-water partition coeff. (20°C) (log K_{ow})	2.13	2.69	3.20	2.77	3.15	3.15
Henry's law constant (25°C) (kPa m^3/mol)	0.55	0.67	0.70	0.50	0.71	0.80
Polarity	Nonpolar	Nonpolar	Nonpolar	Nonpolar	Nonpolar	Nonpolar
Biodegradability	Aerobic	Anaerobic and Aerobic	Aerobic	Aerobic	Aerobic	Aerobic
Maximum contaminant level (MCL) (mg/L)	0.005	1	10[a]	10[a]	10[a]	0.7

[a] MCL = 10 mg/L for the total concentration of the three species of xylene.

The name *ethylbenzene* (C_8H_9) is a precise description—an ethyl group ($-CH_2CH_3$) attached to a benzene ring.

Phenol, also known as carbolic acid, is an aromatic organic compound with the molecular formula C_6H_5OH. It is a white crystalline solid that is volatile. A phenyl group ($-C_6H_5$) is a benzene molecule with one hydrogen atom removed. The phenol molecule is a phenyl group ($-C_6H_5$) bonded to a hydroxyl group (–OH). It is mildly acidic and it has a propensity to cause chemical burns.

Phenol was first extracted from coal tar, but today is produced on a large scale (about 7 billion kg/year) from petroleum. It is an important industrial commodity as a precursor to many materials and useful compounds. Its major uses involve its conversion to plastics or related materials. Phenol and its chemical derivatives are key for building polycarbonates, epoxies, nylon, detergents, herbicides such as phenoxy herbicides, and numerous pharmaceutical drugs.

There is another group of ring compounds known as the *cycloalkanes*. These are saturated hydrocarbons; they have no double bonds. *Cyclopropane* (C_3H_6) is a three-carbon ring and *cyclobutane* is a four-carbon ring. *Cyclohexane* (C_6H_{12}) is a six-carbon ring, but it is not the same as benzene because it does not have the double bonds.

3.9 POLYCYCLIC AROMATIC HYDROCARBONS

The organic chemicals discussed in this and the next few sections tend to be to be persistent, bioaccumulative, and toxic. Examples include *polycyclic aromatic hydrocarbons* (PAHs), PCBs, *dioxins* and *furans*, and *pesticides*.

The PAHs are pure hydrocarbons. Figure 3.7 shows a few that are important. Combining two benzene rings creates *naphthalene* ($C_{10}H_8$), three rings create *phenanthrene* or *anthracene* ($C_{14}H_{10}$), and so on as shown. Their molecular masses (g/mol) are relatively low: anthracene = 178, benz[*a*] pyrene = 252, naphthalene = 128.2, phenanthrene = 178, and pyrene = 202.

The major sources of PAHs are petroleum spills, coking operations, automobile exhaust, and stack emissions from electric power utilities and incineration.

PAHs are persistent in all phases of the environment (air, water, soil, and sediments). A simple way to define persistence is by the half-life of the chemical in the environment. Half-life is the time required for the concentration to decrease by half. The concentration decreases to half the original concentration in one half-life, and to (1/2)(1/2) = 1/4 the original concentration in two half-lives.

Some estimated half-lives are given in Table 3.6. Air environment, water environment, etc. in this context are not precise terms because ambient conditions can vary widely, from freezing cold to very warm, and so on. Therefore, the values given are suggestive more than definitive.

3.10 POLYCHLORINATED BIPHENYLS

PCBs were used in electric capacitors and transformers, hydraulic and lubricating fluids, plasticizers and fireproofing agents, and carbon paper. They were banned in the U.S. in 1979, but by then 2 billion kilograms had been produced worldwide, commencing in the 1960s. Problems remain today from past discharges because of their persistence in the environment.

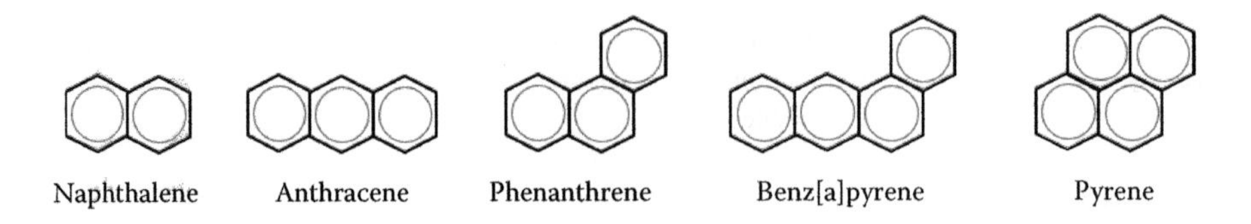

FIGURE 3.7 Example PAHs.

TABLE 3.6
The Persistence of PAHs Classified by Their Average Half-Life

Compound	Air	Water	Soil	Sediment
Indan	~1 day	~1 week	~2 months	~8 months
Naphthalene	~1 day	~1 week	~2 months	~8 months
Biphenyl	~2 days	~1 week	~3 weeks	~2 months
Acenaphthene	~2 days	~3 weeks	~8 months	~2 years
Fluorene	~2 days	~3 weeks	~8 months	~2 years
Phenanthrene	~2 days	~3 weeks	~8 months	~2 years
Anthracene	~2 days	~3 weeks	~8 months	~2 years
Pyrene	~1 week	~2 months	~2 years	~6 years
Fluoranthene	~1 week	~2 months	~2 years	~6 years
Chrysene	~1 week	~2 months	~2 years	~6 years
Benz[a]anthracene	~1 week	~2 months	~2 years	~6 years
Benzo[k]fluoranthene	~1 week	~2 months	~2 years	~6 years
Benzo[a]pyrene	~1 week	~2 months	~2 years	~6 years
Perylene	~1 week	~2 months	~2 years	~6 years
Dibenz[*a,h*]anthracene	~1 week	~2 months	~2 years	~6 years

PCBs are very slightly soluble in water, and also non-biodegradable to a large extent. Compounds of this kind tend to adsorb onto sediment and other particles, and they bioaccumulate throughout the trophic levels of the food chain. For example, the concentration of total PCB increases up the food chain from algae to plankton to planktivores to piscivorous fish to piscivorous birds (gulls and eagles).

The commercial name for PCBs in the U.S. and the U.K. was Aroclor, and the common mixtures were Aroclor 1242, Aroclor 1248, Aroclor 1254, and Aroclor 1260. The last two numbers are the mass percentage of chlorine in the molecule. Aroclor 1254 is 54% chlorine, by mass.

The *biphenyl molecule* shown in Figure 3.8 is composed of two benzene rings. Each ring has six carbon atoms. The hydrogen atom on ten of the carbon atoms can be replaced by chlorine. If there are no chlorine atoms, the molecule is simply *biphenyl*. The carbons are numbered, for example, 2, 3, 2′, and 3′, to indicate the location of attached chlorine atoms.

A *biphenyl* with one attached chlorine is a *monochlorobiphenyl*. There are only three structures that can have one chlorine, that is, with Cl at the 2, 3, or 4 carbon. Because of symmetry, adding Cl to carbons 4, 5, and 6 gives the same structure, and this is true for the carbon atoms that have primed numbers. There are 12 possible dichlorobiphenyls (2 Cl). These 12 molecules have the same molecular mass, but different chemical structures.

There are 209 possible ways to attach from 1 to 10 atoms of chlorine to the biphenyl molecule, as shown in Table 3.7. The different molecular forms are called *congeners*.

In general, bacteria cannot use chlorinated aromatic hydrocarbons as substrate, but some microorganisms are capable of using lower chlorinated PCBs as a *carbon source*. The main degradation pathway is the formation of chlorinated benzoic acids. Benzoic acid is a benzene ring with a hydrogen atom substituted for a carboxylic acid (COOH) group.

These general rules come from Fiedler et al. (1994):

- Increasing the number of chlorine substituents decreases biodegradation of PCBs.
- Ring cleavage occurs preferentially in the unsubstituted ring.
- If there are chlorine substituents on both rings, the ring with fewer chlorine atoms will be hydroxylated first.

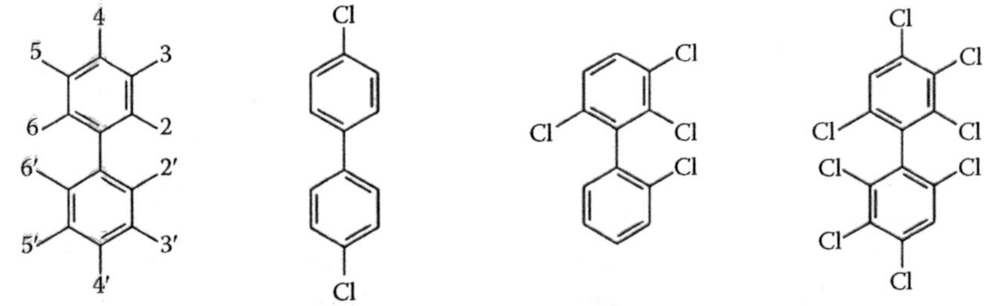

IUPAC nomenclature general formula $C_{12}H_{10-n}Cl_n$

4,4′-dichlorobiphenyl $C_{12}H_8Cl_2$

2,2′,3,6-tetrachlorobiphenyl $C_{12}H_6Cl_4$

2,2′,4,4′,5′,6,6′-octochlorobiphenyl $C_{12}H_2Cl_8$

FIGURE 3.8 PCBs: A chlorine or a hydrogen will be attached at each of the 10 number bonding sites. (Fiedler, H. et al., 1994.)

- PCB metabolism is facilitated when a carbon atom with a chlorine substituent is between two unsubstituted carbon atoms.
- Higher chlorinated congeners having a 2,3,4-trichlorophenyl group are resistant to biological degradation.

PCBs are extremely resistant to conventional aerobic transformation, but they will undergo anaerobic *reductive dechlorination.* Anaerobic degradation of PCBs in Hudson River sediment yields high levels of mono- and dichlorobiphenyls, and markedly lower levels of tri-, tetra-, and pentachlorobiphenyls. The less chlorinated PCBs then are aerobically biodegradable because they are generally less toxic than highly chlorinated PCBs. Highly chlorinated congeners are more readily dechlorinated than lower chlorinated congeners (Fiedler et al., 1994).

Commercial mixtures, as well as the individual PCB congeners, elicit a broad spectrum of biochemical and toxic responses, some of which are similar to those caused by 2,3,7,8-tetrachlorodibenzo-*p*-dioxin.

TABLE 3.7
PCB Congeners and the Commercial Aroclor Mixtures

					Percent of Congeners with Given Number of Chlorine Atoms				
PCB Homolog	**Formula**	**Molar Mass (g/mol)**	**Attached Chlorines**	**Possible Congeners**	**Arochlor 1232**	**Arochlor 1242**	**Arochlor 1248**	**Arochlor 1254**	**Arochlor 1260**
Biphenyl	$C_{12}H_{10}$	154	0	1					
Monochlorobiphenyl	$C_{12}H_9Cl$	188.65	1	3	26	3			
Dichlorobiphenyl	$C_{12}H_8Cl_2$	233.1	2	12	29	13	2		
Trichlorobiphenyl	$C_{12}H_7Cl_3$	257.54	3	24	24	28	18		
Tetrachlorobiphenyl	$C_{12}H_6Cl_4$	291.99	4	42	14	30	40	49	
Pentachlorobiphenyl	$C_{12}H_5Cl_5$	326.43	5	46		22	36	34	12
Hexachlorobiphenyl	$C_{12}H_4Cl_6$	360.88	6	42		4	4	34	38
Heptachlorobiphenyl	$C_{12}H_3Cl_7$	395.32	7	24				6	41
Octachlorobiphenyl	$C_{12}H_2Cl_8$	429.77	8	12					8
Nonachlorobiphenyl	$C_{12}H_1Cl_9$	464.21	9	3					1
Decachlorobiphenyl	$C_{12}Cl_{10}$	498.21	10	1					

Due to the fact that dioxin-like compounds normally exist in environmental and biological samples as complex mixtures of congeners, the concept of toxic equivalents (TEQs) has been developed to simplify risk assessment and regulatory control (U.S. EPA, 2008, 2010; Birnbaum, 1994).

3.11 DIOXINS AND FURANS

Dioxins and *furans* are chemically related compounds that are *persistent organic pollutants (POPs)*. They are found in the environment throughout the world, and they accumulate in the food chain, mainly in the fatty tissue of animals. The higher an animal is in the food chain, the higher the concentration of dioxins tends to be.

More than 90% of human exposure is through food, mainly meat and dairy products, fish, and shellfish. Once dioxins enter the body, they stay a long time because of their chemical stability and their ability to be absorbed by and stored in fatty tissue. Their half-life in the body is estimated to be 7–11 years.

Dioxins are highly toxic and can cause reproductive and developmental problems, damage the immune system, interfere with hormones, and cause cancer. The typical background exposure is not expected to affect human health, but, due to the highly toxic potential, efforts have been made to reduce current background exposure by strict control of industrial processes to reduce formation of dioxins. The major sources are incineration and paper mill pulp waste.

The chemical name for dioxin is 2,3,7,8-*tetrachlorodibenzo-para-dioxin* (TCDD). The name "dioxins" is often used for the family of structurally and chemically related *polychlorinated dibenzo-para-dioxins* (PCDDs) and *polychlorinated dibenzofurans* (PCDFs). Certain dioxin-like PCBs with similar toxic properties are also included under the term *dioxins*. Some 419 types of dioxin-related compounds have been identified, but only about 30 of these are considered to have significant toxicity, with 2,3,7,8-TCDD being the most toxic. These are complex chlorinated ring structures. Four examples are shown in Figure 3.9.

3.12 PESTICIDES

Pesticides are strictly regulated chemicals because of their toxicity. Many have been banned (the most famous of these is DDT) and many are *severely restricted*, which means that virtually all registered uses have been prohibited. A few are shown in Figure 3.10.

Aldicarb was one of the "dirty dozen" pesticides that the environmental group Pesticide Action Network North America targeted in 1985. The U.S. EPA banned Aldicarb in 2010 and required an end to distribution by 2017. Use on citrus and potatoes was banned beginning in 2012, with a complete ban in place by 2018.

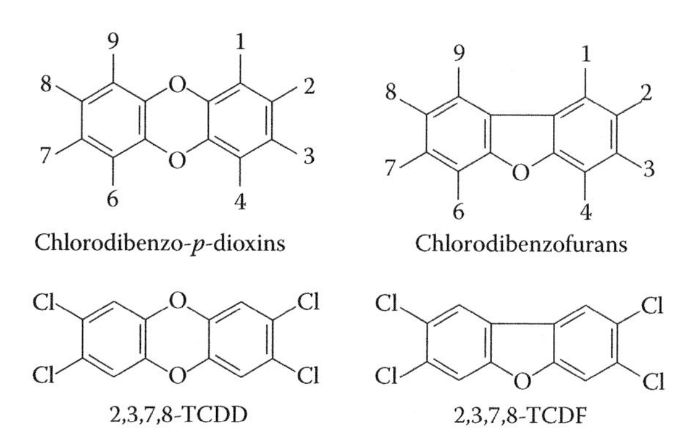

FIGURE 3.9 Four examples of dioxins and furans.

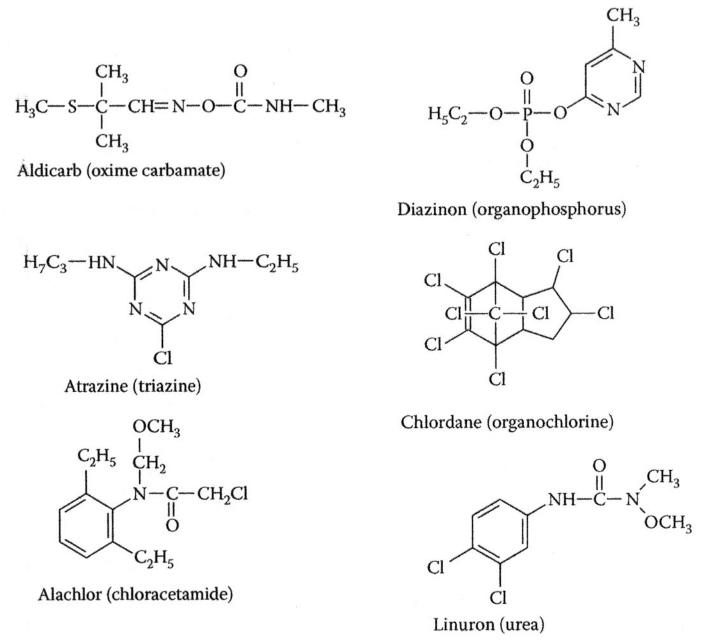

Aldicarb (oxime carbamate)

Diazinon (organophosphorus)

Atrazine (triazine)

Chlordane (organochlorine)

Alachlor (chloracetamide)

Linuron (urea)

FIGURE 3.10 Example pesticides.

Residential use of *Diazinon* was outlawed in the U.S. in 2004, but it is still approved for agricultural uses.

Atrazine was banned in the European Union in 2004 because of persistent groundwater contamination. It is one of the most widely used herbicides in U.S. and Australian agriculture. As of 2001, atrazine was the most commonly identified pesticide in drinking water in the United States. Studies suggest it is an endocrine disruptor, an agent that may alter the natural hormonal system in animals. In 2006, the EPA stated that "the risks associated with the pesticide residues pose a reasonable certainty of no harm," and in 2007 the EPA said that atrazine does not adversely affect amphibian sexual development and that no additional testing was warranted. The EPA opened a new review in 2009 that concluded that "the agency's scientific bases for its regulation of atrazine are robust and ensure prevention of exposure levels that could lead to reproductive effects in humans." However, the EPA review has been criticized, and the safety of atrazine remains controversial (Wikipedia, 2014).

Chlordane, one of the "dirty dozen," was banned by the 2001 Stockholm Convention on Persistent Organic Pollutants. It is a recognized carcinogen and is on at least nine U.S. federal regulatory lists. It is ranked in the worst 10% of chemicals in 10 out of 11 ranking systems. Chlordane adheres to soil particles and enters groundwater only slowly due to its low solubility (0.009 ppm). It degrades only over the course of years. Chlordane bioaccumulates in animals and is highly toxic to fish.

The use of *Alachlor* as an herbicide has been banned in the European Union. The U.S. EPA has set the Maximum Contaminant Level Goal (MCLG) for Alachlor at zero to prevent long-term health effects. The Maximum Contaminant Level (MCL) for drinking water is two parts per billion (2 ppb). Lifetime exposure to levels above the MCL may include cancer and potential damage to the liver, kidney, spleen, lining of nose and eyelids.

Linuron is an example of a urea-based herbicide that is not restricted. It is approved for the control of a wide range of weeds in the cultivation of asparagus, carrots, celery, corn, potatoes, and other vegetables.

TABLE 3.8
Four Solvents and Their Relation to Environmental Problems

Chlorinated Solvent	Formula	Exposure Limit (ppm)	Animal Carcinogen	GWP	ODP	Toxic Air Pollutant
Trichloroethylene (TCE)	CCl_2CHCl	50	Suspect	Regulated	No	Yes
Tetrachloroethylene (Perchoroethylene (PERC)	CCl_2CCl_2	25	Suspect	Regulated	No	Yes
Dichloromethane (methylene chloride)	$CHCl_2$	500	Suspect	Exempt	No	Yes
1,1,1-Trichloroethane (TCA) (methyl chloroform)	CCl_3CH_3	350	—	0.2	0.1	Yes

Notes: GWP = global warming potential—the potential of 1 kg of the chemical to cause global warming relative to the potential of 1 kg of CFC-11, which has a defined GWP = 1.0. Flammable and combustible solvents do not have a GWP because they do not contain halogens. ODP = ozone depletion potential—the potential for ozone depletion of 1 kg of a chemical relative to the potential of 1 kg of CFC-11, which has a defined ODP = 1.0. The higher the chlorine content, the higher the ODP.

3.13 SOLVENTS AND VOLATILE ORGANIC CHEMICALS

Solvents have many important uses in manufacturing, but they are troublesome and toxic. Table 3.8 lists some solvents that have been widely used in the past despite their toxicity and harmful characteristics. Also listed are possible substitutes that are less harmful. For example, benzene might be replaced with toluene, which is much less toxic (though not nontoxic).

3.14 CONCLUSION

A short discussion of toxic organic chemicals in drinking water was presented as a means of introducing some specific names of pollutants.

Several thousand toxic organic chemicals are listed in environmental regulations. The names of these chemicals can be complex and intimidating. Learning the basic schemes for naming organic chemicals will reduce the intimidation factor, and help one recognize the general chemical families to which common compounds belong.

The tutorial section on naming compounds deals with the alkanes (saturated hydrocarbons), the alkenes (unsaturated hydrocarbons), and the aromatic ring compounds in the benzene family. Polyaromatic hydrocarbons, PCBs, pesticides, and solvents are discussed briefly because they are related to many ongoing environmental problems. The PAHs and PCBs are persistent, bioaccumulative, and toxic. The same is true for many pesticides.

4 Measuring Pollutants

4.1 THE DESIGN PROBLEM

The design of pollution control systems requires knowledge of the mass and mass flow rate of all materials that move through the system. Most pollutants are measured as concentrations, and there are a variety of ways that these quantities can be expressed, depending on whether the material is a gas, liquid, or solid.

4.2 THE FUNDAMENTAL UNITS OF MEASUREMENT

The basic SI units that are used to measure mass, length, volume, and time are the kilogram (kg), meter (m), cubic meter (m^3) or liter (L), and second (s). The U.S. conventional units are pounds (lb), feet (ft), gallons (gal), and seconds (s). These are listed in Table 4.1, along with some commonly used multiples and a few factors to convert meters to feet, kilograms to pounds, etc. Almost all the calculations in this book will be done with SI units, but a few exceptions use U.S. units.

4.3 MASS CONCENTRATION: PARTS PER MILLION AND MG/L

Concentrations can be expressed as follows:

For liquids	Mass/volume	Usually kg/m^3, mg/L	
	Mass/mass	ppm = parts per million (mg/kg)	
		ppb = parts per billion (μg/kg)	
For particulates in air and gases			
	Mass/volume	mg/m^3 or $\mu g/m^3$	
For mixtures of gases			
	Mass per volume	mg/m^3 or $\mu g/m^3$	
	Volume per volume	ppmv	
For sludge	Mass/wet mass	kg/T, mg/kg sludge	(T = tonne = 1000 kg)
For solids	Mass/dry mass	kg/T, mg/kg dry solids	

Concentrations of pollutants in natural water and most wastewaters are very low and the convenient units are mg/L or μg/L. kg/m^3 is convenient for certain calculations. The conversion is 1 mg/L = 0.001 kg/m^3, from

$$(1\ \text{mg/L})(1000\ \text{L/m}^3)/(1{,}000{,}000\ \text{mg/kg}) = 0.001\ \text{kg/m}^3$$

Occasionally a brine or industrial wastewater will be reported in g/L, which is equivalent to kg/m^3.

When elements are present in minute or trace quantities, it is more convenient to use parts per billion (ppb) or micrograms per liter (μg/L); a microgram is one-millionth of a gram. For example, in studies of steam purity using a specific ion electrode to measure sodium content, values as low as 0.001 ppm are not uncommon. This is more conveniently reported as 1.0 ppb.

Parts per million (ppm) is frequently used to mean mg/L. To be precise, 1 ppm means 1 g in 1,000,000 g, 1 kg in 1,000,000 kg, or 1 lb in 1,000,000 lb. It makes no difference what units are used, as long as both masses are expressed in the same units.

TABLE 4.1
Some Basic Units and Conversion Factors

Quantity	SI Units	SI Symbol	USCS Units	USCS Symbol	Conversion Factor (SI to USCS)
Length	meter	m	foot	(ft)	3.281
Mass	megagram (1000 kg)	tonne (T)	ton	ton (2000 lb)	0.907
	kilogram	kg	pound	lb	2.205
	gram	g			
	milligram	mg			
	microgram	μg			
Flow rate	cubic meter per sec	m^3/s	cubic feet per second	ft^3/s	35.31
	liter per sec	L/s			
Area	square meter	m^2	square feet	ft^2	10.76
Volume	cubic meter	m^3	cubic feet	ft^3	35.31
	liter	L			
Velocity	meter/sec	m/s	feet per second	ft/s	2.237
Density	kilogram/cubic meter	kg/m^3	pound/cubic foot	lb/ft^3	0.06243
	gram/cubic centimeter	g/cm^3			

The 1,000,000 g reference mass can be grams of water, grams of wet sludge, grams of a moist solid, or grams of a dry solid material like soil or refuse or sludge. Unless it is water, the basis should be specified.

The ppm mass ratio is closely related to milligrams per liter (mg/L). For a concentration of 1000 ppm and a solution density of 1.5 kg/L,

$$(\text{ppm})(\text{Solution density}) = (1000\ \text{mg}/1{,}000{,}000\ \text{mg})(1{,}500{,}000\ \text{mg/L}) = 1{,}500\ \text{mg/L}$$

If the solution density is close or equal to 1.00, then 1 L has a mass of 1,000,000 mg, which, for a concentration of 1000 ppm, gives

$$(1000\ \text{mg}/1{,}000{,}000\ \text{mg})(1{,}000{,}000\ \text{mg/L}) = 1000\ \text{mg/L}$$

Water quality analyses are usually conducted without a measurement of a solution's density. For common water samples, this poses no great inaccuracy because the density of the sample is approximately 1.00 for natural water, municipal water, and most industrial water systems. Municipal sewage is 99.99% water. Thus, for most water and wastewater samples, mg/L and ppm are interchangeable. Likewise, ppb and μg/L are interchangeable. This should not be assumed for highly saline water, industrial wastewater, sludge, soil, or sediments.

Example 4.1 Some Useful Concentration Conversions

Convert mg/L to kg/m^3.

$$1\ \text{mg/L} = (1\ \text{mg/L})(1000\ \text{L/m}^3)(\text{kg}/10^6\ \text{mg}) = 0.001\ \text{kg/m}^3$$

$$1\ \text{g/L} = 1\ \text{kg/m}^3$$

For a solution specific gravity of 1.000 (1,000,000 mg/L):

$$1 \text{ mg/L} = 1 \text{ ppm}$$

$$1 \text{ µg/L} = 1 \text{ ppb (part per billion)}$$

$$1\% = (1 \text{ g/100 g})(1{,}000{,}000 \text{ mg/L}) = 10{,}000 \text{ mg/L}$$

Example 4.2 Mercury in Water

A wastewater effluent of 1000 L/h contains an average of 6 µg/L mercury (Hg). Calculate the mass flow of mercury discharged per day (g/day).

$$\text{Volume of water} = (1000 \text{ L/h})(24 \text{ h/day}) = 24{,}000 \text{ L/day}$$

$$\text{Each liter contains } 6 \text{ µg} = 0.006 \text{ mg} = 0.000{,}006 \text{ g Hg}$$

$$\text{Mass of Hg discharged} = (24{,}000 \text{ L/day})(6 \text{ µg Hg/L})(1 \text{ g}/1{,}000{,}000 \text{ µg}) = 0.144 \text{ g Hg/day}$$

Example 4.3 Suspended Solids Concentration

Wastewater entering a treatment process has a suspended solids concentration of 275 mg/L. Calculate the concentration in (a) kg/m³ and (b) pounds per gallon.

a. (275 mg/L)(1000 L/m³)/(1,000,000 mg/kg) = 0.275 kg/m³
b. 1 kg = 1,000,000 mg = 2.205 lb

$$1 \text{ L} = 0.2642 \text{ gal}$$

$$\left(\frac{275 \text{ mg}}{\text{L}}\right)\left(\frac{1 \text{ L}}{0.2642 \text{ gal}}\right)\left(\frac{2.205 \text{ lb}}{1{,}000{,}000 \text{ mg}}\right) = 0.002295 \text{ lb/gal}$$

A more convenient unit would be 2295 lb/million gallons

Example 4.4 Sludge Volume and Mass

An industry is holding 600 m³ of dense industrial sludge that has specific gravity 1.4. Calculate the sludge mass.

$$\text{Mass of } 1 \text{ m}^3 \text{ of water} = 1000 \text{ kg}$$

$$\text{Mass of } 1 \text{ m}^3 \text{ of sludge} = 1.4(1000 \text{ kg}) = 1400 \text{ kg}$$

$$\text{Sludge mass} = (600 \text{ m}^3)(1400 \text{ kg/m}^3) = 840{,}000 \text{ kg} = 840 \text{ metric tons}$$

Example 4.5 Cadmium in Sludge

A volume of 150 m³/year of liquid sludge that is 10% solids by mass is incorporated into the soil of a cornfield. The cadmium (Cd) concentration of the sludge is 8 ppm, based on the dry sludge solids. The density of the liquid sludge is 1.20 kg/m³. Calculate the mass of Cd that is applied to the field in one year of sludge application.

Basis = 1 year of sludge application

8 ppm = (8 mg Cd)/(1,000,000 mg dry sludge solids) = 8 mg Cd/kg dry solids

10% solids = (10 kg dry solids/100 kg wet sludge) = 0.1 kg dry solids/kg wet sludge

Dry solids in the sludge = (150 m^3)(1200 kg/m^3)(0.10 kg/kg) = 18,000 kg dry solids

Cd in the dry solids = (18,000 kg solids)(8 mg Cd/kg solids) = 144,000 mg Cd = 0.144 kg Cd

4.4 MASS PERCENTAGE AND MASS FRACTION

Mass/mass is the mass fraction. The mass fraction multiplied by 100% is the mass percent (often called weight percent).

Concentrations of pollutants in dense slurries, sludge, sediments, soil, and other solids are given as mass ratios. It is important to make clear whether the mass of bulk material is on a dry or wet basis. The best method is a dry mass basis. A concentration of 1 mg/kg means 1 milligram of pollutant in 1 kilogram of dry material; 1 μg/kg means 1 microgram of pollutant in 1 kilogram of dry material. Also, a concentration of 1 mg/kg means 1 ppm and 1 μg/kg means 1 ppb, on a dry basis.

Concentration as a weight percent can be used for solids or liquids. To say that sludge is "4% solids by weight" means that 4% of the total sludge mass is dry solids. The total sludge mass includes the water and the solids. Thus, 4% solids by weight also means 96% water by weight. And it means 0.04 kg dry solids per kg of wet sludge.

Example 4.6 Mass Percentage of Sludge Solids

An industry is holding 600 m^3 of dense industrial sludge that has specific gravity 1.4. The sludge is 12% solids on a wet mass basis. Calculate the mass of solids in the 600 m^3 of sludge.

Density of sludge = 1.4(1000 kg/m^3) = 1400 kg/m^3

Mass of wet sludge = (600 m^3)(1400 kg/m^3) = 840,000 kg = 840 T

Mass of dry solids in the sludge = (0.12)(840 T) = 100.8 T

Example 4.7 Waste Gas Composition

A gaseous mixture, with total mass of 1000 kg, is 5% benzene, 71% nitrogen, and 24% oxygen. These are mass fractions. The gas volume, at 25°C and 760 mm Hg, is 752,915 L = 752.9 m^3. Calculate the mass of each gas in the mixture and the concentration as mass per volume (g/m^3).

The mass of each gas can be calculated from its mass fraction.

Gas component = (mass fraction)(total gas mass)

Benzene = 0.05(1,000,000 g) = 50,000 g

Nitrogen = 0.71(1,000,000 g) = 710,000 g

Oxygen = 0.24(1,000,000 g) = 240,000 g

Concentrations = (mass of gas)/(total volume of gas)

Benzene = (50,000 g)/(752.9 m^3) = 66.4 g/m^3

Nitrogen = (710,000 g)/(752.9 m^3) = 943 g/m^3

Oxygen = (240,000 g)/(752.9 m^3) = 319 g/m^3

Many problems involve some substance in water, with a specification like, "The mixture is 10% S, by mass." or "Ten percent of the total mass is substance S." or "The mass fraction of S is 0.1."

Define $M = S + W =$ total mass:

$$S = \text{mass of substance } (solids,\ chloride,\ etc.)$$
$$W = \text{ mass of water}$$
$$p_S = \text{ \% substance}$$
$$p_W = \text{\% water}$$

$$p_S + p_W = 100\%$$

$$p_S = \frac{100\,S}{S+W} \quad \Rightarrow \quad W = \frac{S(100 - p_S)}{p_S}$$

$$p_W = \frac{100\,W}{S+W} \quad \Rightarrow \quad S = \frac{W(100 - p_W)}{p_W}$$

In terms of mass fractions,

$$f_S = \text{mass fraction of } S$$
$$f_w = \text{mass fraction of } W$$
$$f_w + f_s = 1$$

$$f_S = \frac{S}{S+W} \quad \Rightarrow \quad W = \frac{S(1 - f_S)}{f_S}$$

$$f_W = \frac{W}{S+W} \quad \Rightarrow \quad S = \frac{W(1 - f_W)}{f_W}$$

Example 4.8 Sludge Thickening

In Figure 4.1, the mass fraction of solids in the 1000 kg/h feed to a sludge thickener is 0.03. The thickened sludge underflow has a solids fraction of 0.06. The thickener supernatant carries 1 kg/h of solids.

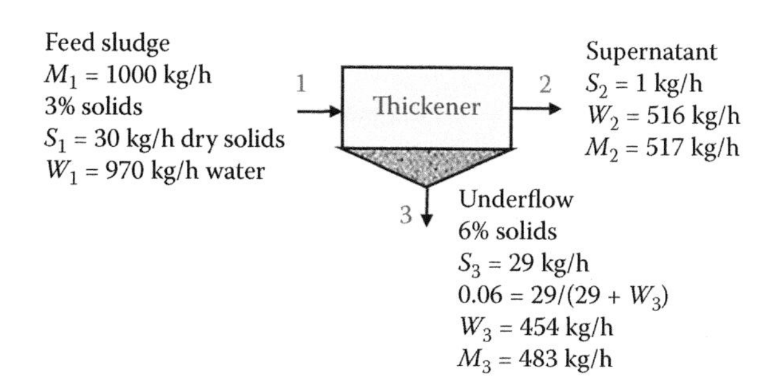

FIGURE 4.1 Sludge thickening.

Using the numbers on the inputs and outputs, we can define

$$M_1 = 1000 \text{ kg/h}$$

$$S_1 = 0.03(1000 \text{ kg/h}) = 30 \text{ kg/h}$$

$$W_1 = (1.0 - 0.03)(1000 \text{ kg/h}) = 970 \text{ kg/h}$$

Also $W_1 = M_1 - S_1 = 970$ kg/h.

Material balance on solids

$$S_3 = S_1 - S_2 = 30 - 1 = 29 \text{ kg/h}$$

Using the definition of mass fraction

$$f_3 = S_3/(S_3 + W_3) = 0.06$$

$$S_3 = 0.06S_3 + 0.06W_3$$

$$W_3 = S_3(1.0 - 0.06)/0.06 = (29 \text{ kg/h})(0.94)/0.06 = 454.3 \text{ kg/h}$$

$$M_3 = S_3 + W_3 = 29 + 454.3 = 483.3 \text{ kg/h}$$

4.5 MASS FLOW RATES

The mass flow rate (i.e., kg/h) of a specific substance is calculated using the concentration of the material and the volumetric flow rate, Q, of the bulk material in which the substance is found. Examples are given in Table 4.2.

Bulk density (ρ) refers to the material as it is delivered or as it flows. For liquids or gases, the units typically are g/L ρ or kg/m^3 (lb/gal or lb/ft^3). The bulk density of solids may be given for a moist or wet material, or it may be given in terms of the dry solids.

The calculation is

$$\text{Mass flow rate of bulk material (kg/h)} = (\rho \text{ kg/m}^3)(Q \text{ m}^3\text{/h})$$

$$\text{Mass flow rate of species } X\text{(kg/h)} = (\text{kg } X\text{/kg})(\rho \text{ kg/m}^3)(Q \text{ m}^3\text{/h})$$

The bulk density of dry air at 0°C is $\rho = 1.293$ kg/m^3. At 70°C, $\rho = 1.029$ kg/m^3, approximately 20% less. This is because air expands when the temperature increases. The calculation is given in the next section.

The bulk density of a 1% solution (by mass) of sodium chloride (NaCl) is 1.0095 kg/m^3. An 8% solution has a bulk density of 1.089 kg/m^3.

TABLE 4.2
Units Commonly Used to Measure Composition and Flow Rate

Mass Units	Concentration (C)	Volume Flow Rate (Q)	Mass Flow Rate (M)
mg	mg/L	L/s	mg/s
kg	kg/m^3	m^3/s	kg/s
kg	kg/m^3	m^3/day	kg/day
lb	lb/gal	gal/min	lb/min

A 10% (by mass) solution of ethanol at 20°C has ρ = 980 kg/m^3. A 50% solution has ρ = 913.8 kg/m^3.

The bulk density of dry loose earth is ρ = 1100 kg/m^3. Dry packed earth has ρ = 1595 kg/m^3 and moist packed earth has ρ = 1600 kg/m^3. This is why one must be clear about whether measurements are based on dry or wet material.

There is a short cut for calculating mass flow rate in U.S. units. When the concentration is measured in mg/L or ppm and the flow rate is measured in million gallons per day (mgd)

$$\text{Mass flow rate (lb/day)} = (8.34\ \text{lb/gal})(\text{Volumetric flow rate, mgd})(\text{Concentration, ppm})$$

The 8.34 lb/gal of water converts the volumetric flow into a mass flow. If the density of the liquid is the same as water, or very nearly so, then 1 ppm = 1 mg/L and lb/day = 8.34(mgd)(mg/L).

Example 4.9 Suspended Solids

Wastewater enters a treatment process at a flow = 4 m^3/s and suspended solids concentration = 275 mg/L. Calculate the mass flow of suspended solids, in kg/day.

$$(275\ \text{mg/L})(1000\ \text{L/m}^3)/(1{,}000{,}000\ \text{mg/kg}) = 0.275\ \text{kg/m}^3$$

$$\text{Mass (kg/day)} = (4\ \text{m}^3\text{/s})(0.275\ \text{kg/m}^3)(86{,}400\ \text{s/day}) = 95{,}040\ \text{kg/day}$$

Example 4.10 Electrostatic Precipitator

Before the installation of an electrostatic precipitator, the stack gas of a power-generating station had a particulate solids concentration of 6 g/m^3. The gas flow rate was 50 m^3/s. The new precipitator removes 24,000 kg/day of particulates. (a) What is the emission rate of particulates, in kg/day, before and after initiating pollution control? (b) Does the new system meet an emission standard of 0.7 g/m^3?

a. Particulate emissions before the electrostatic precipitator was installed
 = (6 g/m^3)(50 m^3/s)(86,400 s/day)(10^{-3} kg/g) = 25,920 kg/day

 Particulate emissions after the electrostatic precipitator was installed
 = 25,920 − 24,000 kg/day = 1920 kg/day

b. After emission control the particulate emissions are
 = (1920 kg/day)/[(50 m^3/s)(86,400 s/day)(10^{-3} kg/g)] = 0.44 g/m^3

 The pollution control system does meet the new standard.

4.6 VOLUME FRACTION AND VOLUME CONCENTRATION

Volume concentrations and volume fraction are used for mixtures of gases. Volume concentrations are the same as volume fractions.

$$\text{Volume percent} = 100(\text{volume/volume}) = 100(\text{volume fraction})$$

Common units are

m^3/m^3
ppmv = parts per million ($L/10^6$ L)
ppbv = parts per billion ($L/10^9$ L)

The volume fraction of a mixture of gases equals the mole fraction. Table 4.3 defines the relation between volume fraction and mole fraction for air and gases.

Concentrations are often reported at standard temperature and pressure (STP). STP is usually 0°C and 1 atmosphere of pressure, but the definition is not uniform for all agencies and technical groups (see Table 4.4).

Air and other gases expand or contract with changes in temperature and pressure. The volume/volume concentration (volume fraction), however, of an individual gas in a mixture does not change as the total gas is compressed or expanded.

A mass/volume concentration does change with temperature and pressure. The volume changes, but the mass contained in that volume is constant.

Design calculations are made at the actual conditions of T °C and P atm. T and P must be stated for the actual volume to have any useful meaning. $(M/V)_{STP}$ and $(M/V)_{TP}$ are related by the Ideal Gas Law:

TABLE 4.3
Volume Fraction and Mole Fraction Definitions for Air and Gases

Units	Concentration (C)	Volume Flow Rate (Q)	Mass Flow Rate (M)
Mass fraction of X	mg X/kg solid mixture	m^3 mixture/h	mg X/h
Volume fraction of Y	m^3 Y/m^3 gas mixture	m^3 mixture/min	mg Y/min
Mole fraction of Z	Moles Z/total moles mixture	m^3 mixture/min	mg Z/min

TABLE 4.4
Standard Conditions for Various Disciplines (Dry Air Only)

Discipline	Temperature	Absolute Temp.	Pressure	Organization
Chemistry/physics (STP)	0°C	273.15 K	100.000 kPa, 0.987 atm	IUPAC
Chemistry/physics (STP)	20°C	293.15 K	101.325 kPa, 1 atm	NIST
Air pollution (new sources)	20°C	293.15 K	101.325 kPa, 1 atm	U.S. EPA
Air pollution (ambient air)	25°C	298.15 K	101.325 kPa, 1 atm	Air pollution
Industrial ventilation	21.1°C (70°F)	529.67°R	29.921 in Hg	
Industrial hygiene	60°F	519.67°R	14.696 psi	U.S. OSHA
ASHRAE[a]	15°C (59.0°F)	288.15 K	101.325 kPa, 1atm	

Source: Wikipedia.

Note: STP indicates standard temperature and pressure for basic science. NTP indicates normal temperature and pressure conditions used in the U.S. for industrial hygiene and air pollution.

[a] ASHRAE, American Society of Heating, Refrigerating, and Air-Conditioning Engineers.

$$PV = nRT$$

where
P = pressure, atm
T = absolute temperature, K
V = gas volume, L
n = number of moles of gas
R = universal gas constant = 0.08205 L atm/mol K

Example 4.11 Venting Toluene

A ventilation airflow of 40,000 m^3/h (at STP) from a printing company contains 2200 ppmv toluene. (a) What is the flow of toluene in m^3/h? (b) The density of toluene is 4.12 kg/m^3. What is the mass flow of toluene in kg/h?

a. Volume fraction of toluene = 2200 ppmv = 22,000 m^3/1,000,000 m^3

Volumetric flow of toluene = (40,000 m^3/h)(2200 m^3/1,000,000 m^3) = 88 m^3/h.

b. Density of toluene = 4.12 kg/m^3(at STP).

Mass flow of toluene = (88 m^3/h)(4.12 kg/m^3) = 362.6 kg/h

Example 4.12 Toluene Emissions

A printing company is limited to emitting 3 kg/h of toluene. The airflow that carries the toluene is 20,000 m^3/h at STP. The density of toluene is 4.12 kg/m^3. What is the allowable concentration (ppmv) of toluene in the emitted air?

Define C_T = allowable concentration of toluene.

Allowed volume of toluene = (3 kg/h)/(4.12 kg/m^3) = 0.728 m^3/h

Actual volume of toluene = (20,000 m^3/h)(C_T m^3/1,000,000 m^3) = 0.728 m^3/h.

C_T = 36.4 ppmv

Example 4.13 Sulfur Dioxide

What is the concentration of sulfur dioxide (SO_2), expressed as ppmv, in combustion gas that is 75% N_2, 7% O_2, 9.85% CO_2, 0.15% SO_2, and 8% H_2O?

Since gas concentrations are provided as volume percent, use the following equation:

$$\text{ppmv for compound} = \frac{\text{Volume \% for compound}}{100\%} \times 10^6$$

$$\frac{0.15\%}{100\%} \times 10^6 = 1500 \text{ ppmv } SO_2$$

4.7 CONVERTING VOLUME AND MASS CONCENTRATIONS IN GASES

Volume and mass concentrations (ppmv and mg/m^3) can be converted using the molar mass (MM) of the pollutant and the Ideal Gas Law. One g-mol of an ideal gas occupies a volume 0.02241 m^3

(22.41 L) at STP (0°C = 273 K and 1 atm). Also, 1 m^3 of an ideal gas contains 1/0.02241 m^3 = 44.623 g-mol of the gas. One lb-mol occupies a volume of 359 ft^3 at STP.

The mass (mg) of a gas occupying 1 m^3 is

$$\frac{\text{Mass of gas (mg)}}{\text{Volume of gas (m}^3\text{)}} = \frac{\text{MM (g/mol)}}{(0.02241)(\text{m}^3\text{/mol})} \times \frac{1000\text{ mg}}{\text{g}} = \frac{\text{MM}(1000)}{0.02241}$$

If a gas mixture contains a pollutant at a concentration of 1 ppmv, or 1 m^3 of pollutant in 1,000,000 m^3 of mixture, the mass concentration will be given by

$$\frac{\text{mg pollutant}}{\text{m}^3\text{ mixture}} = \left(\frac{\text{m}^3\text{ pollutant}}{10^6\text{ m}^3\text{ mixture}}\right)\left(\frac{\text{MM g/mol}}{0.02241\text{ m}^3\text{/mol}}\right)\left(\frac{1000\text{ mg}}{1\text{ g}}\right)$$

$$\frac{\text{mg pollutant}}{\text{m}^3\text{ mixture}} = \text{ppmv}\left(\frac{\text{MM}\times 10^3}{0.02241}\right)\left(\frac{1}{10^6}\right) = \text{ppmv}\left(\frac{\text{MM}}{22.41}\right)$$

Example 4.14 Converting from Volume Concentration to Mass Concentration

The concentration of a gaseous pollutant in air is 50 ppmv. The molar mass of the pollutant is 16 g/mol. Find the concentration as mg/m^3 at standard conditions.

$$\frac{\text{(ppmv)(MM)}}{22.41\text{ L/mol}} = \frac{(50\text{ ppmv})(16\text{ g/mol})}{22.41\text{ L/mol}} = 35.7\text{ mg/m}^3$$

Example 4.15 Mass Concentration in Gases

A gaseous emission has an SO_2 concentration of 25 ppmv. The gas temperature and pressure are 25°C and 1.1 atm. The molar mass of SO_2 is 64 g/mol.

Mass concentration of SO_2 at STP of 1 atm and 0°C.

$$\left(\frac{\text{mg}}{\text{m}^3}\right)_{STP} = \frac{\text{(ppmv)(MM)}}{22.41\text{ L/mol}} = \frac{(25\text{ ppmv})(64\text{ g/mol})}{22.41\text{ L/mol}} = 71.4\text{ mg/m}^3$$

Mass concentration at $T = 25°C$ and $P = 1.1$ atm is

$$\left(\frac{\text{mg}}{\text{m}^3}\right)_{TP} = \left(71.4\frac{\text{mg}}{\text{m}^3}\right)\left(\frac{273\text{ K}}{(273+25)\text{ K}}\right)\left(\frac{1.1\text{ atm}}{1\text{ atm}}\right) = 71.9\text{ mg/m}^3$$

4.8 MOLAR MASS AND MOLAR CONCENTRATION

Mass can also be measured in molar units, such as g-mol, kg-mol, and lb-mol. For example, the MM of calcium (Ca) is 40 g/g-mol, MM of carbon (C) = 12 g/g-mol, and the MM of oxygen (O_2) = 32 g/g-mol. Use these molar masses.

Lime	CaO	40 g/g-mol + 16 g/g-mol = 56 g/g-mol
Carbon dioxide	CO_2	12 + 2(16) = 44 g/g-mol
Calcium carbonate	$CaCO_3$	40 + 12 + 3(16) = 100 g/g-mol

TABLE 4.5
Molar Concentration and Molar Flow Rates

Molar Units	Molar Concentration	Volume Flow Rate	Molar Flow Rate
g-mol	g-mol/L	L/s	g-mol/s
kg-mol	kg-mol/m^3	m^3/day	kg-mol/day
lb-mol	lb-mol/gal	gal/min	lb-mol/min

If masses are expressed as kilograms (kg), molar masses are Ca = 40 kg/kg-mol, C = 12 kg/kg-mol, and O = 32 kg/kg-mol. The molar masses of the compounds are also kg/kg-mol.

Solubility and equilibrium calculations use molar concentrations (mol/L). Often mol/L will be converted to mg/L (or kg/m^3). Mass and molar units are converted using the molar mass of an element or compound. For example, the molar mass of calcium is 40 g/g-mol. A concentration of 1 g-mol per liter (g-mol/L) is equivalent to 40 g Ca/L. A solution with a mass concentration of 0.1 g/L (100 mg/L) has a molar concentration of

$$(0.1 \text{ g/L})/(40 \text{ g/g-mol}) = 0.0025 \text{ mol/L}$$

The mass percentage (mass %) of an element in the compound is the fraction of the compound's mass contributed by that element, expressed as a percentage

$$\text{Mass \% of element } X = \frac{(\text{Atoms of } X \text{ in compound})(\text{molar mass of } X)}{\text{Molar mass of compound}} \times 100$$

Mass flow rates can be expressed in molar units, as shown in Table 4.5. Molar units are most convenient in chemical processes where the chemical species are known and the concentrations are high. A concentration of 1 g-mol/L of calcium carbonate ($CaCO_3$) is 100 g/L. In natural water the concentration of $CaCO_3$ is in the range of 100–400 mg/L, or 0.1–0.4 g-mol/L.

Example 4.16 Freon-12

The chlorofluorocarbon known commercially as Freon-12 has the formula CCl_2F_2.

Atomic masses (g/g-mol): C = 12 Cl = 35.5 F = 19

Molar mass of CCl_2F_2= [12 + 2(35.5) + 2(19)] = 121

Mass fraction of Cl = 2(35.5)/121 = 0.587

Mass % of Cl = 58.7%

Example 4.17 Waste Gas Composition

A gaseous mixture with total mass of 1000 kg is 5% benzene, 71% nitrogen, and 24% oxygen. These are mass fractions. The ambient temperature and pressure are 25°C and 760 mm Hg. The molecular mass of benzene (C_6H_6) is 78 g/g-mol. Calculate the mass of each gas in the mixture, the number of g-moles of each gas, the total gas volume (at STP), the mole fraction and volume fraction of each gas, and the concentration as mass per volume (mg/m^3).

The mass of each gas can be calculated from its mass fraction.

Gas component = (mass fraction)(total gas mass)

$$C_6H_6 = 0.05(1{,}000{,}000\ g) = 50{,}000\ g$$

$$N_2 = 0.71(1{,}000{,}000\ g) = 710{,}000\ g$$

$$O_2 = 0.24(1{,}000{,}000\ g) = 240{,}000\ g$$

Molar mass

$$C_6H_6 = 78\ g/g\text{-mol}$$

$$N_2 = 28\ g/g\text{-mol}$$

$$O_2 = 32\ g/g\text{-mol}$$

Number of gram moles of each gas is

$$C_6H_6 = (50{,}000\ g)/(78\ g/g\text{-mol}) = 641\ g\text{-mol}$$

$$N_2 = (710{,}000\ g)/(28\ g/g\text{-mol}) = 25{,}357\ g\text{-mol}$$

$$O_2 = (240{,}000\ g)/(32\ g/g\text{-mol}) = 7500\ g\text{-mol}$$

Total number of moles of gas = 641 + 25,357 + 7500 = 33,498 g-mol

Mole fractions

$$C_6H_6 = (641\ mol)/(33{,}498\ g\text{-mol}) = 0.01936$$

$$N_2 = (25{,}357\ g\ mol)/(33{,}498\ g\text{-mol}) = 0.7570$$

$$O_2 = (750\ g\ mol)/(33{,}498\ g\text{-mol}) = 0.2239$$

Volume of gas 1 g-mol = 22.465 L at 25°C and 760 mmHg.

Gas volumes

$$C_6H_6 = (641\ g\text{-mol})(22.465\ L/g\text{-mol}) = 14{,}400\ L = 14.4\ m^3$$

$$N_2 = (25{,}537\ g\text{-mol})(22.465\ L/g\text{-mol}) = 559{,}600\ L = 569.6\ m^3$$

$$O_2 = (7500\ g\text{-mol})\ (22.465\ L/g\text{-mol}) = 168{,}500\ L = 168.5\ m^3$$

Total has volume = (33,515 g-mol)(22.465 L/g-mol) = 752,500 L = 752.5 m^3

Volume fractions

$$C_6H_6 = (14.4\ m^3)/(752.5\ m^3) = 0.019$$
$$N_2 = (559.6\ m^3)/(752.5\ m^3) = 0.7570$$
$$O_2 = (168.5\ m^3)/(752.5\ m^3) = 0.2239$$

This shows that the volume fraction = the mole fraction

Concentrations = (mass of gas)/(total volume of gas)

$$C_6H_6 = (50{,}000\ g)/(725.5\ m^3) = 0.0664\ g/m^3 = 66.4\ mg/m^3$$

$$N_2 = (710{,}000\ g)/(725.5\ m^3) = 0.943\ g/m^3 = 944\ mg/m^3$$

$$O_2 = (240{,}000\ g)/(725.5\ m^3) = 0.319\ g/m^3 = 319\ mg/m^3$$

4.9 EQUIVALENT WEIGHTS

Equivalent weight is the mass of a given substance that will

- supply or react with one mole of H^+ cations (or one mole of OH^- anions) in an acid–base reaction
- supply or react with one mole of electrons (e^-) in an oxidation–reduction reaction, or
- combine or displace 1.008 parts of hydrogen by mass, or 8 parts of oxygen by mass

Milligrams per liter (mg/L) and parts per million (ppm) are the simplest forms of expressing concentrations in water and wastewater, and mg/L is the accepted standard basis of reporting a water analysis. In special situations, mass concentrations can be reported as *equivalents per liter* (eq/L) or *milliequivalents per liter* (meq/L). Another possibility is *equivalents per million* (epm).

A milliequivalent is 1/1000 of an equivalent, just as a milligram is 1/1000 of a gram. Because concentrations in water and wastewater are almost always measured as mg/L, the unit of meq/L is more useful than eq/L.

Example 4.18 Equivalent Weight of Ferric and Ferrous Iron

Ferrous iron is Fe^{2+}. The 2+ ionic charge indicates that it can accept 2 moles of electrons (e^-) or 2 moles of OH^- ions. The equivalent weight is the molar mass divided by 2.

$$\text{Molar mass of } Fe^{2+} = 55.8\ g/mol$$

$$\text{Equivalent weight of } Fe^{2+} = (55.8\ g/mol)/(2\ equivalents/mol) = 27.9\ g/eq = 27{,}900\ mg/eq$$

Ferric iron is Fe^{3+}. It can accept 3 moles of electrons or 3 moles of OH^- ions.

$$\text{Molar mass of } Fe^{3+} = 55.8\ g/mol$$

$$\text{Equivalent weight of } Fe^{3+} = (55.8\ g/mol)/(3\ equivalents/mol) = 18.6\ g/eq = 18{,}600\ mg/eq$$

Table 4.6 gives the equivalent weights for common cations and anions (a more complete table including compounds normally used in pollution prevention and control is in Appendix B). They were calculated by dividing the molecular mass of the compound by the number of positive or negative electrical charges that result from the dissolution of the compound.

Calculating the equivalents in oxidation–reduction reactions is more complicated because one needs to know how many electrons are being exchanged. These reactions are used in Chapter 9.

Electroneutrality in a solution means that there must be one positive charge for each negative charge. The total number of positive charges must equal the total number of negative charges. The number of negative meq/L must equal the number of positive meq/L.

TABLE 4.6
Table of Equivalent Weights for Ions

Ion	Formula	Number of ± Charges	Molar Mass (g/mol)	Equivalent Weight (g/eq)
Cations				
Aluminum	Al^{3+}	3	27.0	9.0
Ammonium	NH_4^+	1	18.0	18.0
Calcium	Ca^{2+}	2	40.0	20.0
Copper	Cu^{2+}	2	63.6	31.8
Hydrogen	H^+	1	1.0	1.0
Ferrous iron	Fe^{2+}	2	55.8	27.9
Ferric iron	Fe^{3+}	3	55.8	18.6
Magnesium	Mg^{2+}	2	24.4	12.2
Manganese	Mn^{2+}	2	55.0	27.5
Potassium	K^+	1	39.1	39.1
Sodium	Na^+	1	23.0	23.0
Anions				
Bicarbonate	HCO_3^-	1	61.0	61.0
Carbonate	CO_3^{2-}	2	60.0	30.0
Chloride	Cl^-	1	35.5	35.5
Fluoride	F^-	1	19.0	19.0
Iodide	I^-	1	126.9	126.9
Hydroxide	OH^-	1	17.0	17.0
Nitrate	NO_3^-	1	62.0	62.0
Phosphate (tribasic)	PO_4^{3-}	3	95.1	31.7
Phosphate (dibasic)	HPO_4^{2-}	2	96.0	48.0
Phosphate (monobasic)	$H_2PO_4^-$	1	97.0	97.0
Sulfate	SO_4^{2-}	2	96.0	48.0
Bisulfate	HSO_4^-	1	97.1	97.1
Sulfite	SO_3^{2-}	2	80.0	40.0
Bisulfite	HSO_3^-	1	81.1	81.1
Sulfide	S^{2-}	2	32.0	16.0

Starting with a reasonably complete water analysis, total dissolved solids (TDS) may be calculated as meq/L. Where there is an excess of negative ion meq/L, the remaining positive ion meq/L is likely to be sodium or potassium (or both). For the sake of convenience, it is generally assumed to be sodium. If there is a shortage of negative meq/L, nitrate is added to correct the balance.

Example 4.19 Equivalent Weights and Total Dissolved Solids

Table 4.7 gives the analysis of a water sample as reported in mg/L (column 6). The TDS of this sample is the sum of the ionic concentrations:

$$\text{TDS (mg/L)} = 39 + 69 + 88 + 30 + 385 + 66 + 100 + 12 = 699 \text{ mg/L}$$

can also be expressed as meq/L. The equivalent mass of each ion is its molar mass divided by its charge. The concentration in meq/L is the mass concentration, mg/L, divided by the equivalent weight.

TABLE 4.7
Calculation of Milliequivalent Weights of Anions and Cations in a Water Sample

Constituent	Ion	Molar Mass (g/mol)	Charge	Equivalent Weight (mg/meq)	Concentration (mg/L)	Concentration (meq/L)
Potassium	K^+	39	1	39	39	1.0
Sodium	Na^+	23	1	23	69	3.0
Calcium	Ca^{2+}	40	2	20	88	4.4
Magnesium	Mg^{2+}	24	2	12	30	2.5
						Total Cations = 10.89
Bicarbonate	HCO_3^-	61	1	61	385	6.31
Sulfate	SO_4^{2-}	96	2	48	66	1.38
Chloride	Cl^-	35.5	1	35.5	100	2.82
Phosphate	PO_4^{3-}	95	3	31.7	12	0.38
						Total Anions = 10.89

The equivalent concentration of the cations is 10.89 meq/L. The equivalent concentration of the anions is also 10.89 meq/L. The sum of the cations should equal the sum of the anions, and they do.

The TDS, expressed as meq/L, is

$$\text{TDS (meq/L)} = \text{Total meq/L of cations} + \text{Total meq/L of anions}$$
$$= 10.89 + 10.89 = 21.78 \text{ meq/L}$$

There is no way to convert total meq/L to total mg/L, except to work with the individual ionic concentrations.

The following rules outline where epm can be used and where mg/L (ppm) must be used.

- Either may be used when an exact chemical formula is known. When such knowledge is lacking, mg/L must be used.
- The concentration of all dissolved salts of the individually determined ions must be in mg/L.
- Two or more ions of similar properties whose joint effect is measured by a single determination (e.g., total hardness, acidity, or alkalinity) may be reported in either mg/L or epm.
- The concentration of undissolved or suspended solids should be reported only in mg/L.
- The concentration of organic matter should be reported only in mg/L.
- The concentration of dissolved solids (by evaporation) should be expressed only as mg/L.
- TDS by calculation may be expressed in either mg/L or epm.
- Concentration of individual gases dissolved in water should be reported in mg/L.

4.10 CONCLUSION

Concentrations may be expressed using several choices of units, and the choice may change if the concentration refers to water or wastewater, sludge, soil, or gas.

The unit of choice for water or wastewater is mg/L because mg/L is independent of the density of the solution. If the solution density is 1.00 or very close to it, as it is for fresh water and most wastewaters, mg/L and ppm are interchangeable. This approximation is less valid as the solution density deviates from 1.00.

Because of this, the concentration of a dense sludge or slurry is usually reported as a mass concentration, that is, mg/kg wet sludge or kg/kg dry sludge solids.

The most convenient concentration measures for a mixture of gases are the volume fraction (vol/vol) and ppmv. For particulate solids in air the units typically will be g/m^3 or mg/m^3. The volume fractions for gases in a mixture do not change when the gas expands or contracts. This is not true for mass (e.g., g/m^3) concentrations.

5 Stoichiometry

5.1 THE DESIGN PROBLEM

The design problem is to identify the reactants and products of a reaction and then to calculate their masses.

Stoichiometry is the science of deciphering the qualitative and quantitative information that is given by a *balanced stoichiometric equation*. A balanced stoichiometric equation shows the chemical composition of the reactants and products and the molar proportions in which they react.

This balanced stoichiometric equation is for the neutralization of hydrochloric acid with sodium hydroxide.

$$\underset{\text{Acid}}{HCl} + \underset{\text{Base}}{NaOH} \rightarrow \underbrace{Na^{+} + Cl^{-}}_{\text{Dissolved Salt}} + H_2O$$

It shows what reacts (HCl and NaOH) and what is produced (Na^+, Cl^-, and H_2O). It shows how many moles of HCl and NaOH react (one of each), and how many moles of Na^+, Cl^-, and H_2O are produced (one of each).

It also specifies the material balance. Each atom of the elements that make up the reactants must appear on the products side of the equation. The total mass of the reactants must equal the total mass of the products. The masses of the compounds are calculated using the molar masses.

5.2 ELEMENTS AND COMPOUNDS

A molecule is formed when two or more atoms chemically join together. If the atoms are different from each other, a compound is formed. Not all molecules are compounds, since some, such as hydrogen gas or ozone, consist only of one element or type of atom.

Molecules and compounds are merged and split and rearranged in chemical reactions. Before any process analysis can be done, and before any engineering decisions can be made, we need to know which compounds react with each other and what products are formed. We need to know how much of each reactant is consumed in the reaction and the yield of each product.

Figure 5.1 shows four kinds of reactions. The simplest are addition and decomposition. Slightly more complex are the substitution reactions.

These reactions can be combined into a synthesis pathway. For example,

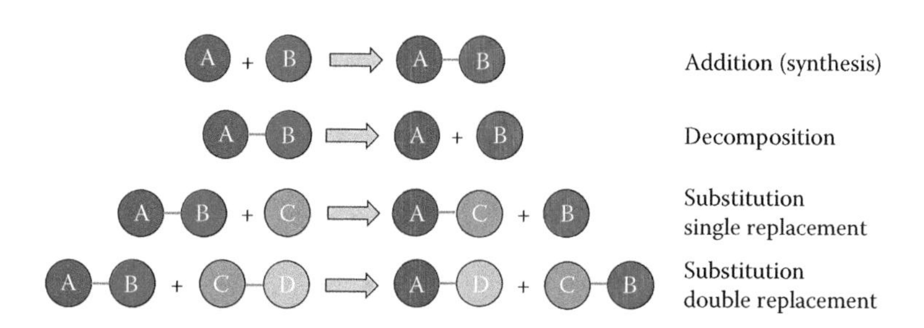

FIGURE 5.1 Four kinds of reactions.

$$A + B \rightarrow AB$$

$$AB + C \rightarrow AC + B$$

$$\text{or} \quad A + B \rightarrow AB$$

$$C + D \rightarrow CD$$

$$AB + CD \rightarrow AC + BD$$

The most desirable kind of reaction is addition because there is no by-product or waste. The other three kinds have by-products or waste products. This is often unavoidable, which is why chemistry is such an important component of pollution prevention and control.

5.3 ATOMIC AND MOLECULAR MASSES

Table 5.1 presents an abbreviated list of the atomic masses of the elements from which the molar mass of a compound can be assembled merely by summing the atomic masses of its constituents. The complete list of elements is given in Appendix A. Compounds that are commonly found or used in pollution prevention and control are listed in Table 5.2, along with their chemical formulas and molar mass.

TABLE 5.1
Atomic Numbers and Atomic Masses

Element		Atomic Number	Atomic Mass	Element		Atomic Number	Atomic Mass	Element		Atomic Number	Atomic Mass
Aluminum	Al	13	26.98	Helium	He	2	4.00	Potassium	K	19	39.10
Antimony	Sb	51	121.75	Hydrogen	H	1	1.01	Radium	Ra	88	226.02
Argon	Ar	18	39.95	Indium	In	49	114.82	Radon	Rn	86	(222)
Arsenic	As	33	74.92	Iodine	I	53	126.90	Selenium	Se	34	78.96
Barium	Ba	56	137.33	Iridium	Ir	77	192.22	Silicon	Si	14	28.09
Beryllium	Be	4	9.01	Iron	Fe	26	55.85	Silver	Ag	47	107.87
Bismuth	Bi	83	208.98	Krypton	Kr	36	83.80	Sodium	Na	11	22.99
Boron	B	5	10.81	Lanthanum	La	57	138.90	Strontium	Sr	38	87.62
Bromine	Br	35	79.90	Lead	Pb	82	207.20	Sulfur	S	16	32.06
Cadmium	Cd	48	112.41	Lithium	Li	3	6.94	Tellurium	Te	52	127.60
Calcium	Ca	20	40.08	Magnesium	Mg	12	24.30	Thallium	Tl	81	204.37
Carbon	C	6	12.011	Manganese	Mn	25	54.94	Thorium	Th	90	232.04
Cerium	Ce	58	140.12	Mercury	Hg	80	200.59	Tin	Sn	50	118.69
Cesium	Cs	55	132.91	Molybdenum	Mo	42	95.94	Titanium	Ti	22	47.90
Chlorine	Cl	17	35.45	Neon	Ne	10	20.18	Tungsten	W	74	183.85
Chromium	Cr	24	51.97	Nickel	Ni	28	58.70	Uranium	U	92	238.03
Cobalt	Co	27	58.93	Niobium	Nb	41	92.90	Vanadium	V	23	50.94
Copper	Cu	29	63.55	Nitrogen	N	7	14.01	Zinc	Zn	30	65.38
Fluorine	F	9	19.00	Oxygen	O	8	16.00	Zirconium	Zr	40	91.22
Gallium	Ga	31	69.72	Palladium	Pd	46	106.40				
Gold	Au	79	196.97	Phosphorus	P	15	30.97				

TABLE 5.2
Formulas for Chemicals Commonly Encountered in Pollution Prevention and Control

Chemical Name	Formula	Molar Mass	Comments
Activated carbon	C	12	Odor control, VOC removal, BOD and COD removal
Aluminum sulfate	$Al_2(SO_4)_3 \cdot 14H_2O$	594	Coagulation/flocculation
Aluminum hydroxide	$Al(OH)_3$	78	Coagulation/flocculation
Aluminum phosphate	$AlPO_4$	122	Phosphorus removal
Ammonia	NH_3	17	Nitrogen removal, eutrophication, toxicity
Ammonium fluorosilicate	$(NH_4)_2SiF_6$	178	Fluoridation of drinking water
Ammonium hydroxide	NH_4OH	35	Acidity regulator, household cleaner
Ammonium nitrate	NH_4NO_3	80	Fertilizer, eutrophication
Ammonium sulfate	$(NH_4)_2SO_4$	132	Fertilizer
Calcium bicarbonate	$Ca(HCO_3)_2$	162	Alkalinity, buffering
Calcium carbonate	$CaCO_3$	100	Corrosion control, water softening
Calcium fluoride	CaF_2	78.1	Fluoridation, neutralization of HF
Calcium hydroxide	$Ca(OH)_2$	74.1	Neutralization, water softening, precipitation
Calcium hypochlorite	$Ca(OCl)_2 \cdot 2H_2O$	179	Disinfection
Calcium oxide (lime)	CaO	56.1	Neutralization, water softening, precipitation
Calcium phosphate	$CaPO_4$	135	Phosphorus removal
Carbon dioxide	CO_2	44	Neutralization, combustion
Chlorine	Cl_2	71	Disinfection
Chlorine dioxide	ClO_2	67	Oxidations
Copper sulfate	$CuSO_4$	160	Algae control
Ferric chloride	$FeCl_3$	162	Coagulation, sludge conditioning
Ferric hydroxide	$Fe(OH)_3$	107	Coagulation, sludge conditioning
Ferrous sulfate	$FeSO_4$	400	Coagulation, sludge conditioning
Fluorosilicic acid	H_2SiF_6	144	Fluoridation
Hydrochloric acid	HCl	36.5	Neutralization
Hydrogen cyanide	HCN	27	Toxic chemical
Hydrogen sulfide	H_2S	34	Odor control, metal removal
Magnesium hydroxide	$Mg(OH)_2$	58.3	Water softening
Nitric acid	HNO_3	63	pH control, metal cleaning
Oxygen	O_2	32	Aeration
Potassium permanganate	$KMnO_4$	158	Oxidizing agent
Sodium aluminate	$NaAlO_2$	82	Coagulation
Sodium bicarbonate	$NaHCO_3$	75	pH Control, cleaning agent
Sodium bisulfite	$NaHSO_3$	105	Reducing agent
Sodium carbonate	Na_2CO_3	106	Water softening
Sodium chloride	NaCl	58.4	Ion exchange regenerant
Sodium fluoride	NaF	42	Fluoridation
Sodium hydroxide	NaOH	40	Neutralization
Sodium hypochlorite	NaOCl	74.4	Disinfection
Sodium silicate	Na_4SiO_4	184	Coagulation aid, drying agent, anticorrosion
Sodium fluorosilicate	Na_2SiF_6	188	Fluoridation
Sodium thiosulfate	$Na_2S_2O_3$	158	Dechlorination
Sulfur dioxide	SO_2	64.1	Air pollution, dechlorination
Sulfuric acid	H_2SO_4	98.1	Neutralization
Ions			
Ammonium	NH_4^+	18	Nitrogen removal, eutrophication, toxicity
Bicarbonate	HCO_3^-	61	Neutralization, water softening, precipitation
Bisulfate	HSO_4^-	97	Reducing agent
Bisulfite	HSO_3^-	81	Reducing agent
Carbonate	CO_3^{2-}	60	Neutralization, water softening, precipitation

(*Continued*)

TABLE 5.2 (*Continued*)
Formulas for Chemicals Commonly Encountered in Pollution Prevention and Control

Chemical Name	Formula	Molar Mass	Comments
Chromate	CrO_4^{2-}	116	Toxicity
Cyanide	CN^-	26	Toxicity
Cyanate	OCN^-	42	Toxicity
Dichromate	$Cr_2O_7^{2-}$	216	Toxicity
Hydroxide	OH^-	17	Neutralization, precipitation
Hypochlorite	OCl^-	51.5	Disinfection, oxidizing agent
Nitrite	NO_2^-	46	Nitrification
Nitrate	NO_3^-	62	Nitrification, nitrogen removal
Orthophosphate	PO_4^{3-}	95	Phosphorus removal, eutrophication
Orthophosphate	HPO_4^{2-}	96	Phosphorus removal, eutrophication
Orthophosphate	$H_2PO_4^-$	97	Phosphorus removal, eutrophication
Sulfate	SO_4^{2-}	96	Odors, corrosion

5.4 STOICHIOMETRY

Analysis of stoichiometry begins with a *balanced chemical equation* that shows the molecules and compounds that react and the products of the reaction. It shows the proportions in which they react and it allows calculation of the masses of the reactants and the products.

By convention, the reactants are on the left-hand side and the products are on the right. All products are shown, even those that are inert.

Inert substances that may be carried along with the reactants are not shown in the equation. They do not react and they produce nothing. They are ignored *for the purpose of the stoichiometry.* They do, however, have bulk and mass, and they are not ignored in the material balance of the reactor system.

The meaning of *balanced* is explained by an example. This reaction for burning heptane explains that heptane reacts with oxygen to produce carbon dioxide and water.

$$\underset{\text{heptane}}{1C_7H_{16}} + \underset{\text{oxygen}}{11O_2} \rightarrow \underset{\text{carbon dioxide}}{7CO_2} + \underset{\text{water}}{8H_2O}$$

Read the equation like a sentence: 1 mole of heptane reacts with 11 moles of oxygen to form 7 moles of carbon dioxide and 8 moles of water. The coefficients that quantify the number of molecules of each species refer to molecules, or to moles of material. They do not refer to mass.

The equation is *balanced* because the number of reacting carbon atoms on the left side is balanced by the same number of carbon atoms in the products on the right side. There are 7 carbon atoms on the left and 7 carbon atoms on the right. The same balance must exist for hydrogen and oxygen atoms. There are 16 hydrogen atoms on each side and 22 oxygen atoms on each side. They have been rearranged, but they have not disappeared.

The reaction equation gives the definite molecular proportions of the reaction. The stoichiometric coefficients (1 for heptane, 11 for oxygen, and so on) refer to molecular units. Eleven oxygen molecules react with each heptane molecule to produce 7 carbon dioxide molecules and 8 water molecules. More precisely, it shows that the 7 carbon atoms from the heptane molecule appear in the 7 molecules of carbon dioxide. Also, (11)(2) = 22 oxygen atoms react and (7)(2) = 14 of them appear in 7 molecules of carbon dioxide and 22 – 14 = 8 of them appear in 8 molecules of water.

Because of conservation of mass, the equation masses are also *balanced*—the mass of reactants equals the mass of products. The 7 carbon atoms that react have the same mass as the 7 carbon atoms that appear in the 7 molecules of CO_2. This is true for each species of atom, and it is, therefore, true for the total mass.

One gram-mole contains about 6×10^{23} molecules (Avogadro's number). One pound-mole contains about 2.7×10^{28} molecules; it has more molecules because one pound mass is 453.6 times greater than one gram mass.

One mole of a substance is defined as the mass of substance divided by its molecular mass.

$$\text{g-mol} = \text{mass in grams per molar mass}$$

$$\text{lb-mol} = \text{mass in pounds per molar mass}$$

and so forth in any mass unit we wish to use.

The mole is a measure of mass. Carbon is used as the reference for atomic mass. The atomic mass of one gram-mole (1 g-mol) of carbon is 12.011 g. Most engineering calculations round this off to 12. The masses of other elements are calculated on that scale.

The key to deciphering the reaction stoichiometry is the conversion of molecular units to mass units. Several examples show how this is done.

Example 5.1 Burning Heptane

The balanced stoichiometric reaction for the burning of heptane is given in the first line of Table 5.3.

TABLE 5.3
Various Interpretations of the Stoichiometric Equation

			Balanced Stoichiometric Equation			
C_7H_{16}	+	$11O_2$	→	$7CO_2$	+	$8H_2O$
			Molecular Basis			
1 molecule of heptane	reacts with	11 molecules of oxygen	to give	7 molecules of carbon dioxide	and	8 molecules of water
$1\ (6.023 \times 10^{23})$ molecules of C_7H_{16}	+	$11\ (6.023 \times 10^{23})$ molecules of O_2	→	$7\ (6.023 \times 10^{23})$ molecules of CO_2	+	$8\ (6.023 \times 10^{23})$ molecules of H_2O
			Molar Basis			
1 g mole C_7H_{16}	+	11 g moles O_2	→	7 g moles CO_2	+	8 g moles O_2
1 kg mole C_7H_{16}	+	11 kg moles O_2	→	7 kg moles CO_2	+	8 kg moles O_2
1 lb mole C_7H_{16}	+	11 lb moles O_2	→	7 lb moles CO_2	+	8 lb moles O_2
1 ton mole C_7H_{16}	+	11 ton moles O_2	→	7 ton moles CO_2	+	8 ton moles O_2
			Molecular Mass of Each Compound			
100 g/g-mol		32 g/g-mol		44 g/g-mol		18 g/g-mol
			Reacting Mass of Each Compound			
1 (100 g) C_7H_{16} = 100 g C_7H_{16}	+	11 (32 g) O_2 = 352 g O_2	→	7 (44 g) CO_2 = 308 g CO_2	+	8 (18 g) H_2O = 144 g H_2O
100 kg C_7H_{16}	+	352 kg O_2	→	308 kg CO_2	+	144 kg H_2O
100 lb C_7H_{16}	+	352 lb O_2	→	308 lb CO_2	+	144 lb H_2O
100 ton C_7H_{16}	+	352 ton O_2	→	308 ton CO_2	+	144 ton H_2O

Converting from a molar to a mass basis uses the molar masses of the compounds, which are as follows:

$$C = 12,\ H = 1,\ \text{and}\ O = 16$$

$$C_7H_{16} = 7(12) + 16(1) = 100\ \text{g/g-mol} = 100\ \text{lb/lb-mol} = 100\ \text{ton/ton-mol}$$

$$CO_2 = 1(12) + 2(16) = 44\ \text{g/g-mol} = 44\ \text{lb/lb-mol} = 44\ \text{ton/ton-mol}$$

$$O_2 = 2(16) = 32\ \text{g/g-mol} = 32\ \text{lb/lb-mol} = 32\ \text{ton/ton-mol}$$

$$H_2O = 2(1) + 16 = 18\ \text{g/g-mol} = 18\ \text{lb/lb-mol} = 18\ \text{ton/ton-mol}$$

The mass, in grams of substance reacting, is calculated by multiplying the molar mass (g/g-mol) of each substance by the number of moles. The proportions of reactants and products hold for any mass units.

Notice that all the mass is accounted for. The mass of reactants equals the mass of products formed: 452 g of reactants yield 452 g of products.

Example 5.2 Stoichiometric Air–Fuel Ratio of Octane

Emissions from motor vehicles are a substantial source of air pollution. Incomplete combustion produces carbon monoxide, nitrogen oxides, and unburned hydrocarbons. A key factor in combustion efficiency is the air–fuel ratio (AFR). The *stoichiometric AFR* will be calculated for octane (a major component of gasoline).

The combustion reaction for pure octane (C_8H_{18}) is

	C_8H_{18} +	$12.5O_2$ →	$8CO_2$ +	$9H_2O$
Molar mass (g)	114	32	44	18
Reacting mass (g)	114	400	352	162

The oxygen for this reaction is supplied in the form of air. Air is approximately 79% nitrogen and 21% oxygen by volume, which means 79/21 = 3.76 moles of nitrogen accompany each mole of oxygen.

$$\text{Moles of oxygen}(O_2) = 12.5\ \text{mol}\ O_2$$

$$\text{Mass of oxygen} = (12.5\ \text{mol}\ O_2)(32\ \text{g/mol}) = 400\ \text{g}\ O_2$$

$$\text{Moles of nitrogen} = (3.76\ \text{mol}\ N_2/\text{mol}\ O_2)(12.5\ \text{mol}\ O_2) = 47\ \text{mol}\ N_2$$

$$\text{Molar mass of}\ N_2 = 28\ \text{g/mol}$$

$$\text{Mass of nitrogen} = (47\ \text{mol})(28\ \text{g}\ N_2/\text{mol}) = 1316\ \text{g}\ N_2$$

$$\text{Mass of air per mole of octane} = 400\ \text{g}\ O_2 + 1316\ \text{g}\ N_2 = 1716\ \text{g air}$$

The *stoichiometric AFR* is

$$\text{AFR} = (1716\ \text{g air})/(114\ \text{g fuel}) = 15.1$$

The masses of air and fuel include all constituents of the air and fuel, whether they are combustible or not. If the fuel is a gas, CO_2 and N_2 would be included in the fuel mass.

The AFR calculated in Example 5.2 is not the stoichiometric AFR for gasoline and diesel fuel, which are mixtures of heptane, octane, some other alkanes, plus additives including detergents and

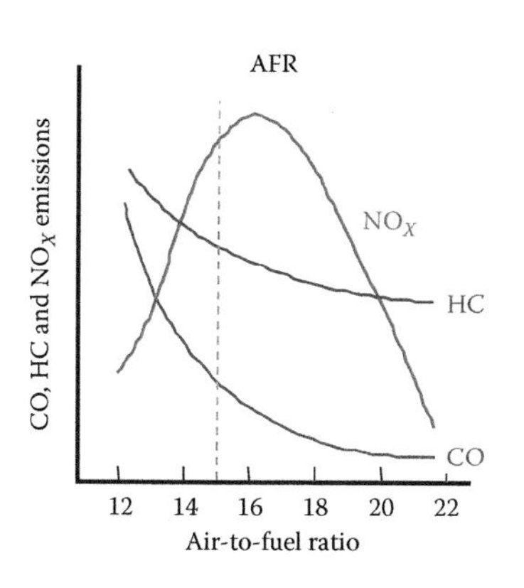

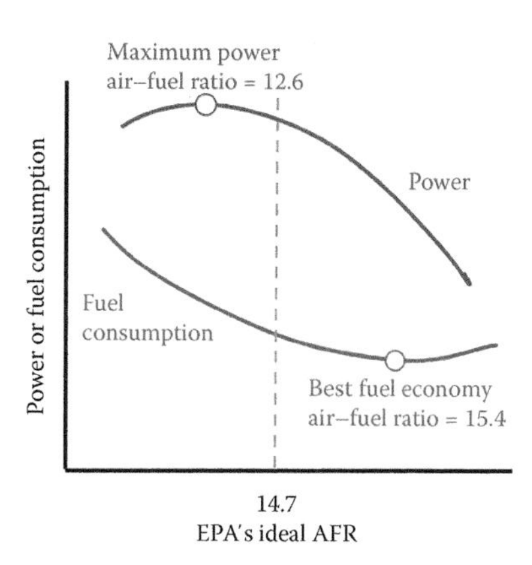

FIGURE 5.2 AFR for automobiles is defined as AFR = 14.7 by the U.S. EPA.

possibly oxidants such as methyl *tert*-butyl ether (MTBE). These compounds alter the stoichiometric ratio, generally downward. For MTBE-laden fuel the stoichiometric ratio can be as low as 14.1:1.

The U.S. Environmental Protection Agency (EPA) has defined an AFR = 14.7 as ideal, based on the function of catalytic converters in automobiles. This is shown in Figure 5.2.

Example 5.3 Combustion of Municipal Refuse

The empirical chemical composition of dry municipal refuse is $C_{65}H_{105}O_{30}N$. Write a balanced stoichiometric equation and find the mass of oxygen consumed in the combustion of 1000 kg/h of this material.

Find the stoichiometric coefficients that balance the reaction equation

$$C_{59}H_{93}O_{37}N + xO_2 \rightarrow aCO_2 + bH_2O + cNO_2$$

where

$$a = 59$$
$$b = 93/2 = 46.5$$
$$c = 1$$
$$2x + 37 = 2a + b + 2c = 2(59) + 46.5 + 2 = 166.5$$

Calculate x, the moles of oxygen required.

$$x = (166.5 - 37)/2 = 129.5/2 = 64.75$$

Balanced reaction:	$C_{59}H_{93}O_{37}N$	+	$64.75\ O_2$	$\rightarrow$	$59\ CO_2 + 46.5\ H_2O + NO_2$
Molar mass (kg)	1,407		32		
Reacting mass (kg)	1,407		32(64.75) = 2,072		

Stoichiometric O_2 requirement = (2072 kg O_2)/(1407 kg dry refuse) = 1.47 kg O_2/kg dry refuse
Mass O_2 per 1000 kg dry refuse = (1000 kg dry refuse/h)(2072 kg O_2)/(1407 kg dry refuse)
= 1470 kg O_2/h
Mass of air (air = 23.2% O_2) = (1470 kg O_2/h)/(0.232 kg O_2/kg air) = 6336 kg air/h
Excess air will be needed for complete combustion.

Example 5.4 Fluoridation

A concentration of 1.0 mg/L of fluoride (F^-) in drinking water is beneficial in preventing tooth decay. Where the natural fluoride concentration is below 1 mg/L, fluoride can be added, usually as the last step in the water treatment process. The European Union Drinking Water Directive specifies a maximum value of 1.5 mg/L. The following three fluoride compounds are commonly used to augment low natural concentrations:

Hexafluorosilicic acid (H_2SiF_6)
Disodium hexafluorosilicate (Na_2SiF_6)
Sodium fluoride (NaF)

All of these dissociate in water to give fluoride ions (F^-).

$$H_2SiF_6 + 2H_2O \rightarrow 6H^+ + 6F^- + SiO_2$$

If a water naturally contains 0.3 mg F^-/L, the mass of hexafluorosilicic acid that must be added to a flow of 16,000 m^3/d to have a concentration of 1.2 mg/L.

$$\text{Mass of } F^- \text{ to be added} = (1.2 - 0.3 \text{ mg } F^-/L)(16{,}000 \text{ m}^3/d)(1000 \text{ L/m}^3)(kg/10^6 \text{ mg})$$

$$= (0.9)(16) = 23.4 \text{ kg } F^-$$

1 mole of hexafluorosilicic acid (H_2SiF_6) provides 6 moles of F^-.
From Table 5.1, molar mass of fluoride (fluorine) = 19 g/g-mol.
From Table 5.1, molar mass of hexafluorosilicic acid (fluosilicic acid) (H_2SiF_6) = 144 g/g-mol.

Using the ratio given by the stoichiometric equation

$$\left(23.4 \text{ kg } F^-\right)\left(144 \text{ g } H_2SiF_6/\text{gmol}\right) / \left(6\left(19 \text{ g } F^-/\text{gmol}\right)\right) = 29.6 \text{ kg } H_2SiF_6$$

Example 5.5 Alkalinity Consumption by Alum

The chemical reaction below shows one mole of aluminum sulfate (alum) reacting with six moles of bicarbonate (HCO_3^-). If 200 mg/L of alum is to be added to water to achieve effective coagulation, how much alkalinity (i.e., HCO_3^-) will be consumed?

$$Al_2(SO_4)_3 \cdot 14H_2O + 6HCO_3^- \rightarrow 2\,Al(OH)_3(S) + 6CO_2 + 14H_2O + 3SO_4^{2-}$$

Molar mass (g)	594	61
Reacting mass (g)	594	366

594 g alum consumes 366 mg bicarbonate.
200 mg alum consumes (366/594)(200 mg) = 123 mg alkalinity as HCO_3^-.
Alkalinity is reported as the equivalent amount of $CaCO_3$.
Molar mass of $CaCO_3$ = 100 g/g-mol.
Alkalinity as $CaCO_3$ = (123 mg HCO_3^-/L) (100 mg $CaCO_3$/61 mg HCO_3^-) = 202 mg/L as $CaCO_3$.

Example 5.6 Struvite Precipitation

Struvite, or magnesium ammonium phosphate (MAP), precipitates according to the following reaction:

$$Mg^{2+} + NH_4^+ + PO_4^{3-} + 6H_2O \rightarrow MgNH_4PO_4 \cdot 6H_2O$$

The formation of struvite in pipes, pumps, and heat exchangers in wastewater treatment plants is a major maintenance problem. On the positive side, struvite is a highly effective source of nitrogen,

TABLE 5.4
The Approximate Composition of Molasses Industry Wastewater

Chemical Constituent	Concentration in Digester Effluent (mg/L)
NH_4^+	1400
Mg^{2+}	21.4
PO_4^{3-}	24
Ca^{2+}	21.2
K^+	2150
COD	3240
pH	7.9

magnesium, and phosphorus for plants and can be used as a slow-release fertilizer at high application rates without damaging plant roots.

The effluent from an anaerobic digester that treats HCO_3^- molasses-based industrial wastewater has the chemical composition given in Table 5.4. The goal is to recover the 1400 mg NH_4^+/L as struvite and sell it as fertilizer. Phosphate and magnesium need to be added in stoichiometric quantities. Use the given stoichiometric equation to calculate these amounts for a 1000 m^3 volume of wastewater.

The stoichiometry is simple:

$$1 \text{ mole Mg} + 1 \text{ mole } NH_4^+ + 1 \text{ mole } PO_4^{3-} \text{ yields 1 mole of struvite}$$

$$Mg^{2+} + NH_4^+ + PO_4^{3-} + 6\,H_2O \rightarrow MgNH_4PO_4 \cdot 6H_2O$$

$$\text{Molar mass} \quad 24.3 \quad 18 \quad 95 \quad 6(18) \rightarrow \quad 245.3$$

Recalling that 1 mg/L = 0.001 kg/m^3

$$\text{Ammonium concentration} = 1400 \text{ mg } NH_4^+/L = 1.4 \text{ kg } NH_4^+/m^3$$

$$\text{Mass of ammonium in } 1000 \text{ m}^3 = (1.4 \text{ kg } NH_4^+/m^3)(1000 \text{ m}^3) = 1400 \text{ kg } NH_4^+$$

$$\text{Magnesium in } 1000 \text{ m}^3 = (0.0214 \text{ kg } Mg^{2+}/m^3)(1000 \text{ m}^3) = 21.4 \text{ kg } Mg^{2+}$$

$$\text{Phosphate in } 1000 \text{ m}^3 = (0.024 \text{ kg } PO_4^{3-}/m^3)(1000 \text{ m}^3) = 24 \text{ kg } PO_4^{3-}$$

$$\text{Moles of ammonium} = (1400 \text{ kg } NH_4^+)/(18 \text{ kg } NH_4^+/\text{kg-mol}) = 77.78 \text{ kg-mol}$$

$$\text{Magnesium required} = (77.78 \text{ kg-mol})(24.3 \text{ kg } Mg^{2+}/\text{kg-mol}) = 1890 \text{ kg } Mg^{2+}$$

$$\text{Phosphate required} = (77.78 \text{ kg-mol})(95 \text{ kg } PO_4^{3-}/\text{kg-mol}) = 7389 \text{ kg } PO_4^{3-}$$

$$\text{Mass of Mg to be added} = 1890 - 21 = 1869 \text{ kg } Mg^{2+}$$

$$\text{Mass of } PO_4 \text{ to be added} = 7389 - 24 = 7365 \text{ kg } PO_4^{3-}$$

There is more to the actual chemistry than is revealed by the stoichiometric equation. Phosphorus can exist as PO_4^{3-} or HPO_4^{2-}, depending on the pH. Ammonia can exist as NH_4^+ or NH_3, also depending on the pH. The minimum solubility of struvite occurs at pH 9–10, so NaOH also will be need for pH control. Equilbirium chemistry is used to understand these chemical shifts and to calculate the amounts of each species that will exist.

5.5 CASE STUDY: AMMONIUM SULFATE FERTILIZER

One way to manufacture ammonium sulfate fertilizer is to combine gypsum ($CaSO_4$) with aqueous ammonia (NH_4OH). We will make a planning evaluation of the technical feasibility of making fertilizer by combining a waste stream that contains $CaSO_4$ with a waste stream that contains NH_4OH.

The gypsum, $CaSO_4$, comes from absorbing SO_2 in the scrubber water from a flue-gas desulfurization process into an aqueous lime solution $Ca(OH)_2$.

Ammonia, NH_3, is available from a coke manufacturing process. Coke is made by carbonizing coal at 1100°C in an oxygen-deficient atmosphere. As the coal is heated there is a release of tar, aromatic hydrocarbon compounds, and other gases. These products are called *coke-oven gas*, and most of the compounds can be recovered as by-products. The composition of coke-oven gas is 48%–55% hydrogen, 28%–30% CH_4, 5%–7.5% CO, 1.5%–2.5% CO_2, and 1%–3% N_2. Ammonia is a minor component in coke-oven gas, but producing one million tons per year of coke will yield 12 T/d of ammonia.

Ammonia is corrosive and it must be removed in coke-oven by-product plants. This can be done in an absorber where a solution of sulfuric acid (about 4% concentration by mass) is sprayed into the gas, or in a saturator in which the gas is bubbled through a bath of acidic solution. This produces ammonium sulfate, which can be crystallized, removed from the solution, dried, and sold. Today the cost of producing ammonium sulfate by this method often outweighs the revenue.

Ammonia can also be scrubbed out of the gas with water. Ammonia is stripped from the scrubbing solution by injecting steam. The ammonia vapors can be used to form a strong ammonia solution or to make ammonium sulfate.

Ammonium salts are very soluble. The solubility of ammonium sulfate is 74.5 g/100 g H_2O at 20°C. On a mass percentage basis, that is,

$$\frac{75\,\text{g}\ (NH_4)_2SO_4}{75\,\text{g}(NH_4)_2SO_4 + 100\,\text{g}\ H_2O} = \frac{75}{175} = 42.86\%\ \text{by mass}$$

The solubility at 30°C is 78.2 g/100 g H_2O, or 78.2/(78.2 + 100) = 43.88%. At 40°C the values are 81.2 g/100 g H_2O, or 81.2/(81.2 + 100) = 44.81%.

Crystallization will begin when the ammonium sulfate solution has been concentrated (by evaporation) to the saturation concentration.

The relevant chemical reactions are

$$NH_3 + H_2O \rightarrow NH_4OH$$

$$2NH_4OH + CO_2 \rightarrow (NH_4)_2CO_3 + H_2O$$

$$(NH_4)_2CO_3 + CaSO_4 \rightarrow (NH_4)_2SO_4 + CaCO_3$$

The overall reaction is

$$2NH_3 + H_2O + CO_2 + CaSO_4 \rightarrow (NH_4)_2SO_4 + CaCO_3$$

The material balance will be made for an input of 1000 kg/h of gypsum.

The mass of $(NH_4)_2CO_3$ and the products are calculated using

$$(NH_4)_2CO_3 + CaSO_4 \rightarrow (NH_4)_2SO_4 + CaCO_3$$

	$(NH_4)_2CO_3$	$CaSO_4$	$(NH_4)_2SO_4$	$CaCO_3$
Molar mass (kg)	96	136	132	100

An input of 1000 kg/h $CaSO_4$ requires (1000)(96/136) = 706 kg/h $(NH_4)_2CO_3$.
Production of $(NH_4)_2SO_4$ = (1000)(132/136) = 971 kg/h.
Production of $CaCO_3$ = (1000)(100/136) = 735 kg/h.
Aqueous ammonia required

$$2NH_4OH + CO_2 \rightarrow (NH_4)_2CO_3 + H_2O$$

	$2NH_4OH$	CO_2	$(NH_4)_2CO_3$	H_2O
Molar mass (kg)	35	44	96	18
Reacting mass (kg)	70	44	96	18

An output of 706 kg/h $(NH_4)_2CO_3$ requires (706 kg/h)(70/96) = 515 kg/h NH_4OH.
CO_2 required = (706 kg/h)(44/96) = 325 kg/h.
Ammonia required

$$NH_3 + H_2O \rightarrow NH_4OH$$

	NH_3	H_2O	NH_4OH
Molar mass (kg)	17	18	35

To make 515 kg/h NH_4OH requires (515)(17/35) = 250 kg/h NH_3.

The overall material balance based on the ideal stoichiometry is shown in Figure 5.3. The actual material balance requires a 10% excess of ammonia, additional water to make the solution, and the removal of water to crystallize the ammonium sulfate. Also, the recovery of ammonium sulfate is about 98%.

The process flowchart is shown in Figure 5.4. The overflow from the gypsum converter contains all of the $(NH_4)_2SO_4$ in the form of a 35% solution of ammonium sulfate. A vacuum evaporator–crystallizer will produce a supersaturated solution so crystals with an average size of 2.4 mm are formed. The diagram shows one crystallizer, but usually two are used in series to shorten the retention time. The crystals are dried and screened, with the undersize crystals being dissolved and recycled to the crystallizer. The $CaCO_3$-laden stream leaving the gypsum converter can be recycled to the SO_2 scrubbing process.

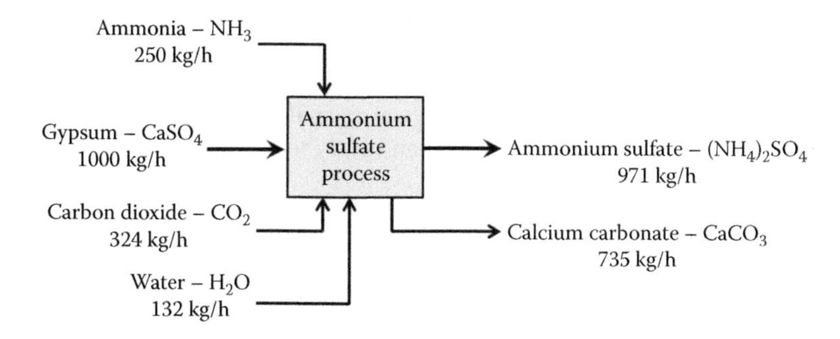

FIGURE 5.3 Material balance for ideal ammonium sulfate stoichiometry.

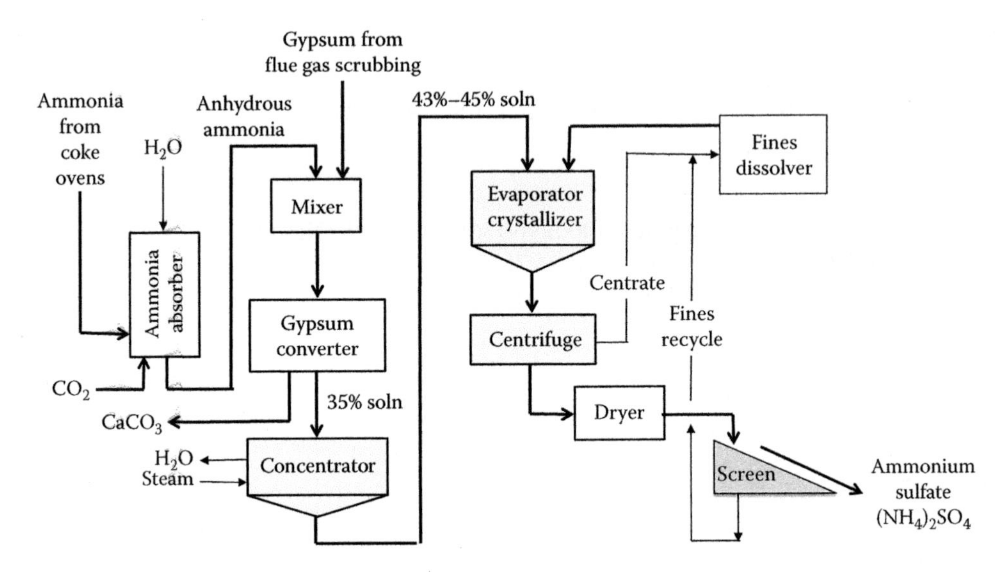

FIGURE 5.4 Process flowchart for ammonium sulfate fertilizer process.

5.6 EMPIRICAL CHEMICAL FORMULAS

The empirical formula of a chemical compound is the simplest positive integer ratio of atoms present in a compound. The empirical formula of hydrogen peroxide, H_2O_2, is simply HO.

Glucose ($C_6H_{12}O_6$), ribose ($C_5H_{10}O_5$), acetic acid ($C_2H_4O_2$), and formaldehyde (CH_2O) all have different molecular formulas but the same empirical formula: CH_2O. This is the actual molecular formula for formaldehyde, but acetic acid has double the number of atoms, ribose has five times the number of atoms, and glucose has six times the number of atoms.

The balanced equations for combustion of these compounds have the same proportions of reactants and products.

$$CH_2O + O_2 \rightarrow CO_2 + H_2O$$

$$C_2H_4O_2 + 2O_2 \rightarrow 2CO_2 + 2H_2O$$

$$C_5H_{10}O_5 + 5O_2 \rightarrow 5CO_2 + 5H_2O$$

$$C_6H_{12}O_6 + 6O_2 \rightarrow 6CO_2 + 6H_2O$$

The chemical compound *n*-hexane has the structural formula $CH_3CH_2CH_2CH_2CH_2CH_3$, which shows that it has 6 carbon atoms arranged in a chain, and 14 hydrogen atoms. Hexane's molecular formula is C_6H_{14}, and its empirical formula is C_3H_7, both showing a C:H ratio of 3:7.

Ammonium nitrite is NH_4NO_2, with the empirical formula NH_2O or $N_2H_4O_2$. This shows the correct proportion of atoms, but it can mislead when we try to guess products when the compound dissociates or ionizes. It is

$$NH_4NO_2 \rightarrow (CO_3^{2-}) + NH_4^+$$

not

$$N_2H_4O_2 \rightarrow N_2 + 2H_2O$$

Ammonium nitrite is the ammonium salt of nitrous acid. When dissolved in water it yields ammonium ion (NO_3^-) and nitrite (NO_2^-). It is used as a rodenticide, micro-biocide, and agricultural pesticide, and is acutely toxic to both humans and aquatic organisms.

Example 5.7 Deriving the Empirical Formula of a Compound

A 100-g sample of a solvent commonly used in paints, inks, and adhesives was analyzed and found to have this composition: 48.64 g carbon (C), 8.16 g hydrogen (H), and 43.20 g oxygen (O). Determine the empirical molecular formula, which will be of the form $C_aH_bO_c$.

Convert the amount of each element in grams to its amount in moles.

$$(48.64\text{ g C})/(12.01\text{ g C/mol}) = 4.049\text{ mol C}$$

$$(8.126\text{ g H})/(1.008\text{ g H/mol}) = 8.095\text{ mol H}$$

$$(43.20\text{ g O})/(16.00\text{ g O/mol}) = 2.7\text{ mol O}$$

The values of *a, b,* and *c* in the empirical formula are written with the smallest possible number of molecules. One of these values will equal 1, and the others will be in proportion to the number of moles in the sample. To find these values, divide each by the smallest of the found values. That is, divide by 2.7.

$$\text{Carbon: } a = (4.049\text{ mol})/(2.7\text{ mol}) = 1.5$$
$$\text{Hydrogen: } b = (8.095\text{ mol})/(2.7\text{ mol}) = 3$$
$$\text{Oxygen: } c = (2.7\text{ mol})/(2.7\text{ mol}) = 1$$

This gives $C_{1.5}H_3O$.

It is conventional to write the formula with integers, when possible. If necessary, multiply each number by the same integer to get whole numbers.

$$a = 1.5(2) = 3$$
$$b = 3(2) = 6$$
$$c = 1(2) = 2$$

Thus, the empirical formula is $C_3H_6O_2$. This formula describes methyl acetate, which is a solvent that is commonly used in paints, inks, and adhesives.

Example 5.8 Methane Formation

A food industry wants to discharge a strong wastewater to a nearby municipal wastewater treatment plant that has an anaerobic sludge digester. The wastewater for a typical day contains a mixture of 1000 kg carbohydrates ($C_{12}H_{22}O_6$) and 850 kg vegetable oils ($C_{54}H_{106}O_6$). Write an empirical formula for the mixed carbohydrates and oils, and write a balanced empirical stoichiometric reaction for the complete conversion to methane (CH_4) and carbon dioxide (CO_2).

The calculations are summarized in Table 5.5.
The empirical formula for 1 mole of the mixed food waste is $C_{99.8}H_{190}O_{28.9}$.
The decimal points give a false sense of precision about the empirical formula, so rounding simplifies it to $C_{100}H_{190}O_{29}$.

The empirical reaction is

$$C_{100}H_{190}O_{29} + xH_2O \rightarrow aCH_4 + bCO_2$$

From the material balance

$$\text{on C}: 100 = a + b$$
$$\text{on H}: 190 + 2x = 4a$$
$$\text{on O}: 20 + x = 2b$$

TABLE 5.5
Calculation of Empirical Formula for Mixed Food Waste

	Carbohydrates $C_{12}H_{22}O_6$	Vegetable Oils $C_{54}H_{106}O_6$	Mixed Food Waste
Mass (kg)	1000	850	1850
Molar mass (kg/kg-mol)	12(12) + 22(1) + 6(16) = 144 + 22 + 96 = 262	54(12) + 106(1) + 6(16) = 648 + 106 + 96 = 850	
Mole fraction C	144/262 = 0.5496	648/850 = 0.7624	
Mole fraction H	22/262 = 0.08397	106/850 = 0.1247	
Mole fraction O	96/262 = 0.3664	96/850 = 0.1129	
Mass of C (kg)	(0.5496)(1000) = 549.6	648.0	1197.6
Mass of H (kg)	(0.0840)(1000) = 84.0	106.0	190.0
Mass of O (kg)	(0.3664)(1000) = 366.4	96.0	462.4
Moles of C/mole of compound	12	54	1197.6/12 = 99.8
Moles of H/mole of compound	22	106	190.0/1 = 190.0
Moles of O /mole of compound	6	6	462.4/16 = 28.9

Solving gives: $a = 68.75$, $b = 31.25$, and $x = 42.5$.

$$C_{100}H_{190}O_{29} + 42.5H_2O \rightarrow 68.75CH_4 + 31.25CO_2$$

5.7 CONCLUSION

Stoichiometry describes the chemical reactions that we use to destroy harmful substances, to convert soluble ions into particulate solids that can easily be removed by settling or filtration, and to neutralize acids and bases.

A balanced chemical reaction shows the chemical composition of the reactants and the products and the molar proportions in which they react. The equation also specifies the material balance. The total mass of the reactants must equal the total mass of the products. Each molecule of the elements that make up the reactants must appear on the products side of the equation. The masses of the compounds are calculated using the molar masses.

In many processes one or more of the products must be separated from the bulk of the material. For example, metal-laden precipitate solids are removed from the carrier water by settling, with the result that a slurry of solids (i.e., sludge) is formed. Estimating the mass of solids is an important calculation. The quantity calculated using the ideal stoichiometry will usually be wrong, perhaps too large and perhaps too small, because wastewater is a complex mixture of substances, many of which are never identified, even though they do involve themselves in the reactions. This creates the need to use empirical stoichiometry, which is the subject of Chapter 6.

6 Empirical Stoichiometry

6.1 THE DESIGN PROBLEM

Chapter 5 explained how stoichiometric calculations are used to estimate the behavior of processes when we know the chemistry and the chemical reaction equation.

Sometimes there is no theoretical stoichiometry. The design problem in such a situation is to discover a useful *empirical stoichiometry* for a complex chemical mixture so that the necessary quantities may be calculated.

Empirical stoichiometry is based on simple experiments that are designed, not to discover fundamental chemical details, but simply to provide the information needed to answer important design questions. The question might be, for example, "How much lime must be added to bring the pH of a complex mixture to pH 7?" or "What mass of settleable solids will form as a result of the lime that is added?"

Turbidity and color have no standard chemical composition. There are no moles of turbidity, moles of colloids, moles of Nephelometric Turbidity Unit (NTU), or moles of color. They are measured in arbitrary units of intensity—NTU in the case of turbidity.

There are no moles of suspended solids (SS), biochemical oxygen demand (BOD), or chemical oxygen demand (COD). They are measured in mass units, but the mass cannot be expressed as moles because the measurements lump, or aggregate, a mixture of perhaps hundreds of different materials.

Turbidity, color, suspended solids, BOD, and COD can be measured, and they can be removed from wastewater by chemical processes. The design problem is to discover the empirical chemistry that will tell us how much can be removed, and what condition will give the best removal. *Jar testing* is the way we do this.

Sometimes the species to be removed are known and quantified, but the chemical composition of the wastewater is not known precisely, and the only chemical equations we can write will be incomplete. The calculations may be worth doing, as a rough approximation, but an experimental check or confirmation may be needed.

In many processes one or more of the products must be separated from the bulk of the material. For example, metal-laden precipitate solids are removed from the carrier water by settling, with the result that a slurry of solids (i.e., sludge) is formed. Estimating the mass of solids is an important calculation. Very often the quantity calculated from the ideal stoichiometry will be wrong because wastewater is a complex mixture of substances, many of which are never identified, even though they do involve themselves in the reactions. This creates the necessity to use empirical stoichiometry.

Sometimes the reactions are known, but the products must be observed to determine particle size, settling velocity, filterability, etc. These characteristics cannot be calculated; they must be observed.

Empirical chemistry is used when actual wastewater obtained from an existing system can be used in experiments that are designed to simulate the full-scale process.

If the industrial production plant has yet to be built, actual wastewater cannot be obtained. In these cases, the designer must rely on experience with similar problems in other cities or factories, or reports in the literature.

6.2 EMPIRICAL STOICHIOMETRY BY EXPERIMENT: JAR TESTS

The most widely used method for learning the empirical chemistry is the *jar test*. This is a simple but powerful tool. It is simple enough to be used by operators and powerful enough to be used in research. It yields information that cannot be calculated precisely from fundamental theory or that can be calculated only approximately and then with great effort. We can learn about stoichiometry, equilibrium, and kinetics without any advance knowledge of the physical and chemical mechanisms that operate in the process. (In this way, it is like the material balance. We can make a material balance without knowing what goes on inside the process boundary if we know what goes in and what comes out.)

Figure 6.1 shows a jar test apparatus that can study six different treatments. The "jars" hold the liquid (e.g., wastewater or river water) to be treated. The jars can be dosed with different chemicals in different concentrations, mixed, aerated, heated, or otherwise manipulated to simulate a full-scale process. After a period of time the contents can be settled or filtered. These conditions are varied systematically until an acceptable treatment is found. Which treatment is best may depend on the clarity, purity, toxicity, etc., of the treated liquid, the quantity of solids produced (sludge), the thickening or dewatering properties of the sludge, and so on.

Figure 6.2 shows a remarkable jar test result. Large settleable floc particles have formed in 15 s. The coagulating agent (an organic polymer) is effective in agglomerating the smaller particles and producing a clear surface layer of liquid.

Mixing times to build a large strong floc are more often in the range of 10–20 min. The only way to learn what will be effective is to experiment with the water to be treated and a variety of chemicals and operating conditions.

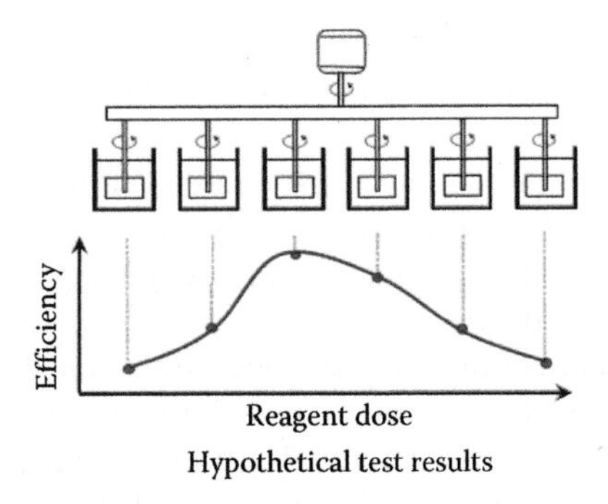

FIGURE 6.1 Jar test to evaluate six treatment conditions, showing a hypothetical, but typical, result.

FIGURE 6.2 Jar test demonstration of a fast-acting flocculant. (Courtesy of Integrated Engineers, Inc.)

TABLE 6.1
Operating Factors That Can Be Studied in Jar Tests and Process Responses That Can Be Observed

Operating Factors That Can Be Studied	Process Responses That Can Be Observed
Chemical dose for coagulation	Effluent quality (turbidity, SS, pH, metals, etc.)
Chemical dose for precipitation	Particle formation time
Acid or base addition	Particle size
Reaction time	Particle settling rates
Mixing time	Sludge volume
Mixing intensity	Sludge thickening behavior
	Mass of sludge solids formed
	pH

Jar tests are used to investigate turbidity, color, and solids removal, precipitation, emulsion breaking, and neutralization. The process responses that can be observed are listed in Table 6.1. None of these can be calculated from theory with sufficient accuracy and confidence to design without some experimental confirmation. The jar test provides the empirical chemistry for coagulant dose, chemical requirements for pH control, and sludge production. It also provides useful empirical information about mixing requirements, floc formation, settling efficiency, and sludge thickening.

Visual results are important. Turbidity and suspended solids removal are obvious at a glance. The time of first visual floc formation and relative settling rates are available without making any physical measurements.

The volume of sludge produced can be seen. The mass of sludge solids can be measured by drying and weighing a sample. The actual sludge is available for dewatering and toxic leaching tests. In metals removal, it is sometimes possible to infer the kind of metal precipitate by its color and the extent of metal removal from pH and sludge quantity.

The empirical jar test has some limitations. It is useless unless representative water or wastewater exists to be tested. If the process that produces the wastewater is to be changed to reduce pollution, the jar test results may not predict how the future wastewater should be treated. The jar test examines small quantities of water or wastewater in a batch system. The real process will use continuous flow through large tanks. Therefore, the designer has to scale up from the jar to the full-scale reactor. Experienced designers give the operator flexibility in controlling the rate of chemical addition and the mixing intensity.

6.2.1 Turbidity Removal by Coagulation/Flocculation and Settling

Clarification of river water for drinking water or use as process water requires removing colloidal particles. *Colloids* are $\leq 10^{-3}$ to 1 μm in diameter. They do not settle under the force of gravity, and they are too small to be removed by ordinary filtration.

Colloids can be encouraged to flocculate by adding a coagulating chemical, usually a positively charged multivalent metal ion like iron (Fe^{3+}) or aluminum (Al^{3+}), which will neutralize the surface charge of the colloid so the individual small particles can be collected into large floc particles that are easy to remove. Organic polymers can also be used.

The process of neutralizing ionic charges is called *coagulation*, and the process of gentle mixing to form a large floc is called *flocculation*. Flocculated particles can be quite large (100–500 μm). They will settle rapidly and are easy to remove by filtration. The amount of coagulant to use cannot be calculated from theory. Some of the added iron or aluminum forms a hydroxide precipitate, some associates with the colloid to neutralize the surface charge, and some remains in solution. What happens depends on the pH and alkalinity of the water. Even if these aspects of the chemistry could be calculated, the time for floc to form, the size of the floc, and the settling properties and sludge volume cannot. Jar tests are used to empirically determine these quantities.

6.2.2 Precipitation of Metals

It is possible to estimate the solubility of the precipitated compounds in water, but this is the not value that is needed to design a real process. The theory applies to a system in a state of equilibrium. Real systems are not in a state of equilibrium. The theory predicts the concentration of soluble metal ion. Effluent standards are written for total metal. Total metal is the sum of soluble metal in all forms (ionic and complexed) plus particulate metal in all forms (as pure metal crystal, amorphous metal compounds, and adsorbed forms). Furthermore, there is usually more than one metal in wastewater. Buffering, complexing, and stabilizing agents are added in the manufacturing process and these interfere with the chemistry of metals removal in waste treatment. Chemical theory is useful to identify precipitation processes that might solve the problem, but only empirical chemistry will tell us whether a promising idea will actually work.

6.2.3 Breaking Emulsions of Oil and Grease

An *emulsion* is a stable mixture of oil and water. Emulsions are ideal for lubrication in machining because they have better cooling properties (by a factor of ten) than liquid oil. Unfortunately, the ideal manufacturing conditions are the opposite of the ideal conditions for waste treatment. Free (non-emulsified) oil can be removed from wastewater by flotation or filtration. Emulsified oil can be removed by these methods if the emulsion is destabilized (broken) so the oil will coalesce into globules. Many emulsions can be broken by chemical coagulation using salt ($NaCl$), sulfuric acid (H_2SO_4), ferric sulfate ($Fe_2(SO_4)_3$), aluminum sulfate ($Al_2(SO_4)_3$), ferric chloride ($FeCl_3$), calcium oxide (CaO), or combinations of these.

6.2.4 pH Control and Neutralization

A special kind of jar test is used to determine chemical requirements for pH control and neutralization and the amount of sludge produced in processing. Test results are plotted in the form of a titration curve. Jar tests are used because simple acid–base equilibrium chemistry falls short of predicting the behavior of real wastewater mixtures because wastewater contains substances other than acids and bases that will react with the added reagent.

TABLE 6.2
Jar Test Results for Dosing Iron and Manganese Waste with Permanganate

Jar #	$KMnO_4$ (mg/L)	Color
1	0.1	Not pink
2	0.15	Not pink
3	0.20	Not pink
4	0.25	Not pink
5	0.30	Pink
6	0.35	Pink

Example 6.1 Jar Test: Potassium Permanganate

Potassium permanganate ($KMnO_4$), a strong oxidizing agent, will be added to a drinking water purification process as a means of oxidizing iron and manganese so they can be removed by filtration. Six stirred jars, each containing 1000 mL, were dosed with $KMnO_4$ as given in Table 6.2.

As the iron and manganese begin to oxidize, the sample will turn varying shades of brown, indicating the presence of oxidized iron and/or manganese. A residual brown or yellow color indicates that the oxidation is incomplete and a higher dosage of $KMnO_4$ is required. The required amount of added permanganate occurs when the solution turns pink and remains pink for at least 10 min. How much potassium permanganate per day is needed to treat a flow of 1 m³/s?

$$\text{Flow rate} = 1\ \text{m}^3/\text{s} = 1000\ \text{L/s} = 60{,}000\ \text{L/min} = 86{,}400{,}000\ \text{L/day}$$

$$KMnO_4\ \text{dose} = (0.30\ \text{mg/L})(86{,}400{,}000\ \text{L/day})/(1{,}000{,}000\ \text{mg/kg}) = 25.92\ \text{kg/day}$$

Example 6.2 Arsenic Removal by Water Softening

Arsenic in drinking water is a problem in many parts of the United States and other countries of the world. Lime softening is one method of removing arsenic. This example is about a test to investigate whether the presence of Cu^{2+} or Zn^{2+} will enhance the adsorption of arsenic onto the calcium carbonate softening precipitate.

These are the test conditions:

- Groundwater was spiked with pentavalent arsenic, As^{5+}, to give an initial concentration of 87.0 parts per billion (ppb).
- Each jar was dosed with 10 mg/L CaO to pH = 8.7.
- Two control tests were run with no Cu^{2+} or Zn^{2+} added.
- Five jars (runs 3–7) were dosed with copper sulfate ($CuSO_4$) to give Cu^{2+} concentrations of 0.5–3.0 mg/L.
- Five jars (runs 8–12) were dosed with zinc nitrate ($ZnNO_3$) to give Zn^{2+} concentrations of 0.5–3.0 mg/L.

TABLE 6.3
Concentration Variations of Calcium Oxide and Divalent Metal Salt and Residual As^{5+}

Experiment	CaO (mg/L)	Cu^{2+} (mg/L)	Zn^{2+} (mg/L)	Initial As^{5+} (ppb)	Final As^{5+} (ppb)	As^{5+} Removed (ppb)	% As^{5+} Removed
Run 1 (control)	10.0	0.0	0.0	87.0	41.0	46.0	52.9
Run 2 (control)	10.0	0.0	0.0	87.0	47.5	39.5	45.4
Average					44.2	42.8	49.1
Run 3	10.0	0.5	0.0	87.0	40.4	46.6	53.6
Run 4	10.0	1.0	0.0	87.0	36.4	50.6	58.2
Run 5	10.0	2.0	0.0	87.0	32.5	54.5	62.6
Run 6	10.0	2.5	0.0	87.0	33.5	53.5	61.5
Run 7	10.0	3.0	0.0	87.0	33.0	54.0	62.1
Run 8	10.0	0.0	0.5	87.0	38.4	48.6	55.9
Run 9	10.0	0.0	1.0	87.0	29.3	57.7	66.3
Run 10	10.0	0.0	2.0	87.0	31.8	55.2	63.4
Run 11	10.0	0.0	2.5	87.0	31.8	55.2	63.4
Run 12	10.0	0.0	3.0	87.0	29.4	57.6	66.2

Source: Data from U.S Patent US 6802980 B1; Sandia Corporation.

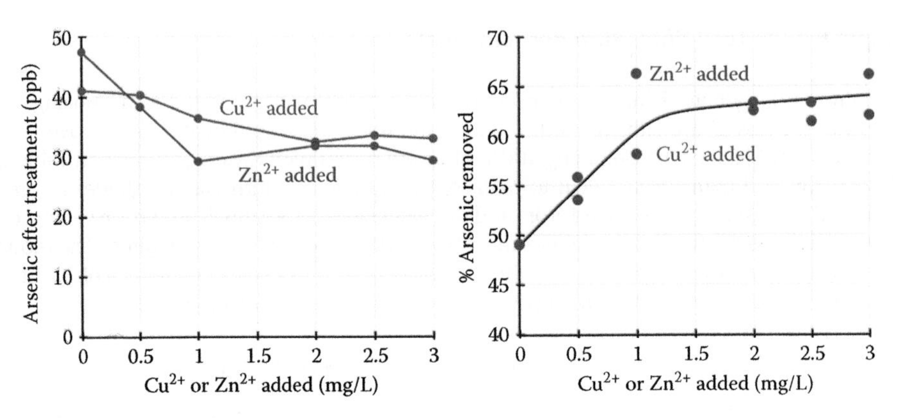

FIGURE 6.3 Removal of arsenic from drinking water by lime softening enhanced by the addition of copper or zinc ions.

- There were no tests with both copper and zinc present.
- The contents of each jar were rapidly mixed for 3 min, slowly mixed for 45 min, and filtered (0.2 μm filter).

The test conditions and results are summarized in Table 6.3 and Figure 6.3.

The two control tests show the experimental variation. The percentages of arsenic removed are 52.9% and 45.4%, with an average of 49.1%. The test results must be interpreted with this in mind.

Adding copper or zinc to the lime-softening process does increase the removal of arsenic from this water. We interpret Figure 6.3 that the two metals are equally effective and that little is gained by adding more than 1 or 1.5 mg/L of Cu or Zn.

6.3 COAGULATION AND FLOCCULATION

Coagulation and flocculation are used to assist the removal of turbidity and color by settling and filtration. The coagulating chemical, mixing conditions, pH, and chemical dosages are determined by jar testing.

The most used coagulating chemicals are aluminum sulfate $Al_2(SO_4)_3 \cdot 14\ H_2O$, ferric chloride $FeCl_3$, or ferrous sulfate $FeSO_4$. Polyaluminum chloride (PACl) is seeing increased use in drinking water treatment.

A stable colloidal solution behaves more like a true solution than like a suspension of solids. "Stable" means that the colloidal particles tend to retain their individual identity instead of coalescing into larger particles. They do not coalesce because each particle has an electrostatic surface, and like-charged particles repel each other. The natural materials found in water and wastewater typically have a negative surface charge. If the electrostatic surface charge is neutralized, the colloids can be encouraged, by gentle mixing, to collide and agglomerate to form larger settleable particles. The surface charge density of colloids can be measured by electrophoretic mobility or zeta potential.

Coagulation is the chemical process of adding chemicals to neutralize the electrostatic surface charge of colloids and very fine particles. Colloids typically have a negative surface charge, so charge neutralization is done by adding multivalent metal cations, usually Al^{3+}, Fe^{3+}, or Fe^{2+}.

The positively charged metal ions collect around the colloid to counteract the particle's natural negative charge. With the proper coagulant dosage, the net charge is reduced to near zero, and the repulsive electrostatic force is too weak to prevent particle collisions and agglomeration. Vigorously mixing the chemicals into the water rapidly disperses the cations and causes them to associate with the colloids before they become involved in side reactions. This rapid mixing step takes only 30–60 s.

Flocculation is the physical process of agglomerating fine particles and coagulated colloids into large settleable particles (floc) by gentle mixing with large paddles that turn at a few revolutions per minute (1–3 rpm). Faster mixing speeds may destroy the floc. Flocculation basins have a hydraulic detention time of 15–45 min. Some of the added chemical forms a hydroxide precipitate that helps sweep up the smaller particles.

As the coagulant reacts to provide neutralizing cations, other reactions occur. If aluminum sulfate (also known as alum) is added to water, this reaction occurs:

$$\underset{\text{aluminum sulfate}}{Al_2(SO_4)_3 \cdot 14H_2O} + \underset{\text{calcium bicarbonate}}{3Ca(HCO_3)_2} \rightarrow \underset{\text{aluminum hydroxide}}{2Al(OH)_3} + \underset{\text{calcium sulfate}}{3CaSO_4} + \underset{\text{carbon dioxide}}{6CO_2} + \underset{\text{water}}{14H_2O}$$

Commercial-grade alum has 14 H_2O attached to the $Al_2(SO_4)_3$. This is water of hydration. Sometimes the alum is shown with 14 waters of hydration, but more often the equation is written without it.

$$Al_2(SO_4)_3 + 3Ca(HCO_3)_2 \rightarrow 2Al(OH)_3 + 3CaSO_4 + 6CO_2$$

Chemically, the water of hydration is not important; it does not enter the reaction. It does, however, add mass, and it is important when you purchase alum, because the molar masses are 342 for $Al_2(SO_4)_3$, 594 for $Al_2(SO_4)_3 \cdot 14H_2O$, and 667 for $Al_2(SO_4)_3 \cdot 18H_2O$. The bulk density of ground commercial alum is 1.04 kg/m^3.

The charge-neutralizing ions of alum are Al^{3+}, $Al(OH)^{2+}$, and $Al(OH)_2^+$. Alum is effective in the range of its minimum solubility, pH 6–8. The precipitate $Al(OH)_3$ assists the flocculation process by sweeping up small particles and contributing weight. The hydroxide–colloidal agglomerates quickly grow large enough to be easily removed by settling and filtration. The aluminum sulfate reacts with the natural alkalinity in the water (represented by $Ca(HCO_3)_2$) and produces CO_2 in the form of carbonic acid. This destruction of alkalinity and production of acid tends to depress the pH of the solution. It may make it necessary to add alkalinity to promote the desired reactions. Alum coagulation is effective in the pH range of 5–8.

Similar reactions occur when iron salts are used. $Fe(OH)_2$ or $Fe(OH)_3$ floc will form, alkalinity will be destroyed, and the pH will tend to drop. The solubility of Fe^{3+} is lower than Al^{3+}. The minimum solubility is near pH 8, but the range of maximum effectiveness is pH 5–9, where more positively charged species exist and where the negative charge on colloids and natural organic matter (NOM) is less.

The solubility of aluminum and ferric iron is discussed in more detail in Chapter 8. Figures 8.5 and 8.6 show the solubility of $Fe(OH)_3$ and $Al(OH)_3$ as a function of pH.

Example 6.3 Color Removal from an Industrial Waste

Treatment of 20,000 m^3/day of colored industrial wastewater was investigated by doing batch laboratory jar tests using alum [$Al_2(SO_4)_3 \cdot 14H_2O$] as a coagulant and lime [$Ca(OH)_2$] for pH control. The best alum dose depended on the alkalinity of the wastewater, the pH, and the nature of color-causing material. The untreated wastewater characteristics were: pH = 7.65, total alkalinity = 100 mg/L as $CaCO_3$, color = 400 units. The quality of water needed for reuse was: pH = 6.0–8.0, color ≤ 30 units. The jar test results are given in Table 6.4.

Figure 6.4, a plot of the data, shows that

- If the lime dose is zero, color removal is sensitive to alum dose and a high dose is needed. The residual color standard of 30 units is met at a dose of 300 mg/L alum, but the pH is below the desired minimum of 6 units.
- If the lime dose = 100 mg/L, color target is met with 200 mg/L alum and the pH = 7.55 is acceptable.
- If the lime dose = 200 mg/L, color removal over the full range of alum doses is less than 30, but the pH exceeds 8 units unless the alum dose is at least 300 mg/L. In this case, alum is being used as an acid to neutralize the wastewater, as well as for coagulation of the color compounds.

TABLE 6.4
Color Reduction by Coagulation Using Lime and Alum

Alum Dose (mg/L)	Lime Dose (mg/L)	pH (units)	Color (units)
0	0	7.65	400
100	0	6.91	170
200	0	6.38	60
300	0	5.97	30
400	0	4.95	37
100	100	8.47	130
200	100	7.55	30
300	100	6.98	32
400	100	6.02	52
100	200	9.45	29
200	200	8.32	23
300	200	7.87	25
400	200	7.15	30

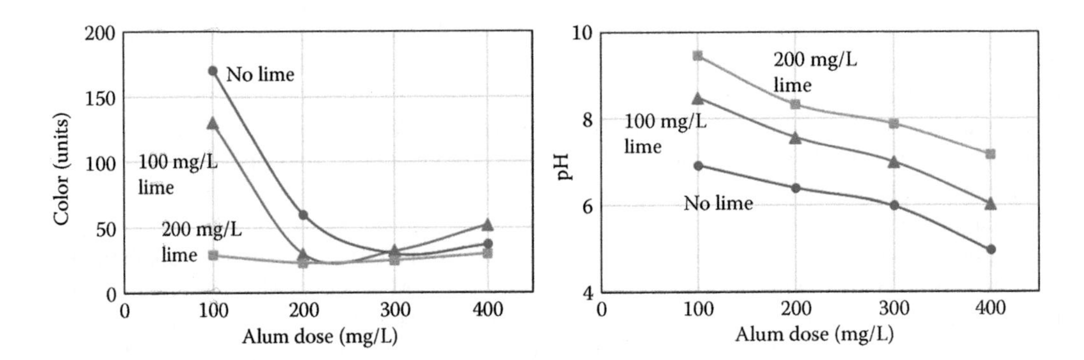

FIGURE 6.4 Plots of color and pH vs alum dose for coagulation data.

- No cost data are given, but lime is cheaper than alum, so we prefer to keep the alum dose low. The best combination seems to be 200 mg/L alum and 100 mg/L lime.
- There might be a better lime–alum dose that was not tested. It would be a good idea, if possible, to try tests in the region of lime = 100–200 mg/L and alum = 50–150 mg/L.

Sludge production data are not available, but suppose that it had been measured and it was found that a 200 mg/L alum dose yielded 200 mg/L settled solids that will need removal and disposal. The mass of solids (M) produced will be

$$M = (200\ \text{mg/L})(20{,}000\ \text{m}^3/\text{day})(1000\ \text{L/m}^3)\ (\text{kg}/10^6\ \text{mg}) = 4000\ \text{kg/day}$$

With a solids concentration of 3% and a specific gravity of 1.02, the sludge flow, Q, will be

$$Q = \frac{4000\ \text{kg solids/day}}{(3\ \text{kg solids/100 kg sludge})(1.02\ \text{kg sludge/L})} = 130{,}719\ \text{L/day} = 131\ \text{m}^3/\text{day}$$

Options for sludge treatment include dewatering by filtration or centrifugation. Blending with sludge produced in other waste treatment processes at the plant may improve its dewatering characteristics.

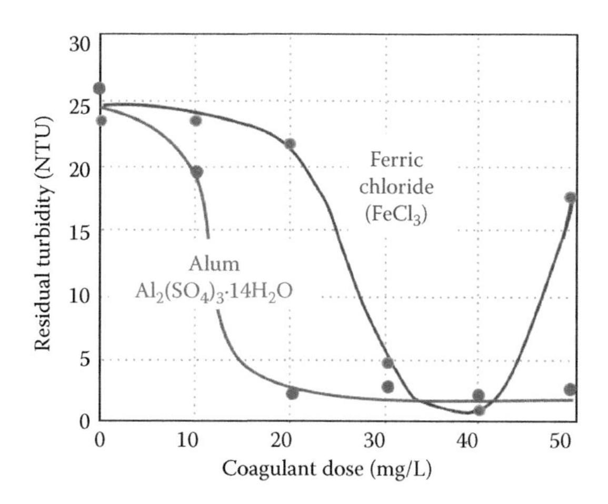

FIGURE 6.5 Jar test results for turbidity removal with alum and ferric chloride.

Example 6.4 Turbidity Removal by Coagulation–Settling

A 6-jar test machine (see Figure 6.1) was equipped with beakers (jars) that could hold 1 L of wastewater. Tests were done at coagulant concentrations of 0, 10, 20 30, 40, and 50 mg/L. Stock solutions of alum and ferric chloride were available in a strength of 1000 mg/L. A 10 mg/L concentration is obtained by adding 10 mL of stock solution to the 1 L sample.

This is the procedure:

1. Start the stirring machine at 80 rpm.
2. Add the chemicals and mix at 80 rpm for 1 min.
3. Reduce the speed to 20 rpm and continue stirring for 20 min.
4. Record a description of the floc in each beaker every 5 min.
5. Stop the stirring and allow the floc to settle for 30 min.
6. Decant a sample and measure the turbidity.

Do this with alum as the coagulant, and repeat the procedure with ferric chloride as the coagulant.

The results are shown in Figure 6.5. The minimum turbidity is about 2 NTU for both coagulants. The most notable observations are that alum is effective at 20 mg/L, and it is effective over a wide range of doses. The dose for ferric chloride must be carefully controlled at about 40 mg/L. This knowledge is relevant to process-control stability and to cost considerations.

6.4 EMPIRICAL STOICHIOMETRY: ESTIMATING SOLID REACTION PRODUCTS

Stoichiometry tells us nothing about the size, density, or settling velocity of the resulting precipitate solids. It tells us nothing about the concentration of the resulting sludge. It cannot because these depend on the sludge composition, which comprise organic or inorganic suspended solids that have nothing to do with the precipitation reaction. Sludge concentration also depends on the design and operation of the physical treatment equipment.

This section will discuss the problems of estimating solid reaction products for a coagulation/flocculation process and for chemical precipitation of phosphorus from wastewater. This discussion continues in Chapter 8—Precipitation Reactions.

Coagulation/flocculation followed by settling and/or filtration is often an effective and economical treatment process. A key part of the design is to estimate the mass and volume of sludge that must be handled. In most real cases, the ideal stoichiometry will not give an accurate prediction.

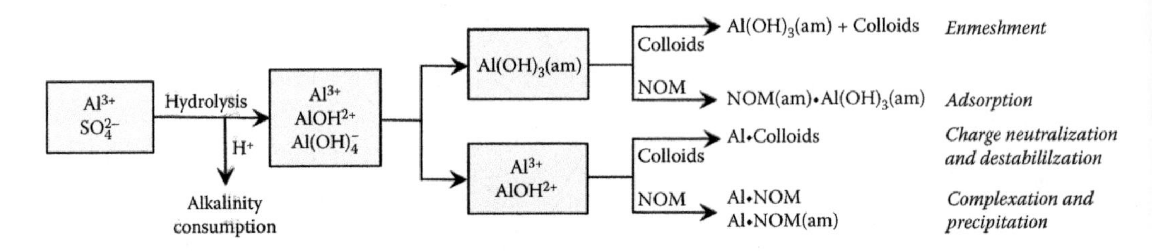

FIGURE 6.6 The creation of solids when aluminum sulfate $(Al_2SO_4)_3$ is added to water that contains natural organic matter (NOM) and colloidal solids. (Pernitsky, D.J., 2003; Pernitsky, D.J. and Edzwald, J.K., 2003.)

Water and wastewater treatment chemistry often involves colloids and suspended-solids products. The colloids may be the target of the treatment process, as in the removal of turbidity from river water, or enhancement of the removal of finely divided suspended solids in a settling tank or a filter. Or, they may coincidentally come into the picture when another species is the target for removal, such as in precipitation of phosphorus from wastewater, or in the precipitation of heavy metals. Either way, whatever happens to the colloids happens simultaneously and in addition to the desired reaction.

Figure 6.6 attempts to explain these complications in the context of alum coagulation to remove turbidity from water.

What we can calculate from a balanced chemical equation is the hydrolysis and formation of aluminum hydroxide and the consumption of alkalinity.

$$Al_2(SO_4)_3 + 6HCO_3^- \rightarrow 2Al(OH)_3(s) + 6CO_2 + 3SO_4^{2-}$$

Other ionized aluminum species depend on the pH, and calculations are unreliable.

$$Al_2(SO_4)_3 \rightarrow 2Al^{3+} + 3SO_4^{2-}$$

$$Al^{3+} + H_2O \rightarrow AlOH^{2+} + H^+$$

$$Al^{3+} + 2H_2O \rightarrow Al(OH)_2^+ + 2H^+$$

$$Al^{3+} + 4H_2O \rightarrow Al(OH)_4^- + 4H^+$$

These happenings cannot be calculated from the known stoichiometry:

$Al(OH)_3$(amorphous solid)	enmeshes colloids and suspended solids
$Al(OH)_3$(amorphous solid)	adsorbs natural organic matter (NOM)
$Al^{3+} + AlOH^{2+} + Al(OH)_4^-$	destabilize colloids so they coagulate
$Al^{3+} + AlOH^{2+} + Al(OH)_4^-$	form complexes and precipitate NOM

We have no way of showing colloids in a balanced chemical equation. We have no way of isolating them from the water to measure their mass, so we measure them indirectly as turbidity.

NOM is a collective measurement. We assume it measures dissolved organics, but we do not measure the organics directly. Often total organic carbon (TOC) or UV absorbance are used as a surrogate for NOM. TOC is another "lumped" measurement.

Suspended solids are not shown in Figure 6.6, but they will be incorporated into the aluminum hydroxide floc particles along with the colloids and NOM.

Removing phosphorus from wastewater by chemical precipitation with ferric iron (Fe^{3+}) yields large quantities of sludge, which is a mixture of precipitated phosphorus compounds, metal hydroxides, and the suspended solids that are coincidentally removed. Sludge disposal typically accounts for about half the cost of a wastewater treatment plant, so it is worth the effort to make the best possible estimates of sludge quantities.

The mass of particles produced could be calculated if the yield were the pure stoichiometric products. That means, if the stoichiometry says the product will be $FePO_4$, you actually do get $FePO_4$. The ideal stoichiometry is an approximation of the real reaction. There are several reasons why this may be a good approximation in a clean simple chemical system, but a poor one for wastewater treatment.

Consider this reaction: $Fe^{3+} + PO_4^{3-} \rightarrow FePO_4$

$FePO_4$ will form, but more complex phosphorus compounds may also form, with an empirical formula like $Fe_X(PO_4)_Y$, or $Fe_X(PO_4)_Y \cdot ZH_2O$. Some of the iron may form hydroxides, $Fe(OH)_2$ or $Fe(OH)_3$ and the hydroxides may entrain particulate solids. The ideal stoichiometry will predict the minimum amount of solids yield, but not the maximum, and not the actual mass, which may change from day to day with changing wastewater composition and operating conditions.

Two examples are given about sludge production when phosphorus is removed from municipal wastewater (Jenkins et al., 1971; Jenkins and Hermanowicz, 1991; Snurer, 2008).

Example 6.5 Phosphorus Precipitation with Pickle Liquor

A treatment plant that serves 145,000 people is ordered to reduce effluent phosphorus (P) from 5 mg/L to 0.5 mg/L. The per capita contributions to the treatment plant are 400 L/day and 2.6 g P. Almost all the phosphorus in municipal wastewater is in the form of phosphate, PO_4^{3-}, but concentrations are reported as P, not PO_4^{3-}.

The treatment requirements are as follows:

Wastewater volume = (145,000 persons)(400 L/day) = 58,000,000 L/day = 58,000 m^3/day
Influent P load = (145,000 persons)(2.6 g P/person) = 377,000 g P/day = 377 kg P/day
Influent P concentration = (377,000 g P/day)(1000 mg/g)/(58,000,000 L/day) = 6.5 mg/L
Existing influent concentration = 6.5 mg P/L
Existing effluent concentration = 5 mg P/L
Required effluent concentration = 0.5 mg/L
P removal required = 6.5 − 0.5 = 6.0 mg P/L

There are a number of ways to remove phosphorus. One possibility is precipitation with ferric chloride ($FeCl_3$). Pickle liquor is hydrochloric acid that contains ferric iron in the form of $FeCl_3$. A nearby industry has large quantities of waste pickle liquor that can be obtained for the price of hauling.

The phosphorus is in the form of phosphate (PO_4^{3-}). The chemistry seems simple—one molecule of ferric iron (Fe^{3+}) reacts with one molecule of phosphate (PO_4^{3-}) to form $FePO_4$, which is an insoluble particle.

	Fe^{3+} +	PO_4^{3-}	→	$FePO_4$
Molar masses	55.85 g	94.97 g		150.82 g

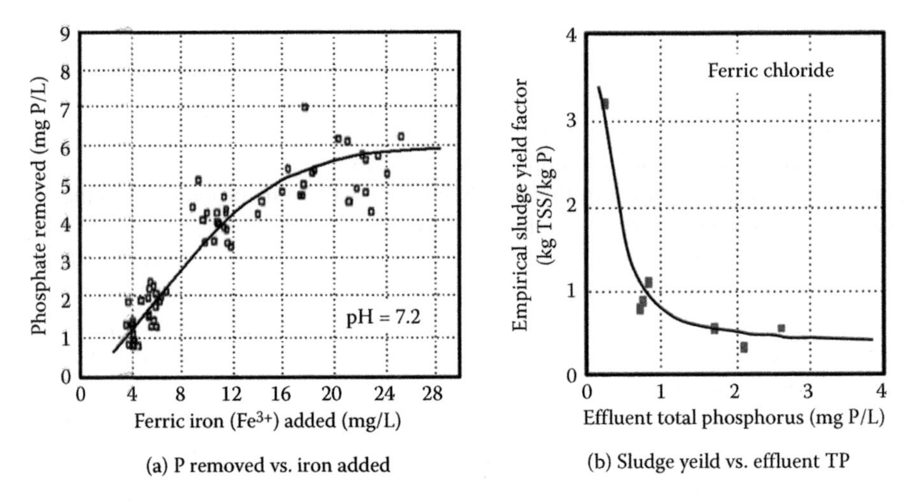

FIGURE 6.7 Experimental data relating (a) phosphorus removal to iron dose and (b) sludge production to effluent total phosphorus concentration. *Total = soluble + particulate forms of P.*

The molar mass of P is 30.97 g. By this stoichiometry, removing one kg of P generates (150.82/30.97) = 4.87 kg of iron phosphate ($FePO_4$).

It seems simple. But—and it is a big but—it takes more than one mole of iron to precipitate one mole of phosphorus. Some reports say it is 2 Fe to 1 P removed.

The excess iron reacts with non-P compounds to form non-P precipitates. And the iron phosphate probably is not pure $FePO_4$. It could be something like $Fe_{1.5}(PO_4)_2 \cdot$Gunk. And, all the iron in the pickle liquid may not be in the ferric form (Fe^{3+}). Ferrous iron (Fe^{2+}) will form a precipitate with phosphorus, but the stoichiometry is different. Also, Fe^{2+} and Fe^{3+} form hydroxides, $Fe(OH)_2$ and $Fe(OH)_3$.

The two graphs in Figure 6.7 show data for the addition of ferric chloride to the aeration tank of an activated sludge process. They are different studies, so be careful with the interpretation.

The influent P concentration is not shown. The typical value when these studies were done was 7–8 mg P/L, and within 10 years it had declined to 5–6 mg/L because of changes in the formulation of household chemicals, laundry detergents in particular.

Graph (a) Experiments were done with a controlled effluent pH = 7.2.
This is a typical pH for domestic wastewater.
P removal per unit of Fe added decreases at higher levels of removal.

Graph (b) Sludge yield increases sharply when effluent P is forced below about 1 mg/L.
This is also shown by the flattening of the curve in graph (a) at higher Fe doses.
Formation of $FePO_4$ is stoichiometric (linear) down to about 1 mg/L.
At lower levels of effluent total phosphorus, $FeCl_3$ is being used to depress the pH.

Excess sludge production is due to formation of non-$FePO_4$ solids.

Estimated solids yield based on graph (b)

P removed = (377 kg P/day)(6/6.5) = 348 kg P/day

Sludge yield coefficient at effluent TP = 0.5 mg P/L = 1.3 kg TSS/kg TP removed

Sludge production = (348 kg P/day)(1.3 kg TSS/kg TP) = 452 kg TSS/day

Example 6.6 Chemical Treatment of Wastewater

A chemical coagulant will be added to raw wastewater as it enters a primary settling tank. This will remove phosphorus and suspended solids and perhaps colloidal solids. Empirical models for estimating sludge production were developed from a series of jar tests by Snurer (2008).

$$S_{Fe} = 27.41 + 2.65\,(P_{In} - P_{Out}) + 1.78\,Fe^{3+}_{added} + (TSS_{In} - TSS_{Out})$$

$$S_{Al} = 21.34 + 2.77\,(P_{In} - P_{Out}) + 3.07\,Al^{3+}_{added} + (TSS_{In} - TSS_{Out})$$

S_{Fe} and S_{Al} are total sludge solids formed from iron or aluminum-based coagulants.
S_{Fe} and S_{Al} are measured as mg dry solids/L.
Fe_{Added}, Al_{Added}, TSS_{In} and TSS_{Out} are measured as mg/L.

The wastewater flow is 20,000 m^3/day with a suspended solids concentration of TSS = 350 mg/L. The TSS removal will be 85% if the coagulant is $FeCl_3$. The removal will be 80% if the coagulant is $Al_2(SO_4)_3$. The influent phosphate (PO_4^{3-}-P) concentration is 8 mg P/L. The effluent concentration will be 1 mg P/L for both coagulants.

Assume ideal stoichiometry for phosphate precipitation to calculate the amount of aluminum and ferric iron added.

The ideal chemistry is

	Fe^{3+} +	PO_4^{3-}	→	$FePO_4$
Molar mass	55.85 g	94.97 g		150.82 g

	Al^{3+} +	PO_4^{3-}	→	$AlPO_4$
Molar mass	26.98 g	94.97 g		121.95 g

1 mole of PO_4^{3-}-P reacts with 1 mole of Fe^{3+} or 1 mole of Al^3. Likewise, 1 mole of P reacts with 1 mole of Fe^{3+} or 1 mole of Al^{3+}.

$$P_{In} - P_{Out} = 8 \text{ mg P/L} - 1 \text{ mg P/L} = 7 \text{ mg P/L}$$

For Fe^{3+} $\quad TSS_{In} - TSS_{Out} = 0.85(350 \text{ mg TSS/L}) = 298 \text{ mg TSS/L}$

For Al^{3+} $\quad TSS_{In} - TSS_{Out} = 0.80(350 \text{ mg TSS/L}) = 280 \text{ mg TSS/L}$

Molar mass of P = 30.97 g

Stoichiometric doses:

$$Fe = (8 \text{ mg P/L} - 1 \text{ mg P/L})(55.85 \text{ mg Fe}/30.97 \text{ mg P}) = 12.6 \text{ mg Fe/L}$$

$$Al = (8 \text{ mg P/L} - 1 \text{ mg P/L})(26.98 \text{ mg Al}/30.97 \text{ mg P}) = 6.1 \text{ mg Al/L}$$

Mass of solids calculated from the ideal stoichiometry

$$(7 \text{ mg P/L})(150.82 \text{ mg } FePO_4/30.97 \text{ mg P}) = 34.1 \text{ mg } FePO_4/L$$

$$(7 \text{ mg P/L})(121.95 \text{ mg } AlPO_4/30.97 \text{ mg P}) = 27.6 \text{ mg } AlPO_4/L$$

Mass of solids predicted using the empirical model

$$S_{Fe} = 27.41 + 2.65(7 \text{ mg P/L}) + 1.78(12.6 \text{ mg Fe/L}) + 298 \text{ mg TSS/L}$$
$$= 366 \text{ mg dry solids/L}$$

$$S_{Al} = 21.34 + 2.77(7 \text{ mg P/L}) + 3.07(6.1 \text{ mg Al/L}) + 298 \text{ mg TSS/L}$$
$$= 357 \text{ mg dry solids/L}$$

6.5 CASE STUDY: STORMWATER TREATMENT BY COAGULATION

The Washington State Department of Transportation wanted to remove sediments and the heavy metals they adsorb from highway runoff (Yonge, 2011). Stormwater detention basins rely on gravity settling to capture sediments. They are effective at capturing the larger particles that are a large percentage (mass basis) of the sediment load. Figure 6.8 shows that adsorbed metals are associated with smaller particles. Particulate zinc is approximately 3.5 times greater on 4 μm particles than on 17 μm particles (average approximate spherical diameter). Lead is almost six times greater. This is because the small particles have a higher organic carbon content and a larger ratio of surface area to mass.

Detention basin effectiveness for removing metals will be increased if they are designed and operated to capture fine sediments. These small particles cannot be removed by gravity settling alone, so chemical coagulation becomes an interesting possibility. Jar tests are an effective tool for evaluating coagulation before moving on to more realistic large-scale testing.

It would be convenient to have "typical highway runoff" of a consistent composition over the course of the test program. This is impossible so *typical stormwater runoff* was prepared to have this composition:

Suspended solids	500 mg/L
Lead	1.8 mg/L
Cadmium	0.06 mg/L
Copper	0.18 mg/L
Zinc	1.3 mg/L

The suspended solids were lake sediment particles that were smaller than 0.075 mm.

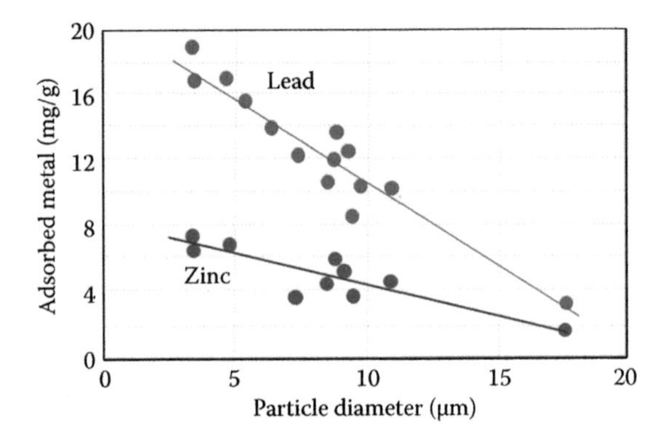

FIGURE 6.8 Lead and zinc adsorption to particles is correlated with particle size. Smaller particles adsorb proportionally more metals.

The experimental factors are as follows:

Jar test procedure:	0.5 min. rapid mix, no slow mix, 4–11 min. settling
Coagulants:	$Al_2(SO_4)_3 \cdot 18H_2O$ or $FeCl_3 \cdot 6H_2O$
Cationic polymers:	SWT 848 or SWT 976

The jar test procedure is unusual because there is no slow mix period. These conditions were thought to be a better simulation of what would happen in a stormwater detention basin treatment system. Chemicals would be added at the inlet, there would be a short but intense mixing period, and then settling for a time that depends on the stormwater flow rate and the volume of the basin.

Notice that the factors are discrete and categorical. There are four discrete choices of coagulating chemicals. The chemicals were not tested in combination. This is not a two-level factorial experimental design, as discussed in Section 6.6.

Preliminary tests showed that $Al_2(SO_4)_3 \cdot 18H_2O$ was ineffective in the 0.5 min rapid mix period. SWT 976 performed well, but not better than SWT 848. $FeCl_3 \cdot 6H_2O$ and SWT 848 showed rapid floc formation and good solids settling. The optimum dosage for $FeCl_3 \cdot 6H_2O$ was 15 mg/L as Fe. Careful dose control was needed because doses 15% more or less than the optimum caused a significant loss in performance. The SWT 848 dosage was 50 mL/L and performance was less sensitive to dosage variations.

The jar test results were enough to justify continuous flow model studies using $FeCl_3$ and SWT 848 as the coagulants. They were not used in combination.

The model detention basins were operated at different levels of surface overflow rate (SOR)

$$\text{SOR (m/s)} = \frac{Q\ (\text{m}^3/\text{h})}{A\ (\text{m}^2)}$$

where Q = stormwater flow rate (m^3/h)
A = area of the stormwater detention basin (m^2)

This can also be defined in terms of the detention basin depth and detention time. Define
θ = hydraulic detention time, h
V = the basin volume, m^3
d = basin depth, m

$$\theta = \frac{V}{Q} = \frac{Ad}{Q} \quad \text{and} \quad Q = \frac{Ad}{\theta}$$

$$\text{SOR} = \frac{Q}{A} = \left(\frac{Ad}{\theta}\right)\frac{1}{A} = \frac{d}{\theta}$$

The SOR can be reduced by reducing the flow through the detention basin, or by increasing the surface area of the basin.

A basin with a larger surface area will capture more small solids (slower settling particles), assuming other basin characteristics are the same. This means that reducing the SOR will increase the efficiency of the detention basin. It will capture more particles and it will capture more small particles and the metals that are adsorbed to them.

Model basin tests were done at four SOR rates (0.52, 0.74, 0.87, and 1.09 m/h). The top left panel of Figure 6.9 shows the lead removal efficiency as a function of SOR for no coagulant (None), $FeCl_3$, and SWT 848 cationic polymer. A portion of the experimental data are reproduced in the plot. The lower SOR removes more of the smaller particles. Both coagulants roughly double the metals removal of a stormwater basin where no coagulant is used.

The remaining panels show the results for copper, zinc, and cadmium. They are similar to the results for lead, but with somewhat lower removal efficiencies.

An unknown fraction of the total metals is soluble and will not be affected by the coagulation process.

Treatment of a storm event can be achieved by (1) treatment of an entire storm event or (2) treatment of the "first flush." First flush treatment is based on time-dependent fluctuations in stormwater characteristics. The first 20% of the total stormwater volume from a drainage basin may contain 80% of the total contaminant load. However, measurement of critical runoff constituents with respect to time during a storm event may allow for the determination of a site-specific first flush volume. If first flush treatment is determined to be appropriate for a given drainage, coagulant cost savings and lower sludge production are the expected advantages.

Chemical costs for a typical storm event can be estimated by applying storm event information for a particular region of interest. For example, using a Type 1A storm event (1.79 in. or 4.55 cm total rainfall) in the Olympia, Washington area, the runoff volume from 1000 ft (305 m) of four-lane highway is approximately 50,000 gal (189 m³).

Assuming that the coagulant dosage defined by this research is applicable to the field stormwater, the storm event would require 2.5 gal (9.5 L) of SWT 848 at a 50 ppmv application. Chemical cost for the entire storm event would be approximately $15 ($6/gal or $1.58/L).

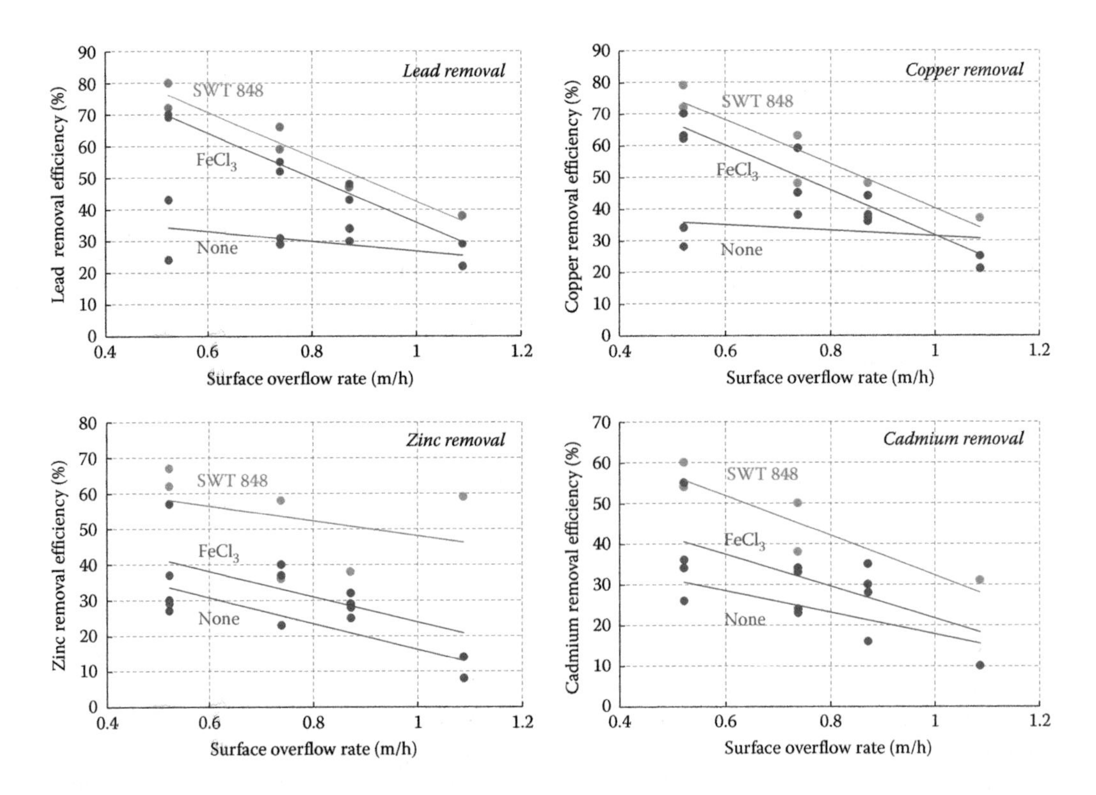

FIGURE 6.9 Removal efficiency of lead and other metals as a function of chemical added and surface overflow rate of the stormwater detention basin.

6.6 STATISTICAL EXPERIMENTAL DESIGN FOR JAR TESTING (AND A CASE STUDY OF EMULSIFIED OIL REMOVAL)

Jar test experiments can yield valuable information for process design. Unfortunately, many jar test experiments are poorly designed because experimenters (engineers and scientists) have been incorrectly taught that varying one factor at a time is good experimental strategy. Data from a real experiment will be used to show that this is a poor strategy and to illustrate an improved design.

Experimental design in this context is the specification of test conditions, the settings of experimental factors such as pH, mixing intensity, mixing time, chemical additions, etc., that can be controlled by the experimenter. The other more obvious aspect of experimental design is the fabrication and assembly of physical equipment.

An industrial wastewater with an initial concentration of 5000 mg/L emulsified oil needed to be cleaned before discharge. The treatment plan was to break the oil emulsion so the oil could be removed by flotation. Jar tests were used to determine the doses of sulfuric acid (H_2SO_4) and ferric chloride ($FeCl_3$) that would give a final oil concentration of less than 100 mg/L.

The first set of five experiments fixed the acid dose at 0.1 g/L and varied the $FeCl_3$ dose from 1.0 to 1.4 g/L. The best residual concentration obtained was 175 mg/L. Based on this result, a second set of tests was done, this time fixing $FeCl_3$ at the apparent best dose of 1.3 g/L in combination with acid doses of 0, 0.1, and 0.2 mg/L. The minimum residual oil concentration was, again, 175 mg/L. The test results for all eight experiments are shown in Table 6.5 and Figure 6.10.

TABLE 6.5
Results of Eight Runs for a One-Factor-at-a-Time Experimental Design

First Set of Experiments: Fix H_2SO_4 Dose and Vary $FeCl_3$ Dose

$FeCl_3$ (g/L)	1.0	1.1	1.2	1.3	1.4
H_2SO_4 (g/L)	0.1	0.1	0.1	0.1	0.1
Residual oil (mg/L)	4200	2400	1700	175	650

Second Set of Experiments: Fix $FeCl_3$ Dose and Vary H_2SO_4 Dose

$FeCl_3$ (g/L)	1.3	1.3	1.3
H_2SO_4 (g/L)	0	0.1	0.2
Residual oil (mg/L)	1600	175	500

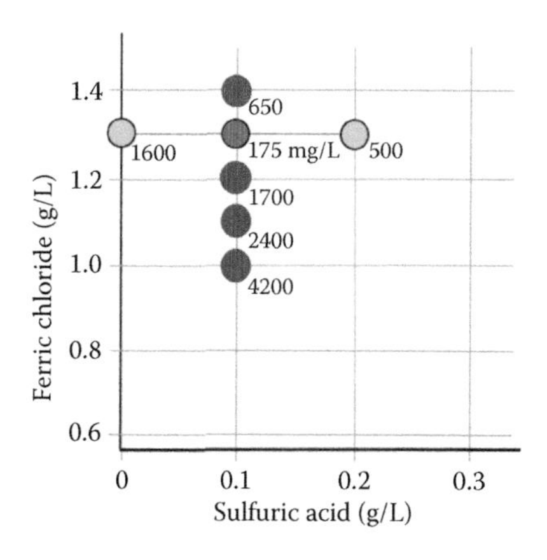

FIGURE 6.10 Results of eight runs on a one-factor-at-a-time experimental design.

This seems to suggest that the best combination of these chemicals, at the massive doses of 0.1 g/L H_2SO_4 and 1.3 g/L $FeCl_3$, could not meet the target of 100 mg/L oil.

This kind of experimental design is called *one-factor-at-a-time* (*OFAT*).

A fundamentally different kind of experimental design simultaneously changes both the acid and the iron reagent, each at two levels. There are two factors (acid and iron reagent) and two levels (high and low) so this is a *2-factor, 2-level design*, for a total of four experimental runs. It is called a 2^2 *experimental design.* (In general, a 2-level design with n factors is a 2^n design.)

The location of the first four runs is based on the experimenter's prior knowledge of the process, which may be a lot or very little. Start at what you believe are best conditions and let the first four runs direct the design of the next four experiments.

Figure 6.11 shows two design cycles of the 2^2 experiment. The first four runs are on the left. It is just luck that this small amount of work has already identified a better operating condition than the seven runs of the OFAT experiment. We accept this good fortune and explore for even better results. The four original experiments suggest a promising direction—downward and to the right—that is, more sulfuric acid and less ferric chloride.

Three new experiments are done, keeping the 100 mg/L result in the pattern. This does indeed bring a better result, as we hoped, but this result was not guaranteed.

Two-level factorial experiments are efficient because a small number of initial tests usually will reveal a hill that can be climbed to maximum performance, or, in our case, a valley that can be followed to lower residual oil concentrations. They can be run with any number of factors.

The strategy is to cycle through the following steps:

$$\text{Design} \rightarrow \text{Experiment} \rightarrow \text{Analysis}$$

Each new design cycle can be expanded, augmented, shifted, or otherwise modified based on what has been learned in the earlier experiments. This is an *iterative experimental design* or *exploratory design*.

In fairness to the experimenters whose data we have borrowed (Pushkarev et al., 1983) to make our case for factorial experimental design, we now show their data in Table 6.6 and Figure 6.12. Twenty-eight data points are reported, but there may have been other experiments for which the oil residual was not measured because the results were obviously uninteresting.

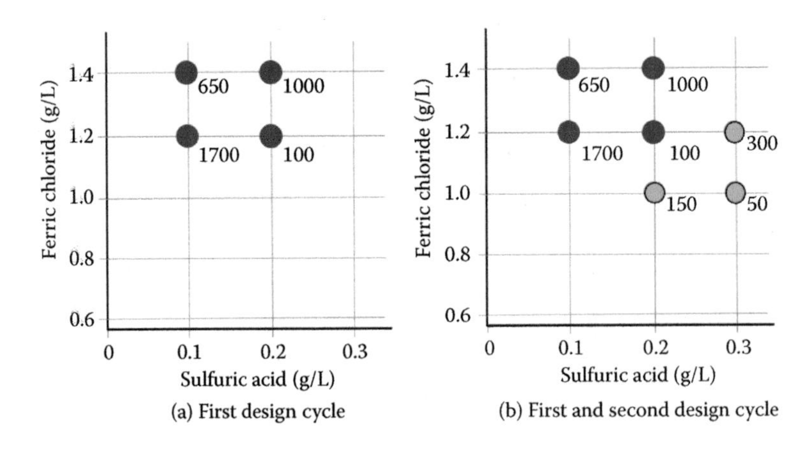

FIGURE 6.11 Two design/experiment cycles of a 2-level, 2-factor experimental design.

TABLE 6.6
Residual Oil (mg/L) after Treatment by Chemical Emulsion Breaking and Flotation

$FeCl_3$ Dose (g/L)	Sulfuric Acid Dose (g/L H_2SO_4) 0	0.1	0.2	0.3	0.4
0.6	—	—	—	—	600
0.7	—	—	—	—	50
0.8	—	—	—	4200	50
0.9	—	—	2500	50	150
1.0	—	4200	150	50	200
1.1	—	2400	50	100	400
1.2	2400	1700	100	300	700
1.3	1600	175	500	—	—
1.4	400	650	1000	—	—
1.5	350	—	—	—	—
1.6	1600	—	—	—	—

Source: Pushkarev, V.V. et al., 1983.

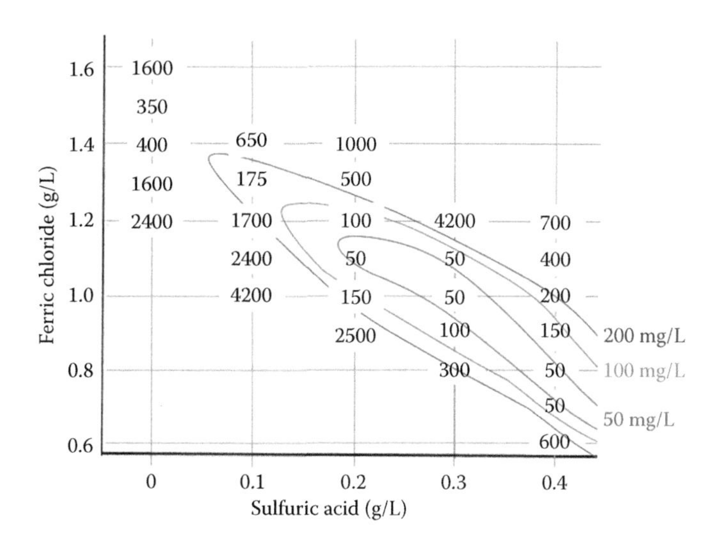

FIGURE 6.12 Data for 28 experimental runs from the oil removal experiments of Pushkarev et al. (1983).

The data show a narrow valley that slants from high $FeCl_3$–low H_2SO_4 to low $FeCl_3$–high H_2SO_4. There is an interaction between $FeCl_3$ and H_2SO_4.

The OFAT experiment cut across the valley, and did so in an unfortunate location where the oil removal efficiency was outside the region where the 100 mg/L target could be met. The fundamental problem with OFAT experiments is that they do not—they cannot—discover the interaction between variables. This is a major weakness, because interactions are common; in fact, they should be expected, and they may be the most interesting result of an experiment.

The interesting result here is that there is not an optimal operating *point*, there is an optimal region. This provides an opportunity to minimize the cost not only by minimizing the chemical doses, but also by finding the minimum combined chemical cost. $FeCl_3$ and sulfuric acid will have different unit costs and the choice will be more iron and less acid, or more acid and less iron.

Finally, we would like to say that this example was not contrived to criticize Pushkarev et al.'s experiment. They may have used a strategy something like we have recommended. Whatever their strategy, the optimal valley was discovered—and they did this with a relatively small number of experimental runs.

Nevertheless, we are justifiably critical of OFAT experiments. They are wasteful and inefficient because they can, and often do, fail to discover important information.

Two-level factorial designs are very efficient. They can be organized with a large number of factors. An initial exploratory design will lead to a better second phase of experiments, and that will lead to a better third round, if more research is indicated.

Two more examples will be given of how statistical experimental design and empirical chemistry have been used to learn about pollution control processes.

The first is decolorization of dye in a textile industry waste. These are *batch tests* in which ozone is fed to a reactor to oxidize the dye. Three factors are investigated. The response is color intensity after treatment.

The second is about discovering operating conditions for polyester manufacturing that will reduce the loss of ethylene glycol into the wastewater. The experiments were done in the full-scale plant under actual manufacturing conditions. The goal was to reduce the wastewater treatment cost.

The emphasis is on experimental strategy—the number of runs and the settings of the independent variables. Statistical calculations will not be shown or explained in any detail. The interested reader can refer to Box, Hunter and Hunter (2005) and Berthouex and Brown (2002).

6.7 CASE STUDY: DECOLORIZATION EXPERIMENT

This experiment deals with decolorization of a complex textile dye in synthetic wastewater. It was expected that ozone (O_3) and hydrogen peroxide (H_2O_2) would be effective when used in combination. Other variables were initial dye concentration, initial pH, and treatment time (Hribernik et al., 2009).

Preliminary tests showed that H_2O_2 addition was not needed and it was dropped in the subsequent experiments. The four factors in Table 6.7 were investigated at low and high level (2 levels). The levels are based on the experimenter's knowledge of the process. The dye levels in the wastewater are estimates for a minimum 5% dye loss and a maximum 20% dye loss, using a known dye in an existing dying process.

TABLE 6.7
High and Low Settings for the Four Factors of a 2^4 Experimental Design to Investigate Decolorization of Textile Dye Wastewater

Factor		Level Low	Level High
A	Initial dye conc. (g/L)	0.1	0.4
B	O_3 flow rate (g/h)	1	2
C	Initial pH	4	12
D	Treatment time (min)	10	40

The experiments were done with ozone gas supplied at the bottom of a 1-L batch reactor. This experiment is a *jar test* in the sense that it tries to simulate a real process with small batch experiments. It is empirical stoichiometry because, while the chemical composition of the dye is known, the stoichiometric reaction with ozone is not. Also, there may be interactions. For example, the effect of pH on the process may be different at high dye concentration than at low dye concentration. Interactions of this kind cannot be captured in a simple stoichiometric equation.

Testing all combinations (treatments) of the four factors at low and high levels requires a minimum of $2^4 = 16$ experimental runs. Four factors at two levels is a 2^4 factorial experimental design. The "2" in the designation is the number of levels and the exponent is the number of factors.

This experiment had duplicate runs of all 16 treatments. The duplicate measurements are averaged to do the calculations. The experimental settings and the results are shown in Table 6.8. The response is absorbance at a wavelength of 525 nm.

Consider the power of this experimental design. There are eight direct comparisons of decolorization for factor A, when it is changed from the low initial dye concentration to the high dye concentration. Runs 1 and 2 make this comparison for O_3 flow rate = 1 g/h, initial pH = 4, and treatment time = 10 min. Runs 3 and 4 make this comparison for O_3 flow rate = 2 g/h, initial pH = 4, and treatment time = 10 min. And so on for runs 5 and 6, 7 and 8, 9 and 10, etc. The 16 runs give eight of these direct comparisons for factor A at all high and low levels of O_3 flow rates, initial pH, and treatment times.

These eight comparisons are averaged to estimate the effect of factor *A* (initial dye concentration), which is +0.450. The effect of factor A is the average difference in absorbance caused by changing a factor from the low level to the high level. The absorbance of the treated wastewater, on average over all treatment levels, increases by 0.45 units when the dye concentration changes from the low to the high concentration.

TABLE 6.8
Replicated 2^4 Factorial Experimental Design to Investigate Four Factors That Influence Decolorization of Textile Wastewater

Factor	*A*	*B*	*C*	*D*	Response, *y*		
Run	Initial Dye Conc. (g/L)	O_3 Flow Rate (g/h)	Initial pH (units)	Time (min)	Absorbance (A) at $\lambda = 525$ nm		Average Response
					Rep. 1	Rep. 2	
1	0.1	1	4	10	0.14300	0.13700	0.14000
2	0.4	1	4	10	0.83500	0.82300	0.82900
3	0.1	2	4	10	0.07062	0.07320	0.07191
4	0.4	2	4	10	0.71600	0.72800	0.72200
5	0.1	1	12	10	0.10611	0.09500	0.10056
6	0.4	1	12	10	0.92000	0.93000	0.92500
7	0.1	2	12	10	0.02100	0.03300	0.02700
8	0.4	2	12	10	0.66200	0.68300	0.67250
9	0.1	1	4	40	0.01588	0.00900	0.01244
10	0.4	1	4	40	0.29100	0.28300	0.28700
11	0.1	2	4	40	0.00500	0.00010	0.00255
12	0.4	2	4	40	0.13900	0.14500	0.14200
13	0.1	1	12	40	0.00010	0.00010	0.00010
14	0.4	1	12	40	0.26828	0.28970	0.27899
15	0.1	2	12	40	0.00010	0.00010	0.00010
16	0.4	2	12	40	0.09900	0.10300	0.10100

The effect of factor *A* can also be computed from the difference between the average response at the high level of A and the average response at the low level of *A*. The effects for factors *B*, *C*, and *D* are calculated in a similar way; that is, by averaging the differences between the high and low levels.

The effect for factor *B* is –0.104. The negative sign indicates that on average the absorbance decreases as factor B (ozone flow rate) is changed from the low level to the high level.

There are 16 independent observations (16 averages of duplicate absorbance measurements), so 16 effects can be estimated. Four of these are the main effects for *A*, *B*, *C*, and *D*. Six are two-factor interactions for *AB*, *AC*, *AD*, *BC*, *BD*, *CD*, and four are three-factor interactions.

Figure 6.13 shows the *main effects* and *interactions* that were estimated from the duplicate measurements of absorbance. (Berthouex and Brown, 2002 explain the calculations.)

Factorial experimental designs will discover important interactions. OFAT experiments cannot do this.

The *two-factor interactions* are the average change in the response when two factors are simultaneously changed from the low to the high levels. If a two-factor interaction is zero, the two factors more than likely act independently of each other in affecting the response. Two-factor interactions are very often important. The two-factor interactions *AB* and *AD* are large enough to be important in this experiment.

It is not common to find that the three-factor interactions are significant. Like the main effects, the interactions can also be positive or negative.

The main effects for *A*, *B*, and *D* are the largest, which means they are the most important in the practical business of removing color from this dye wastewater. To use the jargon of the statistician, they are *significant*. They are larger than we would expect to see if the variation in the response measurements were pure random measurement error. Having duplicate observations allows the random measurement error to be calculated and compared with the magnitude of the effects (Berthouex and Brown, 2002).

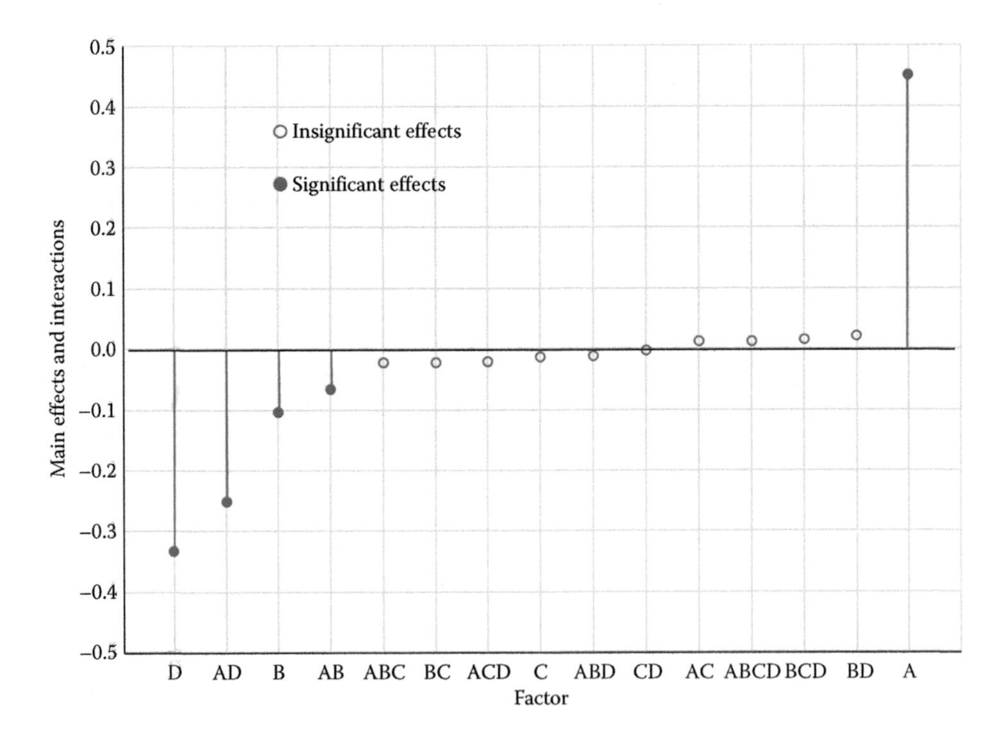

FIGURE 6.13 The *effects* that were estimated from the duplicate measurements of absorbance. The solid red circles with vertical lines identify the significant effects, D, AD, B, AB, and A.

Figure 6.13 shows that 10 of the 15 effects are negative. A negative effect means that increasing the factor reduces the final color (increases the percent color removal). More ozone removes more color. Longer treatment time removes more color. This is a good result for a treatment process that is supposed to remove color (decrease absorbance). The opposite outcome would be a great disappointment.

An unexpected result is that pH appears to have no effect. Its main effect is −0.013, which is so small that it is considered to be zero. The pH range used in the experiment goes from acidic to strongly alkaline. The experimenters must have expected that these levels should be better than pH in a more neutral zone. Or perhaps they already knew the performance at neutral pH and felt that experiments at extreme pH levels were necessary. Or, it is possible that pH might be an active factor at some level between 4 and 12. We have no way to know from the available data.

An empirical model can be fitted to the data. It should be of the form

$$y = \beta_o + \beta_A A + \beta_B B + \beta_D D + \beta_{AB} AB + \beta_{AD} AD$$

where: β are the coefficients of the factors A, B, D, AB, and AD, and
A, B, D, AB, and AD are the numerical levels of those factors.

Only the factor effects that are important are included. Factor C (pH) is not included. The fitted model, obtained using any of the common linear regression programs, is

$$y = -0.1877 + 3.5644A + 0.00637B + 0.002896D - 0.4425AB - 0.05598AD$$

The *coefficients* are the change in y caused by changing a factor by one unit. The *effects* are the change in y caused by changing the level of a factor from low to high.

Figure 6.14 is a response surface plot of the absorbance as a function of the three active factors (initial dye concentration, ozone flow rate, and reaction time). The plot confirms that absorbance decreases rapidly as initial dye concentration decreases, and slowly as ozone flow rate increases.

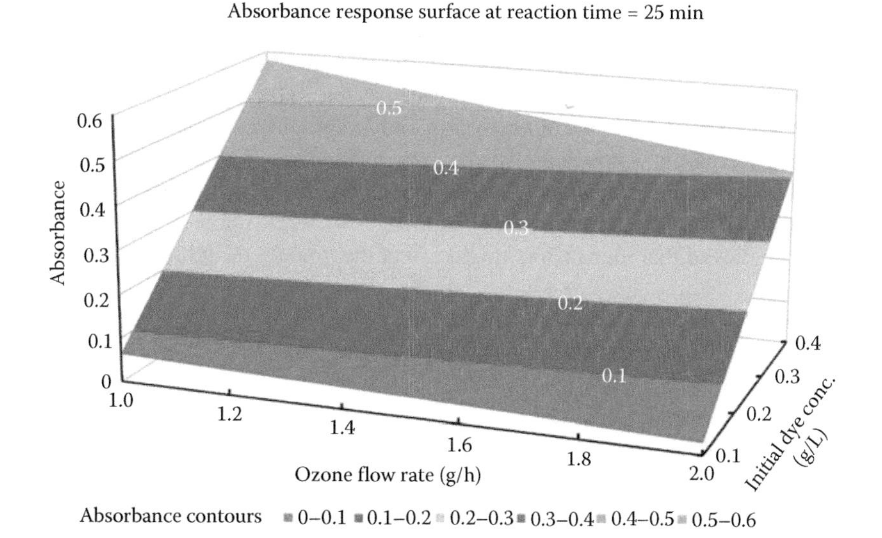

FIGURE 6.14 Response surface of absorbance as a function of initial dye concentration (factor A) and ozone flow rate (factor B) for decolorization process. The reaction time (factor D) is held constant at its average value, $D = 25$ min.

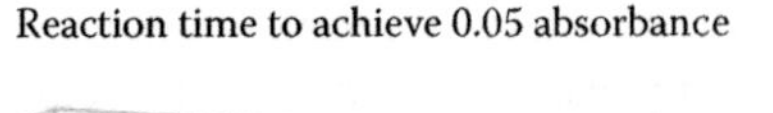

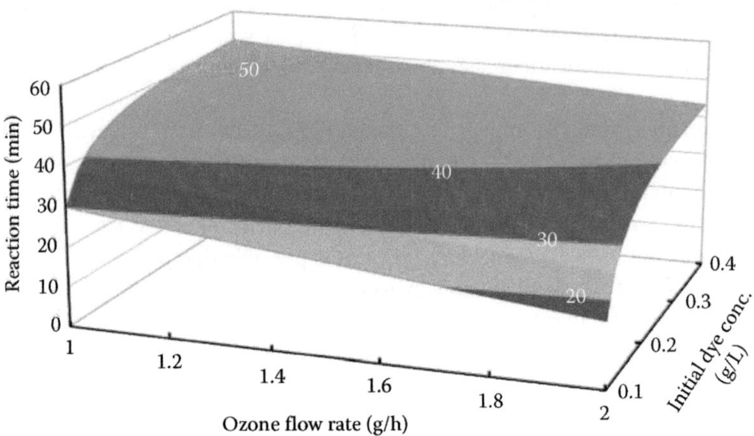

FIGURE 6.15 Model predicted reaction time required to achieve wastewater absorbance of 0.05 as a function of initial dye concentration and ozone flow rate.

The interaction between the two factors causes the decrease in absorbance to be more rapid at high initial dye concentrations than at low concentrations.

The model can also be used to predict any influential factor to achieve a given treatment objective, in this case an absorbance equal to a maximum allowed limit. Figure 6.15 shows the reaction time necessary to reduce the absorbance of the wastewater to 0.05. The figure clearly shows that wastewater decolorization increases significantly with reduced initial dye concentration and modestly with increasing ozone flow rate.

6.8 CASE STUDY: ETHYLENE GLYCOL FACTORIAL EXPERIMENTS

This study focused on the reduction of the wastewater treatment cost during a polyester production process. The volume of wastewater was constant, but the amount of BOD was variable because the amount of ethylene glycol (EG) in the wastewater was variable. The higher the EG, the higher the BOD, and the higher the treatment cost (Akesrisakul and Jiraprayuklert, 2007).

The source of the EG in the wastewater was the esterification reaction in the manufacturing process. The goal was to find the factors that influence EG losses, and to control those factors to minimize EG in the wastewater (and low wastewater treatment cost), while maintaining polyester quality.

Process experts believed that four factors might affect the amount of EG in the wastewater:

T = reaction temperature, which was set at the heat exchanger
S = slurry level in the heat exchanger
F = feed rate of EG
T_{VS} = temperature at vapor separator

These factors were investigated in preliminary experiments. All experiments were performed under actual manufacturing conditions and the level and range of the experimental factors were held within the process specifications.

The response was concentration (mg/L) of EG in the process water effluent.

It was quickly discovered that T_{VS} had a negligible effect over the range of 103°C–109°C that was studied. It was dropped from further studies.

The other three factors were "active" over the specified range and they were studied in a 2^3 factorial experiment that was designed to investigate possible interactions between variables. The settings of these factors, their coded values, and the measured EG concentrations after treatment are given in Table 6.9.

It is common to code the variables in factorial experiments. This makes it easier to see the pattern of the experimental design and it simplifies the calculation of effects and interactions. The coded variables are

$$X_1 = (T - 290)/2 \quad T = 288°C \quad \text{gives } X_1 = (288 - 290)/2 = -1$$

$$T = 292°C \quad \text{gives } X_1 = (292 - 290)/2 = +1$$

$$X_2 = (S - 68.5)/1.5 \quad S = 67\% \quad \text{gives } X_2 = (67 - 68.5)/1.5 = -1$$

$$S = 70\% \quad \text{gives } X_2 = (70 - 68.5)/1.5 = +1$$

$$X_3 = (F - 250)/50 \quad F = 200 \text{ kg/h} \quad \text{gives } X_3 = (200 - 250)/50 = -1$$

$$F = 300 \text{ kg/h} \quad \text{gives } X_3 = (300 - 250)/50 = +1$$

The average level of EG in the wastewater is 67.92 mg/L. Table 6.10 lists the main effects and two-factor interactions calculated from the data. The main effects are large and important. The interactions are statistically "small" and are disregarded.

TABLE 6.9
The 2^3 Factorial Experimental Design to Investigate Ethylene Glycol (EG) Levels in Polyester Manufacturing Wastewater

	Operating Factors			Coded Factors			Response
Order	Temperature (°C)	Slurry Level (%)	Feed Rate (kg/h)	X_1	X_2	X_3	Wastewater EG Conc. (mg/L)
1	288	67	200	−1	−1	−1	65.0
2	292	67	200	1	−1	−1	73.6
3	288	70	200	−1	1	−1	61.5
4	292	70	200	1	1	−1	68.1
5	288	67	300	−1	−1	1	67.0
6	292	67	300	1	−1	1	75.2
7	288	70	300	−1	1	1	62.9
8	292	70	300	1	1	1	70.1

Note: The table shows the actual operating levels and the coded variables.

TABLE 6.10
Main Effects and Interactions Calculated from the 2^3 Factorial Experimental Data

Average	X_1	X_2	X_3	X_1X_2	X_1X_3	X_2X_3	$X_1X_2X_3$
67.92	7.65	−4.55	1.75	−0.75	0.05	−0.05	0.25

The interpretation of the main effects is

- Changing the reaction temperature (X_1) from the low level to the high level increases the EG concentration in the wastewater by 7.65 mg/L
- Changing the slurry level (X_2) from the low level to the high level decreases the EG concentration in the wastewater by 4.55 mg/L
- Changing the EG feed rate (X_3) from the low level to the high level increases the EG concentration in the wastewater by 1.75 mg/L.

The lowest wastewater EG concentration was from Run 3:

Reaction temperature = 288°C
Slurry level in the heat exchanger = 70%
Feed rate of EG = 200 kg/h

A model for EG concentration in the wastewater (y) can be written in terms of the coded variables (the Xs).

$$y = 67.92 + 7.65X_1 - 4.55X_2 + 1.75X_3$$

A model in the original variable metrics can be derived by substituting the expressions for the Xs into this equation.

The trial runs (Figure 6.16) showed increased process stability after implementation of the results of the experiment. The average amount of EG in the wastewater was reduced from 70.5 to 63.3 mg/L. This led to lower BOD in the wastewater, while maintaining the same level of polyester quality.

At the levels shown to be best in the experiments, the EG wastewater concentration was decreased on average 10.15% and the wastewater treatment cost was reduced 27.82%.

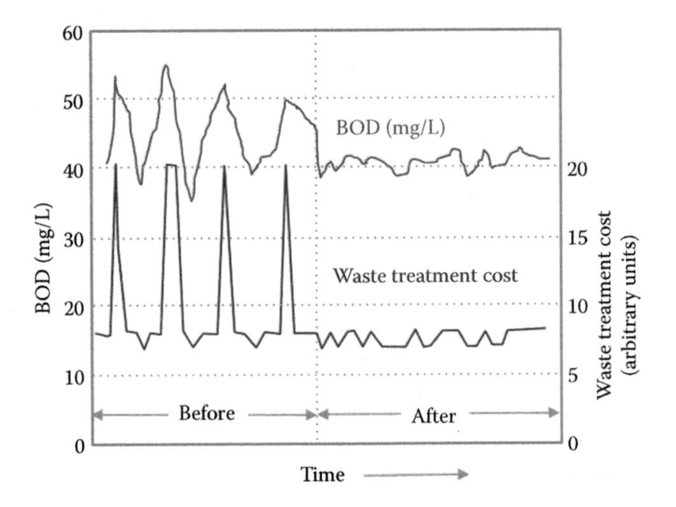

FIGURE 6.16 The before and after results for wastewater treatment at the polyester manufacturing plant. BOD concentrations were reduced and the high variability was removed. The average wastewater treatment costs have been reduced.

6.9 CONCLUSION

Empirical stoichiometry is an experimental method in which a number of interesting observations can be made, including many that we do not know how to calculate. The rate of particle formation, the particle size, and the settling rates can be observed. The sludge produced can be tested for dewatering properties. A variety of chemical reagents can be added, alone or in combination, to discover the best dose, the final solution pH, and the range of effectiveness.

The best-known method of empirical testing is the jar test. A test rig with four or six "jars," which typically hold one liter of test solution, is equipped with variable speed mixers. Chemicals are added to the jars and an interval of mixing, often divided into high-rate and low-rate phases, is followed by an interval of quiescent settling. Samples of the supernatant are collected and analyzed. Sludge may also be tested.

Empirical stoichiometry is also necessary because pollution control work sometimes requires that we work without knowing the chemical names of the reacting species, because they have no chemical names. Turbidity and color are real properties. We can see them and we can measure them, but not in the way we measure the concentration of calcium or lead.

The ideal and the empirical forms of stoichiometry are widely used, sometimes in combination. Sometimes, as for an entirely new facility, the actual wastewater does not exist, so testing cannot be done. Ideal stoichiometry is used to make useful estimates.

7 Chemical Equilibrium for Acids and Bases

7.1 THE DESIGN PROBLEM

When salts, bases, and gaseous substances enter water, they may dissolve and ionize and interact with the water and any other soluble species that may be present. The extent of these interactions is governed by the principles of chemical equilibrium.

Many reactions are *reversible*—they can go forward or backward. The forward reaction converts reactants into products; the reverse reaction converts products back into reactants.

If the chemical mixture (solution or gas) is undisturbed, the reaction will find an equilibrium—a balance. The concentration of all the chemical species will appear to be constant. No net change occurs in the amounts of products and reactants. The forward and reverse reactions are operating, but the rates of conversion are equal in both directions. This condition is called *equilibrium.*

$$A \leftrightarrows B$$

The design problem is to calculate the concentrations of the ionic species when they have reached their equilibrium position.

7.2 CHEMICAL EQUILIBRIUM

Figure 7.1 shows two conditions of equilibrium for compounds A and B. The concentrations at equilibrium do not have to be equal. Regardless of how the experiment is begun, the ratio of the equilibrium concentrations of compounds A and B will be the same. All reacting species reach equilibrium at the same time.

The material balances are as follows:

$$[A]_{eq} + [B]_{eq} = [A]_0$$

and

$$[A]_{eq} + [B]_{eq} + [C]_{eq} = [A]_0$$

For the reaction in the left panel of Figure 7.1, the chemical equilibrium constant requires that at equilibrium

$$\frac{[B]_{eq}}{[A]_{eq}} = K$$

Equilibrium may occur within a fraction of a second (it does for strong acids), or it may take some measurable time. Equilibrium theory tells us nothing about the time required. It tells us the final balance between the species in solution. And it indicates how the balance may be shifted by adding more reactant or removing product.

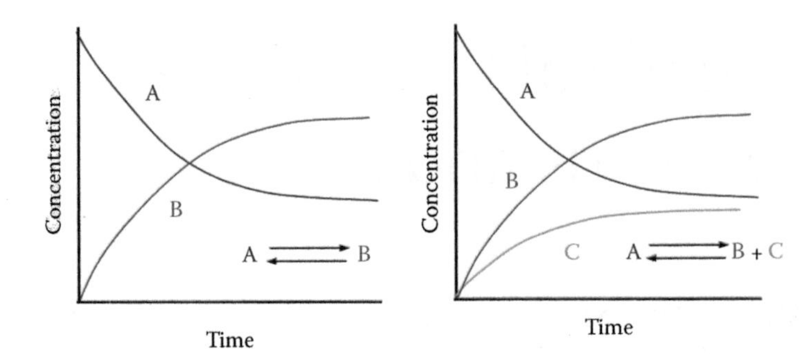

FIGURE 7.1 The equilibrium concentrations do not have to be equal. All reactions reach equilibrium at the same time.

If more of a reactant or a product is added to the system, the equilibrium will shift to a new position. This behavior lets us manipulate, or drive, a chemical reaction toward a desired condition that is safest, most efficient, or yields the desired outcome.

A general reversible chemical reaction is

$$aA + bB \leftrightarrows cC + dD$$

The double arrow indicates that it is reversible. The universally valid equilibrium equation is

$$K = \frac{[C]^c[D]^d}{[A]^a[B]^b}$$

where

[A], [B], [C], and [D] are the chemical activities of species A, B, C, and D

a, *b*, *c*, and *d* are the stoichiometric coefficients.

The rules for defining the chemical activities are as follows:

- For substances in solution, the activity is approximated by the molar concentration (mol/L).
- For gases, activity is approximated by the partial pressure (atm).
- For solvents such as water, the activity is always taken as unity.
- Crystalline solid materials and precipitates also always have activity equal to unity.

The activity does depend on the strength of the solution, but at the low concentrations normally encountered in water and wastewater treatment, molar concentrations can be assumed equal to the activity. That assumption will be used in all the examples of this chapter. This assumption is not true is when working with seawater and marine systems.

When this equation is applied to a mixture of *dissolved* substances, K is called an *ionization constant*. Sometimes a subscript is added for clarity; K_a is an acid ionization constant and K_b is a base ionization constant.

If one of the substances is a *solid* (crystal or precipitate), K is called a *solubility product* and for clarity is indicated by K_{sp}. The convention is to write the reaction as though the solid, A_aB_b, is dissolved.

$$A_aB_b \leftrightarrows aA + bB$$

The activity of the solid $[A_aB_b]$ is one, so the equilibrium reaction becomes

$$K_{sp} = \frac{[A]^a[B]^b}{[A_aB_b]} = [A]^a[B]^b$$

This chapter is about ionization reactions. Chapter 8 is about solubility and precipitation.

7.3 IONIZATION OF WATER

Water molecules act as both acids and bases. One water molecule acting as a base can accept a hydrogen ion from another that acts as an acid.

$$\underset{\text{Acid}}{H_2O} + \underset{\text{Base}}{H_2O} \leftrightarrows \underset{\text{Hydronium ion}}{H_3O^+} + \underset{\text{Hydroxyl ion}}{OH^-}$$

H_3O^+ is the hydronium ion and OH^- is the hydroxyl ion. This is often written in a simplified form

$$H_2O \leftrightarrows H^+ + OH^-$$

The reversible reaction is

$$\text{Forward reaction} \qquad H_2O \rightarrow H^+ + OH^-$$

$$\text{Reverse reaction} \qquad H_2O \leftarrow H^- + OH^-$$

This reaction is so fast that it will always be at equilibrium. Many important reactions in environmental systems are fast and equilibrium can be assumed without serious error.

At 25°C, pure water has pH = 7 and

$$[H^+] = [OH^-] = 10^{-7} \text{ moles per liter (mol/L)}$$

The ionization constant at standard temperature and pressure (25°C, 1 atm) is

$$K_W = [H^+][OH^-] = 10^{-14}$$

This means that the product of hydrogen ion concentration and hydroxyl ion concentration always equals 10^{-14}. If $[H^+]$ is known, $[OH^-]$ is easily calculated, and vice versa. Changing $[H^+]$ by adding an acid will simultaneously change $[OH^-]$, and vice versa. For example,

$$[H^+] = 10^{-2} \text{ and } [OH^-] = 10^{-12}$$

$$[H^+] = 10^{-3} \text{ and } [OH^-] = 10^{-11}$$

$$[H^+] = 10^{-9} \text{ and } [OH^-] = 10^{-5}$$

where $[H^+]$ = hydrogen ion concentration (mol/L)
$[OH^-]$ = hydroxyl ion concentration (mol/L)

K_W is temperature dependent, as shown in Table 7.1.

TABLE 7.1
The Ionization of Water (K_w) Is a Function of Temperature

Temperature (°C)	$K_W = [H^+][OH^-]$
0	0.115×10^{-14}
10	0.293×10^{-14}
20	0.681×10^{-14}
25	1.008×10^{-14}
30	1.471×10^{-14}
40	2.95×10^{-14}

7.4 pH

pH is a fundamental characteristic of chemical solutions. It determines the extent of ionization of soluble compounds and the formation of solids by ions that tend to precipitate. Low pH indicates acidic conditions.

The definition of pH is

$$pH = -\log_{10}[H^+]$$

This convenient "p" notation saves us having to write the concentrations as exponentials. The "p" comes from "potential," and the formula $-\log_{10}[X]$ is used in other cases where logarithmic values are easier to understand.

Taking negative logarithms of the equilibrium expression for water,

$$K_W = [H^+][OH^-] = 10^{-14}$$

gives $$-\log_{10} K_W = -\log_{10}[H^+] - \log_{10}[OH^-] = 14$$

and $$pK_W = pH + pOH = 14$$

The pH scale for aqueous solutions runs from 0 to 14. A change of one pH unit represents a 10-fold change in $[H^+]$.

At $$pH = 2 \quad [H^+] = 10^{-2} \text{ and } [OH^-] = 10^{-12}$$

$$pH = 3 \quad [H^+] = 10^{-3} \text{ and } [OH^-] = 10^{-11}$$

$$pH = 9 \quad [H^+] = 10^{-9} \text{ and } [OH^-] = 10^{-5}$$

Table 7.2 shows how K_W and the neutral pH depend on temperature. Even though the pH of pure water changes with temperature, it is still neutral. In pure water, there will always be the same number of hydrogen ions and hydroxide ions. That means that the pure water remains neutral even if its pH changes.

We are strongly conditioned to the idea of 7 being the pH of pure water, and anything else feels strange. Remember that if the value of K_W changes, then the neutral value for pH changes as well.

TABLE 7.2
K_W and Calculated Neutral pH as a Function of Temperature

Temperature (°C)	$K_W = [H^+][OH^-]$	pH
0	0.115×10^{-14}	7.47
10	0.293×10^{-14}	7.27
20	0.681×10^{-14}	7.08
25	1.008×10^{-14}	7.00
30	1.471×10^{-14}	6.92
40	2.95×10^{-14}	6.67
50	5.476×10^{-14}	6.63
60	9.5×10^{-14}	6.51

Is this temperature effect important in pollution control? Not really. Many industries are required to neutralize their effluents. Neutral is defined in most regulations to be pH 6–8 or 6–8.5. Wastewater treatment is done at ambient temperatures, say between 10°C and 30°C, so the temperature effect is much smaller than the working range of "a neutralized effluent pH."

Example 7.1 Calculating the Neutral pH

At 25°C, $K_W = 1.008 \times 10^{-14}$, or 10^{-14} for our purposes. Neutral pH in pure water exists when

$$[H^+] = [OH^-]$$

$$[H^+][OH^-] = K_W = 10^{-14}$$

and $$[H^+][H^+] = [H^+]^2 = K_W = 10^{-14}$$

$$[H^+] = 10^{-7}$$

$$pH = -\log_{10}[H^+] = 7.0$$

At 10°C, $K_W = 0.293 \times 10^{-14}$. Neutral pH in pure water exists when $[H^+] = [OH^-]$

$$[H^+][H^+] = [H^+]^2 = K_W = 0.293 \times 10^{-14}$$

$$[H^+] = \sqrt{K_W} = \sqrt{0.293 \times 10^{-14}} = 0.541 \times 10^{-7}$$

$$pH = -\log_{10}[H^+] = -\log_{10}\left(0.541 \times 10^{-7}\right) = 7.27$$

Figure 7.2 shows pH values for a variety of common substances and also the pH of the untreated wastewater effluent from some industrial operations.

7.5 pH CONTROL AND NEUTRALIZATION

The success of many chemical manufacturing processes as well as many water and wastewater treatment processes depends on keeping the operating pH at a specified level (which may be acidic, alkaline, or near neutral). Most precipitation processes are pH dependent, as are many oxidation–reduction reactions, disinfection, and most biological treatment processes.

Neutralization is pH control that is designed to achieve a pH near 7. This is necessary for effluents that are discharged to streams (typical pH limit of 6–8.5) or to municipal sewers (typical pH limit of 6–9).

Neutralization is accomplished by adding an acid to a basic solution or by adding base to an acidic solution in an amount that will react to yield a pH near 7.0. The general chemical reaction is

$$\underset{\text{acid}}{HA} + \underset{\text{base}}{MOH} \rightarrow \underset{\text{metal salt}}{MA} + \underset{\text{water}}{H_2O}$$

The three critical decisions in designing a neutralization system are as follows:

- How much neutralizing reagent is needed? This is learned by calculation or from a titration curve.
- Which metal salt will be formed? Will it be soluble or will it form a solid?
- Which reagent should be used? This is determined by cost and by solids production.

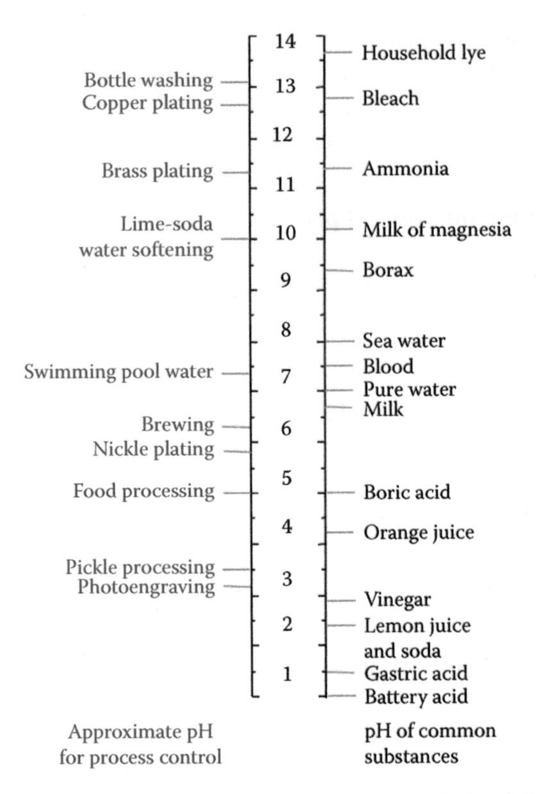

FIGURE 7.2 pH of some common substances and effluents from some industrial operations.

The three general rules for metal salt formation and solubility are as follows:

- HCl, HNO_3, and acetic acid give soluble products when reacted with any base.
- All Na^+ salts and all Cl^- salts are soluble.
- Ca^{2+} may precipitate with H_2SO_4 or phosphoric acid.

Figure 7.3 is the titration curve for an acid that is neutralized by the addition of a base chemical. The *equivalence point* in a titration is the point at which all the acid (or base) molecules in the solution react with an exact stoichiometric amount of the titrant. The pH is equal to 7 at the equivalence point only if a strong acid is used to titrate a strong base (or vice versa). If a strong acid is used to titrate a weak base, the equivalence point will be at a pH lower than 7 because of hydrolysis. For a strong base titrating a weak acid, the equivalence point will be at a pH higher than 7.

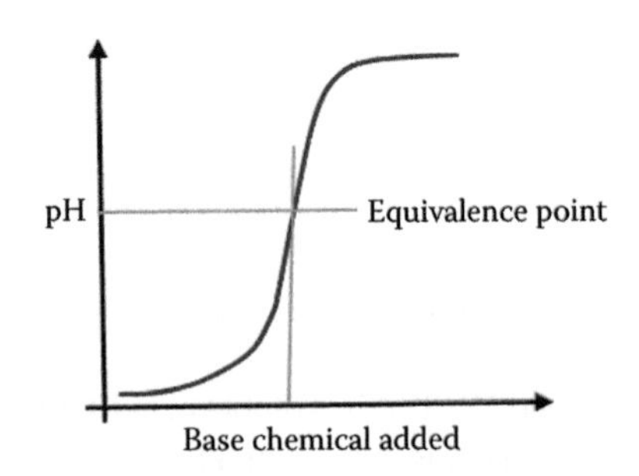

FIGURE 7.3 The equivalence point of a titration curve for neutralization of an acid.

Table 7.3 gives data on neutralization requirements for a variety of bases by four different acids. Table 7.4 gives the approximate pH for dilute solutions of some common acids and alkalis in water. These data are useful to indicate the relative effectiveness of different reagents.

These tables provide a convenient starting point, but wastewater solution chemistry is usually complicated, so a simple stoichiometric calculation does not give the actual chemical dose needed. What's important is to know that sulfuric acid, H_2SO_4, can be neutralized with hydrated lime, $Ca(OH)_2$. The wastewater will contain substances in addition to H_2SO_4. If the acid were used to clean steel parts, we would expect an acidic solution that contains dissolved iron. Whatever neutralizing agent is added, some of it will react with the iron.

Waste pickle liquor is a special case. Pickling is the process of cleaning metal by dipping it in acid to dissolve rust and scale. Hydrochloric acid and sulfuric acid are both widely used in pickling. After some time, the acid must be dumped and replaced. The acid that is dumped will be

TABLE 7.3
Theoretical Requirements to Neutralize Pure Acid

	kg Required to Neutralize 1 kg of 100%			
Basic Reagent	**Sulfuric**	**Nitric**	**Hydrochloric**	**Hydrofluoric**
Anhydrous ammonia	0.347	0.270	0.467	0.851
Calcium hydroxide, $Ca(OH)_2$	0.756	0.588	1.016	1.852
Calcium carbonate, $CaCO_3$	1.020	0.794	1.372	2.501
Calcium oxide, CaO	0.572	0.445	0.769	1.401
Magnesium oxide, MgO	0.411	0.320	0.553	1.007
Magnesium hydroxide, $Mg(OH)_2$	0.595	0.463	0.800	1.458
Potassium carbonate, K_2CO_3	1.409	1.096	1.895	3.453
Sodium carbonate, Na_2CO_3	1.081	0.841	1.453	2.649
Sodium hydroxide, NaOH	1.144	0.890	1.538	2.804
Dolomitic limestone	0.95			
Dolomitic lime, hydrated	0.65			
High calcium limestone	1.06			
High calcium lime, hydrated	0.80			

TABLE 7.4
Approximate pH for Dilute Aqueous Solutions of Acid and Alkaline Solutions

Chemical	Concentration (mass %)	pH	Chemical	Concentration (mass %)	pH
Ammonia	1.7	11.6	Sodium carbonate	0.5	11.6
	0.17	11.1	Sodium hydroxide	3.8	14.0
	0.017	10.6		0.38	13.0
Calcium carbonate	Saturated	9.4		0.038	12.0
Calcium hydroxide	Saturated	12.4	Hydrochloric acid	3.5	0.1
Magnesium hydroxide	Saturated	10.5		0.35	1.1
Potassium hydroxide	5.3	14.0		0.035	2.0
	0.53	13.0	Sulfuric acid	4.7	0.3
	0.053	12.0		1.0	0.9
Sodium bicarbonate	0.8	8.4		0.47	1.2
	1.0	8.6		0.047	2.4

contaminated with iron; the waste pickle liquor composition is primarily $FeCl_2$ in HCl, or $FeSO_4$ in H_2SO_4. In addition to the chloride salt or sulfate salt, the reaction products probably will include precipitated iron compounds. The reactions for neutralizing waste hydrochloric-acid pickle liquor with hydrated lime, $Ca(OH)_2$, are as follows:

$$FeCl_2 + Ca(OH)_2 \rightarrow Fe(OH)_2 + CaCl_2$$
$$2HCl + Ca(OH)_2 \rightarrow CaCl_2 + 2H_2O$$

The $CaCl_2$ will be soluble, but the $Fe(OH)_2$ will precipitate and it may need to be removed before the neutralized acid can be discharged. If sulfuric-acid-based pickle liquor is neutralized with lime, the salt will be $CaSO_4$. If the concentrations are high, this will precipitate along with the iron hydroxide.

$$FeSO_4 + Ca(OH)_2 \rightarrow Fe(OH)_2 + CaSO_4$$
$$H_2SO_4 + Ca(OH)_2 \rightarrow CaSO_4 + 2H_2O$$

The net result is that acid is destroyed, iron is lost, and there probably will be a volume of sludge to dewater and handle. We will see later that acid sometimes can be recovered instead of destroyed by neutralization.

The amount of chemical needed to achieve a target pH is not easy to calculate for a complex mixture, so common practice is to measure the titration curve of the reaction. The shape of the titration curve will depend on the acid and other impurities in the solution and on the neutralizing reagent. Figure 7.4 shows two titration curves for neutralizing waste sulfuric acid from a metal cleaning operation by adding NaOH. The impurities in the acid (iron and other metals) distort the titration curve of pure sulfuric acid in a way that is difficult to predict. The titration curve captures all the complicated chemistry in a simple graph that can be used to calculate the amount of chemical required to neutralize the acid waste.

Both the pH and the flow of an industrial waste can change rapidly. In some industries, the pH of a waste can change from 2 to 12 in less than a minute. Reagent pumps and valves cannot be manipulated fast enough to keep up with such changes, even if we could devise precise mathematical rules for reagent addition. These process control problems are resolved by equalizing the influent pH variations in well-mixed tanks. If the pH variations are large and/or rapid, a series of two or three mixed tanks with pH monitoring and automatically controlled chemical addition are used.

The addition of neutralizing chemicals may be controlled by a feedback or feed-forward strategy, or a combination of these. Feedback control measures effluent pH and makes an empirical adjustment in the rate at which reagent is added to the mixing tank. Feed-forward control measures influent pH and computes a reagent addition in anticipation of the new material entering the reactor. This anticipatory control is good when fluctuations are large and rapid, especially if used in combination with effluent "fine-tuning" by feedback control.

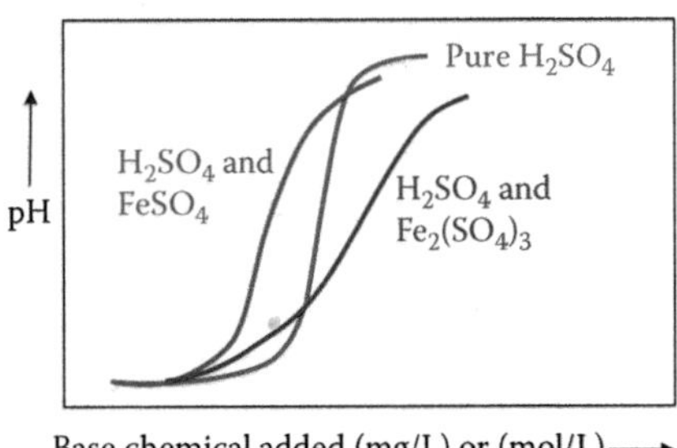

FIGURE 7.4 Titration curves for pure sulfuric acid compared to waste sulfuric acid from a metal cleaning operation.

Example 7.2 Neutralization with Feedback Control

A wastewater flow of 600 L/min is to be neutralized. The problem is difficult because of the pH variability. The pH ranges between 2 and 11 and it can change from one extreme to another in less than 10 s. The titration curve for one sample of the wastewater when it was acidic is shown in Figure 7.5.

Even if the wastewater were always similar to this one sample, the steepness of the titration curve at the target pH makes it difficult to meter the reagent accurately enough to hit the target. When the waste flow rate and composition both change rapidly, it is not possible to sequence metering pumps and valves fast enough to hit the target. The solution is to use two mixed reactors in series, as shown in Figure 7.6.

The first reactor (volume = 800 L) is instrumented to measure pH of the influent and send this information to a controller that will calculate a reagent dose and adjust the reagent feed. This is called "feed-forward control" because the information needed to make the decision is fed ahead of the wastewater coming into the reactor.

Feed-forward control alone will not solve the problem because the pH controller algorithm will not be correct for all situations. The process control is improved by adding feedback control. The pH of the effluent from the first reactor is measured and the reagent dose is adjusted in response. Feedback control by itself cannot solve the problems because it is reacting to what has already happened. The combination of feed-forward and feedback control combines anticipation and reaction and should perform better than either system alone. The second mixed reactor (volume = 4000 L) acts as a blender to smooth out (equalize) pH variations in the effluent from reactor 1.

Figure 7.7 shows the influent and effluent pH levels of a feed-forward pH control system with feedback adjustment. The system consists of a stirred reactor followed by an unstirred blender (mixing is by the fluid turbulence). [Imagine the process shown in Figure 7.6 without the mixer in reactor 2.] The "feed-forward controller" uses the observed pH values to compute a reagent dose, which is added in response to entering pH levels. The feedback controller adds some fine tuning to the feed-forward control. The process can produce an effluent with nearly constant pH, even when there are large and rapid fluctuations in influent pH.

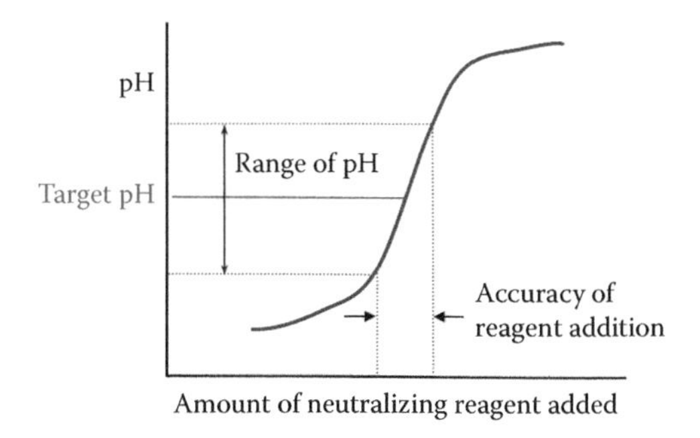

FIGURE 7.5 Steepness of the titration curve makes it difficult to hit the target effluent pH.

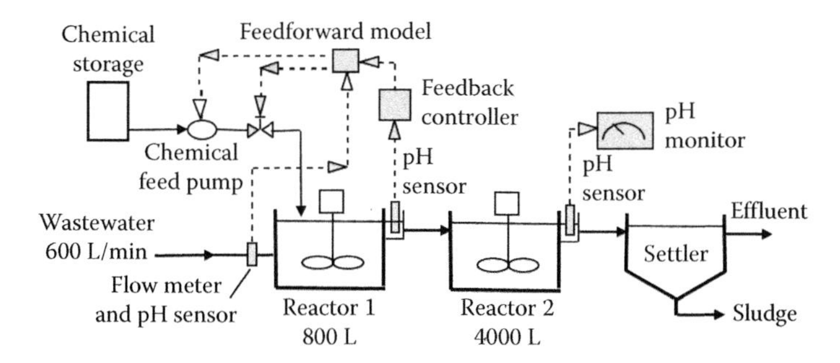

FIGURE 7.6 Neutralization system using two mixed reactors in series with the first instrumented for feed-forward and feedback control, and the second serving as a blender.

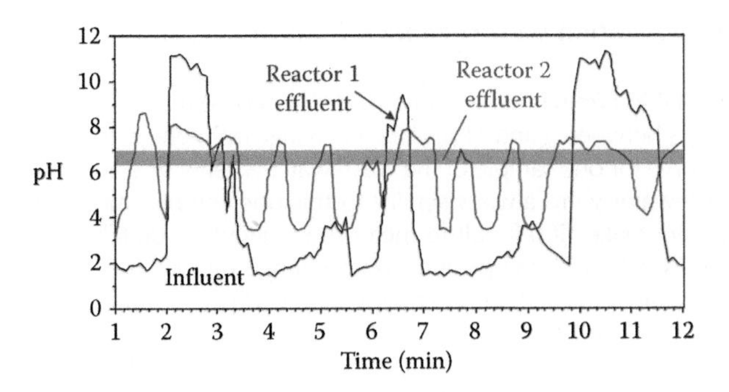

FIGURE 7.7 Influent and effluent pH of reactors 1 and 2. Reactor 1 has mechanical mixing. Reactor 2 is mixed only by the turbulence of the fluid.

7.6 ACID–BASE REACTIONS

The ionization of a strong acid, such as HCl, is not reversible.

$$HCl + H_2O \rightarrow H_3O^+ + Cl^-$$

where H_3O^+ is the hydronium ion. This is often written more simply as

$$HCl \rightarrow H^+ + Cl^-$$

Weak acids and weak bases only partially ionize when they dissolve. There is an equilibrium between the ionic and the molecular forms. Hydrofluoric acid and ammonia are examples

$$HF \leftrightarrows H^+ + F^-$$

$$NH_3 + H_2O \leftrightarrows NH_4^+ + OH^-$$

The equilibrium constants are called ionization constants, because of the ionization that occurs

$$K = \frac{[H^+][F^-]}{[HF]} \text{ and } K = \frac{[NH_4^+][OH^-]}{[NH_3]}$$

The values of K are typically very small, so, as with pH and pK_w, they are often reported as pK values where

$$pK = -\log_{10}(K)$$

Table 7.5 gives ionization constants and their pK values for some common weak acids and bases.

Example 7.3 Ionization Constant of a Weak Acid I

In a 0.12 M solution, a weak monoprotic acid, HA, is 5.0% ionized and 95% unionized. Calculate the ionization constant for the weak acid.

$$HA \leftrightarrows H^+ + A^-$$

$$K_a = \frac{[H^+][A^-]}{[HA]}$$

TABLE 7.5
Ionization Constants for Some Common Acids and Bases at 25°C

Compound	Reaction	Ionization Constant	p*K*
Acetic acid	$CH_3COOH \leftrightarrows CH_3COO^- + H^+$	$K = 1.8 \times 10^{-5}$	4.75
Carbonic acid	$CO_2 + H_2O \leftrightarrows H_2CO_3$	$K = 1.58 \times 10^{-3}$	2.8
	$H_2CO_3 \leftrightarrows HCO_3^- + H^+$	$K_1 = 4.5 \times 10^{-7}$	6.35
	$HCO_3^- \leftrightarrows CO_3^{2-} + H^+$	$K_2 = 4.7 \times 10^{-11}$	10.33
Chromic acid	$H_2CrO_4 \leftrightarrows HCrO_4^- + H^+$	$K_1 = 1.8 \times 10^{-1}$	0.74
	$HCrO_4^- \leftrightarrows CrO_4^{2-} + H^+$	$K_2 = 3.2 \times 10^{-7}$	6.49
Formic acid	$H_2CO_2 \leftrightarrows HCO_2^- + H^+$	$K = 1.8 \times 10^{-4}$	3.75
Hydrocyanic acid	$HCN \leftrightarrows CN^- + H^+$	$K = 6.2 \times 10^{-10}$	9.21
Hydrofluoric acid	$HF \leftrightarrows F^- + H^+$	$K = 6.3 \times 10^{-4}$	3.2
Hydrogen sulfide	$H_2S \leftrightarrows HS^- + H^+$	$K_1 = 8.9 \times 10^{-8}$	7.05
	$HS^- \leftrightarrows S^{2-} + H^+$	$K_2 = \sim 1 \times 10^{-19}$	~19
Hypochlorous acid	$HOCl \leftrightarrows OCl^- + H^+$	$K = 4.0 \times 10^{-8}$	7.40
Phosphoric acid	$H_3PO_4 \leftrightarrows H_2PO_4^- + H^+$	$K_1 = 6.9 \times 10^{-3}$	2.16
	$H_2PO_4^- \leftrightarrows HPO_4^{2-} + H^+$	$K_2 = 6.2 \times 10^{-8}$	7.21
	$HPO_4^{2-} \leftrightarrows PO_4^{3-} + H^+$	$K_3 = 4.8 \times 10^{-13}$	12.32
Ammonium hydroxide	$NH_4OH \leftrightarrows NH_4^+ + OH^-$	$K = 1.8 \times 10^{-5}$	4.75
Ammonium	$NH_4^+ \leftrightarrows NH_3 + H^+$	$K = 5.5 \times 10^{-10}$	9.25

Source: Linde, D.R., ed., 2009; Speight, J., 2017.

Calculate the concentrations of all species in the solution.

5% ionized $[H^+] = [A^-] = 0.05(0.12\ M) = 0.0060\ M = 6 \times 10^{-3}\ M$

95% unionized $[HA] = 0.95(0.12\ M) = 0.114\ M$

Substitute the concentrations in the ionization constant expression to get the value of K_a.

$$K_a = \frac{[H^+][A^-]}{[HA]} = \frac{(6.0 \times 10^{-3})(6.0 \times 10^{-3})}{0.114} = 3.2 \times 10^{-4}$$

Example 7.4 Ionization Constant of a Weak Acid II

The pH of a 0.10 M solution of a weak monoprotic acid, HA, is 2.97. Calculate the value of its ionization constant.

$$HA \leftrightarrows H^+ + A^-$$

$$K_a = \frac{[H^+][A^-]}{[HA]}$$

Calculate the concentrations of all species in the solution.

$$\text{pH} = 2.97 \text{ so} = [H^+] = 10^{-2.97} = 1.1\times10^{-3}\text{ M}$$

$$[H^+] = [A^-] = 1.1\times10^{-3}\text{ M}$$

$$[HA] = \left(0.10 - 1.1\times10^{-3}\right) = 0.0989\text{ M}$$

Substitute the concentrations in the ionization constant expression to get the value of K_a.

$$K_a = \frac{[H^+][A^-]}{[HA]} = \frac{(1.1\times10^{-3})(1.1\times10^{-3})}{0.0989} = 1.22\times10^{-5}$$

Example 7.5 Ionization Constant for Hydrogen Fluoride

A solution with an initial concentration of 0.5 mol/L hydrogen fluoride (HF) is found to have a hydronium concentration of 0.0185 M (pH = 1.73) at equilibrium. The ionization reaction is

$$HF + H_2O \leftrightarrows H_3O^+ + F^-$$

Calculate the acid ionization constant K_a and the molar concentration of fluoride (F^-).

$$K_a = \frac{[H_3O^+][F^-]}{[HF]}$$

The activity of water, the solvent, is 1.0 and does not appear in this expression because pure substances are not included.

$[H_3O^+]$, $[F^-]$, and $[HF]$ are the molar concentrations at equilibrium.

One mole of HF ionizes to give one mole of F^- and one mole of H_3O^+.

$$[H_3O^+] = [F^-] = 0.0185\text{ mol/L}$$

Also $$[HF]_{eq} = [HF]_{initial} - [F^-] = [HF]_{initial} - [H_3O^+]$$

The equilibrium expression can now be written as

$$K_a = \frac{[H_3O^+][F^-]}{[HF]_{initial} - [F^-]} = \frac{(0.0185)(0.0185)}{(0.500) - (0.0185)} = 7.11\times10^{-4}$$

This value is reasonably close to the one in Table 7.5 (6.3×10^{-4}), given solution preparation and pH measurement errors and unknown temperature differences.

Example 7.6 Ionization Constant for Ammonia

An ammonia solution has an initial concentration 0.00062 mol/L (76.5 mg/L) and pH = 10.0. Calculate the base ionization constant, K_b.

$$NH_3 + H_2O \leftrightarrows NH_4^+ + OH^-$$

and $$K_b = \frac{[NH_4^+][OH^-]}{[NH_3]}$$

From the dissociation of water

$$K_W = [H^+][OH^-] = 10^{-14}$$

$$[H^+] = 10^{-pH} = 10^{-10} \text{ mol/L}$$

$$[OH^-] = \frac{K_W}{[H^+]} = \frac{10^{-14}}{10^{-10}} = 10^{-4} \text{ mol/L}$$

Because 1 mol of NH_3 reacts with 1 mol of H_2O to produce 1 mol of NH_4^+

$$[NH_4^+] = [OH^-] = 10^{-4} \text{ mol/L}$$

$$[NH_3]_{eq} = [NH_3]_{initial} - [OH^-] = 0.00062 - 0.0001 = 0.00052 \text{ mol/L}$$

Substitute the concentrations in the ionization constant expression to get the value of K_b.

$$K_b = \frac{[NH_4^+][OH^-]}{[NH_3]_{initial} - [OH^-]} = \frac{(0.0001)(0.0001)}{(0.00062)-(0.0001)} = \frac{1\times10^{-8}}{5.2\times10^{-4}} = 1.9\times10^{-5}$$

Example 7.7 Ammonia Ionization

When you have a sample analyzed for ammonia, the laboratory reports the total ammonia concentration. Total ammonia is the sum of ammonia, NH_3, and ammonium ion, NH_4^+. It can be important to know which species predominates. NH_3 is toxic to fish, while NH_4^+ is not. Compute the ratio of NH_3 to NH_4^+ concentrations in a water sample.

The equilibrium reaction between NH_3 and NH_4^+ is

$$NH_4^+ \leftrightarrows NH_3 + H^+$$

with

$$K = \frac{[NH_3][H^+]}{[NH_4^+]} = 5.5\times10^{-10}$$

The ionization constant defines the balance between the species in this reaction.

The quantities in the brackets are concentrations in moles per liter. The concentration of $[H^+]$ can be expressed as

$$pH = -\log_{10}[H^+] \quad \text{and} \quad [H^+] = 10^{-pH}$$

Thus, we have

$$K = \frac{[NH_3][H^+]}{[NH_4^+]} = \frac{[NH_3]\ 10^{-pH}}{[NH_4^+]}$$

$$\frac{[NH_3]}{[NH_4^+]} = \frac{K}{10^{-pH}} \quad \text{and} \quad [NH_3] = \frac{K}{10^{-pH}}[NH_4^+]$$

Substituting the value of K gives

$$\frac{[NH_3]}{[NH_4^+]} = \frac{5.5\times10^{-10}}{10^{-pH}} \quad \text{and} \quad [NH_3] = \frac{5.5\times10^{-10}}{10^{-pH}}[NH_4^+]$$

As the pH increases, NH_3 increases relative to NH_4^+. For any given concentration of total ammonia, the water becomes more toxic to fish and other aquatic life.

At pH = 7, almost all of the total ammonia is in the ionized (nontoxic) form.

$$[NH_3] = \frac{5.5\times10^{-10}}{10^{-pH}}[NH_4^+] = \frac{5.5\times10^{-10}}{10^{-7}}[NH_4^+] = 5.5\times10^{-3}[NH_4^+] = 0.0055[NH_4^+]$$

This ratio changes by a factor of 10 for a change of 1 pH unit. At pH 8 and 9,

$$[NH_3] = \frac{5.5\times10^{-10}}{10^{-8}}[NH_4^+] = 0.055[NH_4^+]$$

$$[NH_3] = \frac{5.5\times10^{-10}}{10^{-9}}[NH_4^+] = 0.55[NH_4^+]$$

Example 7.8 Fraction of Total Ammonia That Is UnIonized

The total ammonia concentration is 20 mg/L at pH 7.0 and 25°C. Calculate the percentage of total ammonia that is unionized; that is, calculate the NH_3 concentration.

$$\text{Total Ammonia} = [NH_3] + [NH_4^+]$$

Using the result from Example 7.6, the ratio of ammonia to total ammonia is

$$\frac{[NH_3]}{[NH_3] + [NH_4^+]} = \frac{\frac{K}{10^{-pH}}[NH_4^+]}{\frac{K}{10^{-pH}}[NH_4^+] + [NH_4^+]} = \frac{1}{1+(10^{-pH}/K)}$$

Thus, for $K = 5.5 \times 10^{-10}$ and pH = 7.0,

$$\frac{[NH_3]}{[NH_3] + [NH_4^+]} = \frac{1}{1+(10^{-pH}/K)} = \frac{1}{1+(10^{-7}/5.5\times10^{-10})} = \frac{1}{183} = 0.0055 = 0.55\%$$

so that

$$[NH_3] = 0.0055(20\text{ mg/L}) = 0.11\text{ mg/L}$$

The ionization constant changes with temperature. For $T = -2$°C to 40°C,

$$pK = 0.090387 + \frac{273.933}{T}$$

The result is that increasing temperature increases the fraction of unionized ammonia, as shown in Figure 7.8.

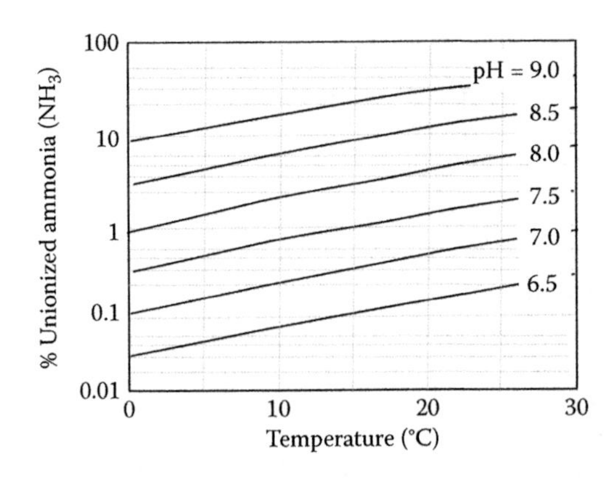

FIGURE 7.8 Percentage of total ammonia that is unionized as a function of temperature and pH.

Example 7.9 Hydrogen Sulfide

Hydrogen sulfide gas (H_2S) is toxic and highly odorous. It is added to natural gas, which is odorless, to provide a warning of gas leaks. The odor threshold is 0.00047 ppm (0.47 ppb), the point at which 50% of humans can detect an odor without being able to identify it. The recognition threshold is 0.0047 ppm. The Occupational Safety and Health Administration permissible exposure limit (8-h weighted average) is 10 ppm. At high concentrations (100–150 ppm) the olfactory nerve is paralyzed after a few inhalations and sense of the smell goes away, along with the awareness of danger. This is why people die from H_2S exposure. The lethal concentration for 50% of humans is 800 ppm for 5 min exposure.

Show that the formation of H_2S is favored by acidic conditions by calculating the percentage of H_2S as a function of pH.

The first ionization of H_2S is

$$H_2S \rightleftarrows HS^- + H^+ \quad \rightarrow \quad K_1 = \frac{[HS^-][H^+]}{[H_2S]} = 8.9 \times 10^{-8}$$

The second ionization

$$HS^- \rightleftarrows S^{2-} + H^+ \quad \rightarrow \quad K_2 = \frac{[S^{2-}][H^+]}{[HS^-]} = 1 \times 10^{-19}$$

Following the derivation in Example 7.7, we can find

$$\%H_2S = \frac{100\,[H_2S]}{[H_2S] + [HS^-] + [S^{2-}]} = \frac{100}{1 + \frac{K_1}{[H^+]} + \frac{K_1}{[H^+]}\frac{K_2}{[H^+]}}$$

which gives these values:

pH	4.0	5.0	5.5	6.0	6.5	7.0	7.5	8.0	8.5	9.0	10.0
% H_2S	99.9	99.1	97.3	91.8	78.0	52.9	26.2	10.1	3.4	1.1	0.1

Example 7.10 Hydrogen Sulfide II

Continue with Example 7.9 by computing the percentage of HS^- and S^{2-} as a function of pH, and plot the results, including H_2S.

The first ionization of H_2S is

$$H_2S \rightleftarrows HS^- + H^+ \quad \rightarrow \quad K_1 = \frac{[HS^-][H^+]}{[H_2S]} = 8.9 \times 10^{-8}$$

The second ionization

$$HS^- \rightleftarrows S^{2-} + H^+ \quad \rightarrow \quad K_2 = \frac{[S^{2-}][H^+]}{[HS^-]} = 1 \times 10^{-19}$$

From Example 7.8, we know

$$\%H_2S = \frac{100[H_2S]}{[H_2S] + [HS^-] + [S^{2-}]} = \frac{100}{1 + \frac{K_1}{[H^+]} + \frac{K_1}{[H^+]}\frac{K_2}{[H^+]}}$$

Likewise

$$\%HS^- = \frac{100[HS^-]}{[H_2S] + [HS^-] + [S^{2-}]} = \frac{100}{\frac{[H^+]}{K_1} + 1 + \frac{K_2}{[H^+]}}$$

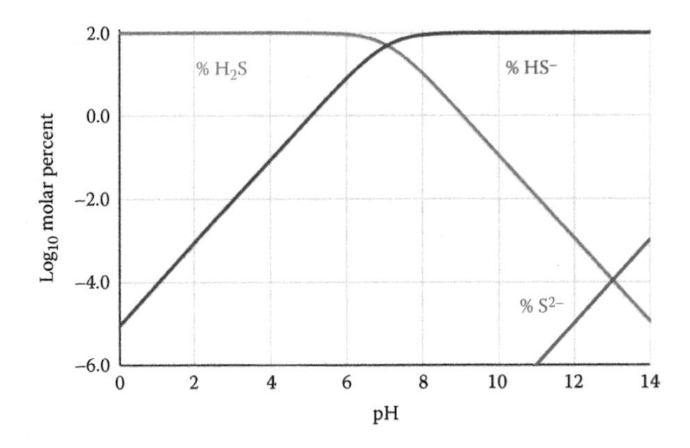

FIGURE 7.9 Distribution of sulfide species as a function of pH.

and

$$\%S^{2-} = \frac{100\,[S^{2-}]}{[H_2S] + [HS^-] + [S^{2-}]} = \frac{100}{\frac{[H^+]\,[H^+]}{K_1\;\;K_2} + \frac{[H^+]}{K_2} + 1}$$

which gives these percentages.

pH	1	2	3	4	5	6	7	8	9	10	11	12	13	14
% H_2S	100	100	100	99.9	99.1	91.8	52.9	10.1	1.1	0.1	0	0	0	0
% HS^-	0	0	0	0.1	0.9	8.2	47.1	89.9	98.9	99.9	100	100	100	100
% S^{2-}	0	0	0	0	0	0	0	0	0	0	0	0	0	0

Over the pH range 0 to 14, only H_2S and HS^- have measurable amounts. The percentages for S^{2-} are all essentially zero. The reason for this is the very small value of the second ionization constant for hydrogen sulfide. Table 7.5 gives a $K_2 \approx 10^{-19}$ for H_2S. This is much smaller than the $K_2 = 10^{-12}$ that is reported in many texts, a value which has been shown to be incorrect (Myers, 1986). Using $K_2 \approx 10^{-19}$ means that the concentration of S^{2-} in aqueous solution will always be negligible. We shall see the importance of this in Example 7.13 and our discussion of metal sulfide solubility in Chapter 8.

The very small values of S^{2-} can be seen by plotting the percent distribution of H_2S, HS^-, and S^{2-} on a log scale, as shown in Figure 7.9. Even at a pH (14) with the largest S^{2-} concentration, it is only 0.001% (one in a hundred thousand) of the total sulfide concentration.

7.7 THE ICE TABLE AND SOLVING EQUILIBRIUM PROBLEMS

The *ICE table* is a simple matrix structure that is used to simplify the calculations in reversible equilibrium reactions (often involving weak acids or weak bases). The letters in ICE stand for

I = Initial molar concentration
C = Change in concentration
E = Equilibrium concentration

The matrix actually has four rows and an alternate name is RICE (Wikipedia), for

R = Reaction
I = Initial molar concentration

C = Change in concentration
E = Equilibrium concentration

The ICE (RICE) table is also known as the ICE Box (RICE Box).

ICE tables set up and organize the variables and constants needed to calculate the unknown quantities in equilibrium reactions. These may be concentrations of reactants, or products, or they may be the equilibrium constant. Molar concentrations *must* be used, because the chemical reaction and equilibrium constant show the balance among moles of reactants and products at equilibrium.

The chemical equation defines the chemical equilibrium. The problem will usually indicate initial or equilibrium concentrations of reactants and/or products. The problem usually gives the value of the ionization (equilibrium) constant, K.

Set up the table by writing the equilibrium equation across the first row. Then create three rows for

I = Initial molar concentration
C = Change in concentration
E = Equilibrium concentration

The empty table is shown for the general reaction:

$$aA + bB \leftrightarrows dD + eE$$

In the chemical equilibrium, a, b, d, and e represent the stoichiometric coefficients of the reaction and can be 1, 3, 4, 2, 1/2, etc. A, B, D, and E represent the reactants and products.

R	aA	+	*b*B	⇆	*d*D	+	eE
I							
C							
E							

Next, all the known concentrations are filled in. Many equilibrium problems give the initial concentrations of reactants and the equilibrium constant, and the final concentrations of reactants and products are to be calculated.

Suppose for the equilibrium above, the initial concentrations of A and B were given as [A] = 0.750 M and [B] = 1.500 M, and the equilibrium constant was given. These initial concentrations should be entered.

R	aA	+	*b*B	⇆	*d*D	+	eE
I	0.750		1.500		0		0
C							
E							

Notice that 0s are entered for the initial concentrations of D and E.

The C row, representing change, is based on noting that the equation can only shift to the right, since there are no products present. The amount that it shifts is x. Write the stoichiometric coefficient in front of x with a minus sign for any species that will have a decrease in concentration. Use a plus sign in front of x of any species for which the concentration increases. The table should now look like the example below.

R	aA	+	bB	⇆	dD	+	eE
I	0.750		1.500		0		0
C	$-ax$		$-bx$		$+dx$		$+ex$
E							

The last line shows the equilibrium concentrations, which are the sum of the concentrations in the *I* row and the *C* row. The table should now read

R	aA	+	bB	⇆	dD	+	eE
I	0.750		1.500		0		0
C	$-ax$		$-bx$		$+dx$		$+ex$
E	$0.750 - ax$		$1.500 - bx$		dx		ex

The equilibrium concentrations are used in the mathematical form of the equilibrium expression.

$$K = \frac{[D]^d[E]^e}{[A]^a[B]^b}$$

becomes

$$K = \frac{[dx]^d[ex]^e}{[0.750 - ax]^a[1.500 - bx]^b}$$

The stoichiometric coefficients *a*, *b*, *d*, and *e* will be known.

If *K* is known, *x* can be calculated. Or, vice versa, if the equilibrium concentrations are known, then *K* can be calculated.

Once *x* is known, the equilibrium concentrations can be calculated using the mathematical relations in the *E* row. This is usually fairly easy compared to the rest of the problem.

Now we set up the ICE table for the case when the initial concentrations of products are known. For [D] = 0.500 M and [E] = 0.850 M, the table will look like:

R	aA	+	bB	⇆	dD	+	eE
I	0		0		0.500		0.850
C	$+ax$		$+bx$		$-dx$		$-ex$
E	ax		bx		$0.500 - dx$		$0.850 - ex$

The equilibrium equation

$$K = \frac{[D]^d[E]^e}{[A]^a[B]^b}$$

becomes

$$K = \frac{[0.500 - dx]^d[0.850 - ex]^e}{[ax]^a[bx]^b}$$

where *a*, *b*, *d*, and *e* are known.

Many times, assumptions and approximations can be made that will simplify the calculations. A summary of the ICE method is as follows:

- Write the chemical equation and set up the *ICE* table.
  Row *I*—Enter the initial *molar* concentration of each species given in the problem, reactants, and products.

Row *C*—Enter the change in concentrations of all species.
Row *E*—Enter the concentration of species when the system is at equilibrium.

- Write the equilibrium expression (the mathematical equation) and substitute the equilibrium concentrations from the E row in the *ICE* table.
- Make appropriate assumptions and approximations to simplify the equations.
- Determine the numerical value of x.
- Use the value of x to determine the equilibrium concentrations of all species. The equations written in row *E* of the table should be used to do this.
- Check any assumptions that were made.

Example 7.11 Hydrocyanic Acid Ionization

A hydrogen cyanide concentration of 3,000 mg/m^3 in air will kill a human in about 1 min. Acidic conditions favor HCN; basic conditions favor the less toxic CN^-. Calculate the concentration of the species in 0.02 M (540 mg/L) hydrocyanic acid solution (HCN). Use $K_a = 4.9 \times 10^{-10}$.

$$HCN + H_2O \leftrightarrows H_3O^+ + CN^-$$

Reaction	HCN	+	H_2O	⇌	H_3O^+	+	CN^-
Initial	0.02				0		0
Change	$-x$				$+x$		$+x$
Equilibrium	$(0.02 - x)$				x		x

$$[H^+] = [H_3O^+]$$

$$K_a = \frac{[H^+][CN^-]}{[HCN]} = \frac{(x)(x)}{(0.02 - x)} = \frac{x^2}{(0.02 - x)} = 6.2 \times 10^{-10}$$

Assume that the acid dissociates only slightly, so that $x = [CN^-] \ll 0.02$ and

$$0.02 - x \approx 0.02$$

$$K = \frac{x^2}{0.02} = 6.2 \times 10^{-10}$$

$$x^2 = 12.4 \times 10^{-12}$$

The simplifying assumption gives

$$x = [H^+] = [CN^-] = 3.5 \times 10^{-6}\ M$$

and $$[HCN] = (0.02 - x) = 0.0199965\ M$$

The calculated value of $x = 3.5 \times 10^{-6}$ M $\ll$ 0.02 M, so the simplification is valid.
Calculate the percent ionization for the 0.02 M acid

$$\%\ \text{Ionization} = \frac{[HCN]_{\text{ionized}}}{[HCN]_{\text{Total}}} \times 100\% = \frac{3.5 \times 10^{-6}}{0.02} \times 100\% = 0.0175\%$$

Example 7.12 Ammonia Buffer Solution

Calculate the pH of a buffer solution that initially consists of 0.0500 M NH_3 and 0.0350 M NH_4^+. Note: K_a for NH_4^+ is 5.5×10^{-10}. The equation for the reaction is as follows:

$$NH_4^+ \leftrightarrows H^+ + NH_3$$

Because the reaction has a 1:1 stoichiometry the x moles NH_4^+ will lose is equal to the x moles that H^+ and NH_3 will gain.

We know that initially there is 0.0350 M NH_4^+ and 0.0500 M NH_3. Before the reaction occurs, no H^+ is present, so it starts at 0.

Reaction	NH_4^+	$\rightleftharpoons$	H^+	+	NH_3
Initial	0.035		0		0.050
Change	$-x$		$+x$		$+x$
Equilibrium	$(0.035 - x)$		x		$0.050 + x$

Apply the equilibrium values to the expression for K_a.

$$5.5 \times 10^{-10} = \frac{[H^+][NH_3]}{[NH_4^+]} = \frac{x(0.050 + x)}{0.035 - x}$$

Assuming x is negligible compared to 0.0500 and 0.0350

$$(0.050 + x) \approx 0.050 \quad \text{and} \quad (0.035 - x) \approx 0.035$$

the equation simplifies to

$$5.5 \times 10^{-10} = \frac{[H^+][NH_3]}{[NH_4^+]} = \frac{x(0.050)}{0.035}$$

$$x = [H^+] = 3.85 \times 10^{-10}$$

$$pH = -\log(3.85 \times 10^{-10})$$

$$pH = 9.41$$

Example 7.13 Ionization of Hydrogen Sulfide (Hydrosulfurous Acid)

Calculate the concentration of all species in 0.01 M hydrogen sulfide solution (H_2S).
First ionization

$$H_2S \leftrightarrows H^+ + HS^- \qquad K_1 = \frac{[H^+][HS^-]}{[H_2S]} = 8.9 \times 10^{-8}$$

Reaction	H_2S	$\leftrightarrows$	HS^-	+	H^+
Initial	0.01		0		0
Change	$-x$		$+x$		$+x$
Equilibrium	$(0.01 - x)$		x		x

$$K_{a1} = \frac{[H^+][HS^-]}{[H_2S]} = \frac{(x)(x)}{0.01 - x} = 8.9 \times 10^{-8}$$

$$x^2 + 8.9 \times 10^{-8} x - 8.9 \times 10^{-10} = 0$$

$$x = \frac{-b \pm \sqrt{b^2 - 4ac}}{2a} = \frac{-8.9 \times 10^{-8} \pm \sqrt{(8.9 \times 10^{-8})^2 - 4(1)(-8.9 \times 10^{-10})}}{2(1)}$$

$$x = 2.98 \times 10^{-5}\ \text{M} \approx 3 \times 10^{-5}\ \text{M}$$

$$x = [H^+] = [HS^-] = 0.00003\ \text{M}$$

$$\text{pH} = -\log_{10}\left(3 \times 10^{-5}\right) = 4.5$$

$$\left[H_2S\right] = 0.01 - x = 0.01 - 0.00003 = 0.00997\ \text{M}$$

Approximate solution:

Assume that $x = [HS^-] \ll 0.01$ and

$$0.01 - x \approx 0.01$$

$$K = \frac{(x)(x)}{0.01 - x} = \frac{x^2}{0.01} = 8.9 \times 10^{-8}$$

$$x^2 = 8.9 \times 10^{-10}$$

The simplifying assumption gives

$$x = [H^+] = [HS^-] = 3 \times 10^{-5}\ \text{M}$$

and $$\left[H_2S\right] = \left(0.01 - x\right) = 0.00997\ \text{M}$$

The calculated value of $x = 3 \times 10^{-5}$ M $\ll 0.01$ M, so the simplification is valid.

Second ionization

$$HS^- \leftrightarrows H^+ + S^{2-} \qquad K_2 = \frac{[H^+][S^{2-}]}{[HS^-]} = 1 \times 10^{-19}$$

Reaction	HS^-	$\leftrightarrows$	S^{2-}	+	H^+
Initial	3×10^{-5}		0		3×10^{-5}
Change	$-y$		$+y$		$+y$
Equilibrium	$(3 \times 10^{-5} - y)$		y		$(3 \times 10^{-5} + y)$

Approximate solution:

Assume that $y = [S^{2-}] \ll 3 \times 10^{-5}$ so that

$$(3 \times 10^{-5} - y) \approx 3 \times 10^{-5} \quad \text{and} \quad (3 \times 10^{-5} + y) \approx 3 \times 10^{-5}$$

$$K_2 = \frac{[H^+][S^{2-}]}{[HS^-]} = \frac{(3 \times 10^{-5} + y)(y)}{(3 \times 10^{-5} - y)} \approx \frac{3 \times 10^{-5} y}{3 \times 10^{-5}} \approx 1 \times 10^{-19}$$

$$y \approx 1.3 \times 10^{-19}\ \text{M}$$

The calculated value of $y = 1 \times 10^{-19}$ M $\ll 3.0 \times 10^{-5}$ M, so the simplification is valid.

$$[S^{2-}] = y \sim 1 \times 10^{-19} \text{ M}$$

This vanishingly small concentration for S^{2-} confirms our earlier observation in Example 7.9 that the second dissociation of H_2S essentially does not occur in aqueous solution.

7.8 CHLORINATION

Implementing the disinfection of drinking water by chlorination may be the single most important public health policy of the twentieth century. Where it is used, it has virtually eradicated waterborne diseases such as cholera and typhoid fever. It has saved millions of lives.

Chlorine is a strong oxidizing agent and that makes it important in pollution control. It reacts with ammonia, hydrogen sulfide, and with many of the compounds that are measured as total organic carbon, chemical oxygen demand, and biochemical oxygen demand.

The use of chlorine as an oxidizing agent is explained in Chapter 8. Here we discuss the importance of chlorine ionization to form hypochlorous acid and hypochlorite.

Chlorine is often supplied to a water or wastewater utility as a liquefied gas. The liquid vaporizes when the pressure is reduced and the chlorine gas is added to water. A hydrolysis reaction occurs within a few seconds at ordinary water temperatures to form hypochlorous acid (HOCl). This is an instantaneous, irreversible reaction.

$$Cl_2 + H_2O \rightarrow HOCl + H^+ + Cl^-$$

or

$$Cl_2 + OH^- \rightarrow HOCl + Cl^-$$

Hypochlorous acid ionizes to form hypochlorite (OCl^-). This is a reversible reaction that is influenced by pH.

$$HOCl \leftrightarrows H^+ + OCl^-$$

Both hypochlorous acid and hypochlorite are oxidizing agents and disinfectants, but HOCl is a much stronger disinfectant, and a much stronger oxidizing agent, than OCl^-. Low pH favors the formation of HOCl and disinfection is more efficient at lower pH.

Hypochlorites and bleaches work in the same general manner as chlorine gas. They react with water to form hypochlorous acid. The reactions of sodium hypochlorite (NaOCl) and calcium hypochlorite ($Ca(OCl)_2$) with water are as follows:

$$Na(OCl) + H_2O \rightarrow HOCl + Na^+ + OH^-$$

$$Ca(OCl)_2 + 2H_2O \rightarrow 2HOCl + Ca^{2+} + 2OH^-$$

Hypochlorous acid also reacts with ammonia to form chloramines. Chloramines are effective disinfectants. They are weaker than hypochlorite, but more stable, which is an advantage when disinfecting protection is needed in long pipelines and water distribution systems.

The formation of monochloramine is

$$\underset{\text{Ammonia}}{NH_3} + \underset{\text{Hypochlorous Acid}}{HOCl} \rightarrow \underset{\text{Monochloramine}}{NH_2Cl} + \underset{\text{Water}}{H_2O}$$

Monochloramine may then react with more hypochlorous acid to form a dichloramine ($NHCl_2$). Then the dichloramine may react with hypochlorous acid to form a trichloramine ($NHCl_3$).

$$NH_2Cl + HOCl \rightarrow NHCl_2 + H_2O$$
$$NHCl_2 + HOCl \rightarrow NCl_3 + H_2O$$

Example 7.14 looks at the effect of pH on the hypochlorite-hypochlorous ionization equilibrium in water.

Example 7.14 Chlorine and Hypochlorous Acid

The ionization reaction is

$$HOCl \leftrightarrows H^+ + OCL^-$$

The ionization constant is

$$K = \frac{[OCl^-][H^+]}{[HOCl]} = 4.0 \times 10^{-8}$$

The quantities in the brackets are concentrations in moles per liter. The concentration of [H^+] can be expressed as pH: pH = $-\log_{10}[H^+]$ and [H^+] = 10^{-pH}. Thus,

$$K = \frac{[OCl^-][H^+]}{[HOCl]} = \frac{[OCl^-]10^{-pH}}{[HOCl]}$$

which gives $$[OCl^-] = K[HOCl]10^{pH}$$

Total Chlorine = [OCl^-] + [HOCl]

Using the result from Example 7.7, the fraction of the total chlorine that is hypochlorite is

$$\frac{[OCl^-]}{[HOCl]+[OCl^-]} = \frac{K[HOCl]10^{pH}}{[HOCl]+K[HOCl]10^{pH}} = \frac{1}{1+(10^{-pH}/K)}$$

Figure 7.10 shows the fractions of HOCl and OCl^- as a function of pH. $K = 4 \times 10^{-8}$ from Table 7.4.

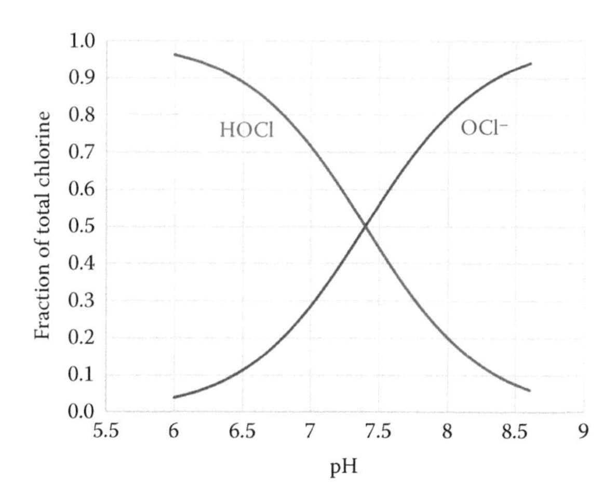

FIGURE 7.10 Fraction ionization of hypochlorous acid and hypochlorite.

7.9 CARBONATES AND ALKALINITY

The pH does not reveal how much acid or base must be added to neutralize a wastewater. This is especially true for wastewaters that have some buffering capacity due to the presence of weak acids and their salts.

Alkalinity is the capacity of a solution to neutralize an acid. Alkalinity is measured by titration with a strong acid down to a reference pH, which is pH 4.5 for total alkalinity and 8.3 for carbonate alkalinity. Alkalinity is equal to the amount of acid added to reach those pH endpoint, and is expressed as mg/L $CaCO_3$. Figure 7.11 is a titration curve for a water sample that has a very high initial pH.

Assuming the alkalinity is due mainly to carbonate species, which is reasonable in most natural water systems and in municipal wastewater,

$$\text{Total Alkalinity} = [HCO_3^-] + 2[CO_3^{2-}] + [OH^-] - [H^+]$$

$$\text{Carbonate Alkalinity} = [CO_3^{2-}] + [OH^-] - [H_2CO_3^*] - [H^+]$$

Figure 7.11 shows that the carbonate (CO_3^{2-}) will be negligible when the pH is less than pH 8.3. Conversely, we can say that below pH 8.3 the alkalinity is predominately in the form of HCO_3^-.

Since carbonates contribute to water and wastewater alkalinity, alkalinity is often reported as mg $CaCO_3$/L.

The equilibrium equations for the *carbon dioxide–bicarbonate–carbonate* system are as follows:

$$CO_2(aq) \leftrightarrows CO_2(g) \qquad K_H = \frac{p_{CO_2}}{[CO_2]_{aq}}$$

$$CO_2(aq) + H_2O \leftrightarrows H_2CO_3 \qquad K = \frac{[H_2CO_3]}{[CO_2]_{aq}}$$

$$H_2CO_3 \leftrightarrows H^+ + HCO_3^- \qquad K_1' = \frac{[H^+][HCO_3^-]}{[H_2CO_3]}$$

$$HCO_3^- \leftrightarrows H^+ + CO_3^2 \qquad K_2 = \frac{[H^+][CO_3^{2-}]}{[HCO_3^-]}$$

Since it is difficult to distinguish between CO_2(aq) and H_2CO_3, it is common practice to combine the two equilibrium equations to get

$$K_1 = \frac{[H^+][HCO_3^-]}{[H_2CO_3]+[CO_2]_{aq}} = \frac{K_1'}{K+1}$$

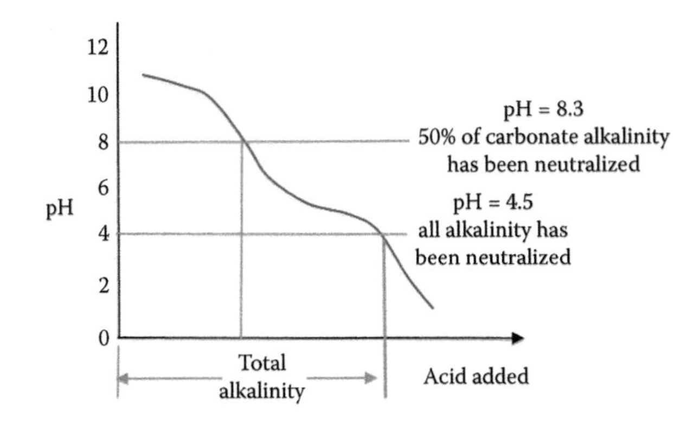

FIGURE 7.11 Titration curve to measure total and carbonate alkalinity.

and to define $[H_2CO_3^*] = [H_2CO_3] + [CO_2]_{aq}$

and

$$K_1 = \frac{[H^+][HCO_3^-]}{[H_2CO_3^*]}$$

where

$$pK_1 = 6.35 \text{ and } pK_2 = 10.33.$$

The equations for a carbonate solution in contact with solid $CaCO_3$ are as follows:

Equilibrium: $K_s = [Ca^{2+}][CO_3^{2-}]$
Electroneutrality (charge balance): $2[Ca^{2+}] + [H^+] = [HCO_3^-] + 2[CO_3^{2-}] + [OH^-]$
Mass balance (assuming all ions come from the solid):

$$C_t = (Ca^{2+}) = (H_2CO_3^*) + (HCO_3^-) + (CO_3^{2-}) + (OH^-)$$

Example 7.15 Alkalinity Destruction in Coagulation

Commercial grade alum has $14H_2O$ waters of hydration attached to the $Al_2(SO_4)_3$ molecule. When it is added to water as a coagulant, it reacts with the natural bicarbonate alkalinity:

$$Al_2(SO_4)_3 \cdot 14H_2O + 6HCO_3^- \rightarrow 2Al(OH)_3 + 3SO_4^{2-} + 6CO_2 + 14H_2O$$

Calculate the mass of alkalinity (as mg/L $CaCO_3$) destroyed per unit mass of added alum.
Molar masses:

Commercial grade alum = 594 g/mol (PMB made changes here)
Bicarbonate: 6(61 g) = 366 g/mol

Bicarbonate destroyed = 366/594 = 0.62 g bicarbonate/g commercial grade alum added
Expressing the bicarbonate as mg/L $CaCO_3$ (multiply by the ratio of the equivalent weights)

$$\text{Bicarbonate alkalinity} = 366 \text{ g } HCO_3^-(50 \text{ g } CaCO_3 / 61 \text{ g } HCO_3^-) = 300 \text{ g as } CaCO_3$$

Bicarbonate alkalinity destroyed = 300/594 = 0.51 g bicarbonate alkalinity/g commercial grade alum added

Example 7.16 Carbon Dioxide and pH in an Anaerobic System

Anaerobic wastewater treatment processes and sludge digestion processes generate methane gas and carbon dioxide from acetic acid and other volatile fatty acids (butyric acid, etc.). These are weak acids, but they do tend to depress the pH in the anaerobic reactor. Carbon dioxide forms H_2CO_3, which also acts as an acid in the system. This is critical because the desired range for the process pH is 6.6–8.5. Lower pH inhibits the microbes that convert the acetic acid to methane. A high concentration of alkalinity is needed to neutralize the volatile fatty acids and carbonic acid formed in the process.

The *CO_2–water system* is controlled by the partial pressure of the carbon dioxide, which is denoted by p_{CO_2}. If the biogas is 35% CO_2 by volume, the value of p_{CO_2} = 0.35 atm. Even a CO_2 level as low as 1%–3% (0.01–0.03 atm partial pressure) has a significant effect on the pH of water.

The relations are

$$[H_2CO_3] = K_H p_{CO_2} \quad \text{and} \quad K_1 = \frac{[HCO_3^-][H^+]}{[H_2CO_3]}$$

Combining these gives

$$[H^+] = \frac{K_1 K_H p_{CO_2}}{[HCO_3^-]} \quad \text{and} \quad [HCO_3^-] = \frac{K_1 K_H p_{CO_2}}{[H^+]}$$

where

$K_H = 0.0246$ mol/L-atm at 35°C.

$K_1 = 4.8 \times 10^{-7}$ at 35°C (for ionic strength of 0.0 M, see notes in the following text)

$[HCO_3^-]$ = the bicarbonate alkalinity if the pH = 7–8

Calculate the alkalinity required to neutralize the carbon dioxide, assuming $p_{CO_2} = 0.3$ atm, pH = 7, and temperature = 35°C (35°C is the typical operating temperature for a mesophilic anaerobic sludge digester).

$$[HCO_3^-] = \frac{K_1 K_H p_{CO_2}}{[H^+]} = \frac{(4.8 \times 10^{-7})(0.0246 \text{ mol/L-atm})(0.3 \text{ atm})}{(10^{-7})} = 0.0354 \text{ mol/L } HCO_3^-$$

The molar mass of $[HCO_3^-]$ is 61 g/mol, so

$$(0.0354 \text{ mol/L})(61 \text{ g/mol}) = 2.159 \text{ g/L} = 2159 \text{ mg/L}$$

Convert this to an equivalent of $CaCO_3$, which is how alkalinity is reported. The molar mass of $CaCO_3$ = 100 g/L

$$(2159 \text{ mg } HCO_3^- /\text{L})(100 \text{ mg } CaCO_3)/(61 \text{ mg } HCO_3^-) = 3539 \text{ mg } CaCO_3/\text{L alkalinity}$$

Notes:

1. The concentrations of dissolved chemicals can be quite high in an anaerobic system, so the ionic strength may be greater than 1.0. This will change the value of K_1. $K_1 = 1 \times 10^{-6}$ for ionic strength of 0.2 M at 35°C. This would increase the amount of alkalinity needed.
2. The dissociation constant $K_1 = 4.7 \times 10^{-7}$ at 30°C and $K_1 = 5.19 \times 10^{-7}$ at 50°C.
3. Henry's constant, K_H, increases with temperature, which means that the solubility of CO_2 in water decreases with an increase in temperature. The solubility at 50°C is about 67% of the solubility at 30°C.
4. The combined effects of (2) and (3) result in the alkalinity requirement being lower at 50° than at 30°C.

7.10 ANOTHER LOOK AT ACID–BASE EQUILIBRIA USING p*K* VALUES

This section expands the interpretation of acid–base equilibria, using the concept of p*K*.

The "p" operator means that some quantity is transformed by taking the negative base 10 logarithm.

Just as

$$pH = -\log_{10}[H^+]$$

$$pK = -\log_{10}[K]$$

where *K* is the ionization constant.

Table 7.5 gave ionization constants for some common acids and bases and their p*K* values.

Most weak bases accept a proton (H^+) from a water molecule to leave an OH^- ion. The ionic (dissociated) form of a weak acid (F^- in the reaction below) is called its conjugate base.

Likewise, the ionic (protonated) form of a weak base (NH_4^+ in the reaction below) is called its conjugate acid.

Ammonia and hydrofluoric acid are examples:

$$\underset{\text{weak base}}{NH_3 + H_2O} \leftrightarrows \underset{\text{conjugate acid}}{NH_4^+} + OH^-$$

$$\underset{\text{weak acid}}{HF} \leftrightarrows H^+ + \underset{\text{conjugate base}}{F^-}$$

The pK values convey useful information about a weak acid or base. When the pH of a solution equals the pK, the concentrations of the weak acid and its conjugate base are equal. At pH values below pK, the acid form dominates. At pH values above pK, the dissociated form (conjugate base) has the higher concentration.

Here's how it works for a weak acid:

$$HA \leftrightarrows H^+ + A^-$$

The equilibrium ionization constant is

$$K = \frac{[H^+][A^-]}{[HA]}$$

Taking negative base 10 logarithms yields

$$-\log K = -\log[H^+] - \log\left(\frac{[A^-]}{[HA]}\right) \quad \text{or} \quad pK = pH - \log\left(\frac{[A^-]}{[HA]}\right)$$

When pK = pH,

$$\log\left(\frac{[A^-]}{[HA]}\right) = 0 \quad \text{or} \quad \frac{[A^-]}{[HA]} = 1$$

Example 7.17 Dissociation of Hypochlorous Acid

Table 7.5 gives pK = 7.4 for hypochlorous acid (HOCl). Therefore, if the pH of a hypochlorous acid solution is 7.4, the concentrations of HOCl and OCl^- are the same: [HOCl] = [OCl^-]. Lowering the pH below 7.4 shifts the equilibrium to favor HOCl. Raising it above 7.4 favors OCl^-. This is shown in Figure 7.10, but let's confirm it with calculations. A drinking water has a pH of 6.8 and total chlorine concentration of 0.5 mg/L. Calculate the concentration of HOCl and OCl^-.

The molar mass of chlorine is 35.45 g/mol, so the total molar concentration of chlorine in solution is

$$\text{Total chlorine} = [OCl^-] + [HOCl] = (0.5\ \text{mg/L})/(35.45 \times 10^3\ \text{mg/mol}) = 0.000014\ \text{M} = 1.4 \times 10^{-5}\ \text{M}$$

From the relation between pH and pK,

$$pK = pH - \log\left(\frac{[A^-]}{[HA]}\right) = pH - \log\left(\frac{[OCl^-]}{[HOCl]}\right)$$

Substituting [OCl^-] = 1.4×10^{-5} M − [HOCl] gives

$$pK = pH - \log\left(\frac{1.4 \times 10^{-5}\ \text{M} - [HOCl]}{[HOCl]}\right)$$

Solving for [HOCl]

$$\log\left(\frac{1.4\times10^{-5}\,\text{M}-[\text{HOCl}]}{[\text{HOCl}]}\right)=\text{pH}-\text{p}K=6.8-7.4=-0.6$$

$$\frac{1.4\times10^{-5}\,\text{M}-[\text{HOCl}]}{[\text{HOCl}]}=10^{-0.6}=0.25$$

$$1.25[\text{HOCl}]=1.4\times10^{-5} \quad\Rightarrow\quad [\text{HOCl}]=1.12\times10^{5}\,\text{M}$$

and

$$[\text{OCl}^-]=1.4\times10^{-5}\,\text{M}-1.12\times10^{-5}\,\text{M}=2.8\times10^{-6}\,\text{M}$$

The molar ratio of [HOCl] to total chlorine is $(1.12 \times 10^{-5})/(1.4 \times 10^{-5}) = 0.8$, which agrees with the value shown in Figure 7.10.

7.11 CONCLUSION

The pH of natural waters is critical for the health of plants and animals. The pH of water used in the home must be within certain ranges to prevent corrosion and scaling in pipes, water heaters, and other utilities. Industrial discharges to municipal sewers are required to maintain a pH between 6 and 8.5 to prevent damage to sewers and pumps. The pH in wastewater treatment processes is manipulated, sometimes toward acidic conditions and sometimes toward alkaline conditions, to control the ionization and precipitation of pollutants.

Neutralization is used by a variety of industries. The chemistry is simple. Add an acid to lower the pH, or add a base to increase the pH. The implementation can be difficult because the pH of industrial wastewater can change from 2 to 10 in a few seconds. A very small amount of neutralizing reagent can shift the pH by several units. The addition of neutralizing chemicals must be managed with instruments that can measure pH and quickly operate control values and feed pumps. Mixed reactors, often in series, are used to blend the incoming wastes and smooth out the sudden variations.

Ionization constants can be used to calculate the concentrations of chemical species in relatively pure or simple solutions. As the wastewater chemistry becomes more complex, stoichiometric and ionization calculations give way to empirical stoichiometry.

The empirical stoichiometry of neutralization is revealed by the titration curve. Titrating the wastewater with different neutralizing reagents will reveal the amount required for neutralization, as well as the tendency for sludge to be produced. If sludge is produced, it can be observed for settling rates, compaction density, filtration rates, and other design information.

Acid–base reactions are the simplest examples of reversible reactions and chemical equilibria, but other ionization reactions have similar chemistry. A reversible reaction will find an equilibrium between the reactants and the products, and this can be calculated using published ionization constants.

The ICE (or RICE) tables are a handy way to organize the calculations. *R* is for the reaction, *I* is for the initial condition, *C* is for the change in the concentrations of the chemical species, and *E* is for the equilibrium concentrations.

Ammonia, hydrogen sulfide, chlorine, and the carbonates are chemical systems that play a special role in pollution control work.

Ammonia is important because it is toxic to fish, and because it is a nutrient in natural waters and in biological wastewater treatment processes. It will stimulate algae blooms, and it will cause oxygen to be consumed, as bacteria convert it to nitrate.

Hydrogen sulfide is important because it is odorous, toxic, and corrosive, and because when it is burned it is converted to sulfur dioxide, which is a controlled air pollutant.

Chlorine is important as a bacterial disinfectant in drinking water and as an oxidizing agent in pollution control systems. The form of the chlorine (hypochlorous acid or hypochlorite) affects both its oxidizing and disinfecting power.

The carbon dioxide–bicarbonate–carbonate system dominates in controlling the buffering chemistry of natural waters, and that makes it important for drinking-water process design.

8 Precipitation Reactions

8.1 THE DESIGN PROBLEM

Precipitation converts dissolved pollutants into solid particles that can be removed by some simple physical separation method like settling or filtration. Controlled precipitation is used to remove metals from industrial wastewater, remove phosphorus from wastewater, and to soften drinking water and boiler feed water. Uncontrolled precipitation causes scaling and clogging in boilers, hot water heaters, coffee makers, and pipes.

The precipitate that is formed must be removed, so precipitation processes are always supported by some kind of a separation process. Effluent limits for metals are given as total metal. If the effluent limit is 0.1 mg/L total copper, and soluble copper can be reduced to 0.05 mg/L, the concentration of copper in the solid form cannot exceed 0.05 mg/L. A useful precipitation reaction will leave only a trace of the objectionable or valuable material in solution. The ultimate effluent quality will depend on the separation of the precipitate from the solution. The separation of solid from liquid may be done by settling if the effluent requirement is not too strict, or by filtration if the allowable effluent concentration is very small.

Precipitation has two disadvantages. The first is that a precipitating reagent (sometimes more than one) must be added. As a general principle, we prefer treatment processes that do not require the addition of a new material. The second is that the precipitate solids must be removed as sludge or as filter backwash water. Sludge processing and disposal is expensive. The sludge will be classified as a hazardous material if it contains heavy metals. This will increase the cost of sludge disposal and the waste generator will have a long-term liability for future problems caused by the sludge even though it may be properly placed in a legal landfill. These factors often discourage precipitation and encourage using technology that can recover the metal and reuse the water. (See Berthouex and Brown, 2014, for examples of pollution prevention.)

8.2 SOLUBILITY RULES: BASIC GUIDELINES

Precipitation reactions are used to remove a soluble (dissolved) substance from solution by transforming it into an insoluble (particulate) form. Table 8.1 outlines how a chemist thinks of the solubility for common salts, bases, and gaseous substances in water. A *soluble* compound exists only in its ionic form. A statement such as "The oxides and hydroxides of all other common metals are insoluble" means that precipitation should be a feasible means of removing the metal from solution.

8.3 SOLUBILITY PRODUCTS

There is a mathematical relationship between the concentrations of soluble ions that will exist in solution when the solid compound formed from these elements is present. The equilibrium of two soluble ions A and B that are created by dissolving a solid A_aB_b

$$A_aB_b \leftrightarrows aA + bB$$

is determined by the equilibrium expression:

$$K_{sp} = \frac{[A]^a[B]^b}{[A_aB_b]} = [A]^a[B]^b$$

TABLE 8.1
Solubility Rules

Salts and Bases

1. Nitrates, chlorates, and acetates of all metals are soluble in water
2. All common sodium, potassium, and ammonium salts are soluble in water
3. The chlorides, bromides, and iodides of all metals except lead, silver, and mercury are soluble in water. The lead salts $PbCl_2$, $PbBr_2$, and PbI_2 are soluble in hot water. The water-soluble chlorides, bromides, and iodides are also soluble in dilute acids
4. The sulfates of all metals except lead, mercury, barium, strontium, and calcium are soluble in water. Silver sulfate is slightly soluble. The water-insoluble sulfates are soluble in dilute acids
5. The carbonates, phosphates, borates, sulfites, chromates, and arsenates of all metals except sodium, potassium, and ammonium are insoluble in water, but are soluble in dilute acids. $MgCrO_4$ is soluble in water; $MgSO_4$ is slightly soluble in water
6. The sulfides of all metals except barium, strontium, calcium, and magnesium, sodium, and potassium are insoluble in water
7. The hydroxides of sodium, potassium, and ammonium are very soluble in water. The hydroxides of calcium, strontium, and barium are moderately soluble. The oxides and hydroxides of all other common metals are insoluble

Gases

1. Compounds that are gases at ordinary temperatures include NH_3, CH_4, HF, HCl, HBr, HI, H_2S, CO_2, SO_2, and CO
2. Solubility of gases in water at 20°C:
   - Very soluble in water (400–700 volumes of gas/volume water): HCl, HF, HBr, HI, NH_3
   - Fairly soluble in water (40 volumes of gas/volume water): SO_2
   - Slightly soluble in water (1–2.5 volumes of gas/volume water): CO_2, Cl_2, H_2S
   - Very slightly soluble in water (0.015–0.03 volumes of gas/volume water): N_2, H_2, CO, O_2, CH_4

K_{sp} is the solubility product of the solid A_aB_b. [A], [B], and $[A_aB_b]$ are the activities of the chemical species in solution. The activity of a solid is always 1.0. In this case, $[A_aB_b] = 1.0$, hence the simplification to

$$K_{sp} = [A]^a[B]^b$$

The activities [A] and [B] are approximated by the molar concentrations (mol/L).

The chemical balance at equilibrium can be shifted to favor the formation of more solids by adding either dissolved species A or B to the solution. Conversely, removing A or B will cause more A_aB_b to dissolve to replace the ions that were removed. This is *Le Chatelier's principle.*

What we said above is valid so long as there is some solid A_aB_b in the system. When the solids are removed from the solution, the equilibrium between the dissolved ions may change, or we may take additional steps to change it. For example, we may want to shift some safe distance from a condition that might cause more solids to precipitate. We do this to avoid mineral scaling in cooling towers, boilers, and municipal water distribution systems.

Tables 8.2 and 8.3 contain the solubility products (K_{sp}) for compounds that are useful or important in pollution prevention and control.

The solubility of metal sulfides is more complex than for hydroxides. Chemistry books and handbooks of chemical data give three definitions for the solubility product of metal sulfides. These are K_{sp}, K^*_{sp}, and K_{spa}. Values for several metal sulfides are given in Table 8.4. These are explained in the following paragraphs.

The first definition (K_{sp}) is based on the dissolution that forms free sulfide ion. It is based on an incorrect value of the second ionization constant of H_2S ($K_2 \approx 10^{-13}$), used prior to about 1980. This has appeared in many older books on pollution control. It is still useful even though this chemistry

TABLE 8.2
Solubility Product (K_{sp}) for Metal Hydroxides in Water at 20°C

Metal Hydroxides	K_{sp}
Manganous hydroxide—$Mn(OH)_2$	1.9×10^{-13}
Ferrous hydroxide—$Fe(OH)_2$	4.9×10^{-17}
Ferric hydroxide—$Fe(OH)_3$	2.8×10^{-39}
Zinc hydroxide—$Zn(OH)_2$	3.0×10^{-17}
Nickel hydroxide—$Ni(OH)_2$	5.5×10^{-16}
Stannous hydroxide—$Sn(OH)_2$	5.5×10^{-28}
Cobalt hydroxide—$Co(OH)_2$	5.9×10^{-15}
Chromium hydroxide—$Cr(OH)_3$	6.3×10^{-31}
Lead hydroxide—$Pb(OH)_2$	1.4×10^{-15}
Cadmium hydroxide—$Cd(OH)_2$	7.2×10^{-15}
Silver hydroxide—AgOH	2.0×10^{-8}
Copper hydroxide—$Cu(OH)_2$	2.2×10^{-20}
Mercuric hydroxide—$Hg(OH)_2$	3.2×10^{-26}

Source: Speight, J., 2017.

TABLE 8.3
Solubility Product in Water at 18°C–25°C for Selected Metal Compounds

Element	Compound	K_{sp}	Compound	K_{sp}	Compound	K_{sp}
Calcium	$CaCO_3$	3.3×10^{-9}	CaF_2	2.7×10^{-11}	$CaSO_4$	4.9×10^{-5}
	$CaHPO_4$	1×10^{-7}	$Ca_3(PO_4)_2$	2×10^{-29}	$Ca_3(AsO_4)_2$	6.8×10^{-19}
Cadmium	$CdCO_3$	1×10^{-12}	$Cd_3(PO_4)_2$	2.5×10^{-33}	$Cd_3(AsO_4)_2$	2.2×10^{-23}
Chromium	CrF_3	6.6×10^{-11}	$CrPO_4$	2.4×10^{-23}	$CrAsO_4$	7.7×10^{-21}
Copper	$CuCO_3$	1.4×10^{-10}	$Cu_3(PO_4)_2$	1.4×10^{-37}		
Iron	$FeAsO_4$	5.7×10^{-21}	$FePO_4$	9.9×10^{-16}		
Lead	$PbCO_3$	7.4×10^{-14}	$Pb_3(PO_4)_2$	8.0×10^{-43}	$PbCrO_4$	2.8×10^{-13}
	$Pb_3(AsO_4)_2$	4×10^{-36}				
Magnesium	$Mg(OH)_2$	1.8×10^{-11}	$MgNH_4PO_4$	2.5×10^{-13}	MgF_2	5.16×10^{-11}
	$MgCO_3$	6.82×10^{-6}	$Mg_3(PO_4)_2$	1.04×10^{-24}		
Mercury	Hg_2CO_3	3.6×10^{-17}	$Hg_2(CN)_2$	5×10^{-40}	Hg_2Cl_2	1.4×10^{-18}
	Hg_2SO_3	1×10^{-27}				
Silver	Ag_2CO_3	8.46×10^{-12}	Ag_3PO_4	8.9×10^{-17}		
Tin	$Sn(OH)_4$	1.0×10^{-56}				
Zinc	$ZnCO_3$	1.46×10^{-10}	$Zn_3(PO_4)_2$	9×10^{-33}		

Source: Speight, J., 2017.

is not strictly correct. And, because it is simple and easy to understand we will use it in some later sections of this chapter.

$$MeS \leftrightarrows Me^{2+} + S^{2-} \qquad K_{sp} = [Me^{2+}][S^{2-}]$$

A second definition (K^*_{sp}) is based on the realization that the true value of the second ionization constant of H_2S is much smaller and uncertain ($K_2 \approx 10^{-19}$), so that free sulfide ion (S^{2-}) cannot exist

TABLE 8.4
Solubility Products for Some Important Metal Sulfides

Metal Sulfides	K_{sp}	K^*_{sp}	K_{spa}
Manganous sulfide—MnS	2.5×10^{-13}	3×10^{-14}	3×10^{7}
Ferrous sulfide—FeS	6.3×10^{-18}	6×10^{-19}	6×10^{2}
Zinc sulfide—ZnS	1.6×10^{-24}	2×10^{-25}	2×10^{-4}
Nickel sulfide—NiS	3.2×10^{-19}	3×10^{-20}	–
Stannous sulfide—SnS	1.0×10^{-25}	1×10^{-26}	1×10^{-5}
Cobalt sulfide—CoS	4.9×10^{-21}	5×10^{-22}	–
Lead sulfide—PbS	8.0×10^{-28}	3×10^{-28}	3×10^{-7}
Cadmium sulfide—CdS	8.0×10^{-27}	8×10^{-28}	8×10^{-7}
Silver sulfide—Ag_2S	6.3×10^{-50}	6×10^{-51}	6×10^{-30}
Copper sulfide—CuS	6.3×10^{-36}	6×10^{-37}	6×10^{-16}
Mercuric sulfide—HgS	1.6×10^{-52}	2×10^{-53}	2×10^{-32}
Chromium	No precipitate		

Sources: K_{sp} from Speight, J. 2017.
K^*_{sp} from Brown, T.E. et al. 2006 and Hill, J.H. and Petrucci, R.H. 1999.
K_{spa} from Myers, R.J. 1986 and Lide, D.R. 2009.

in aqueous solution because it hydrolyzes immediately to HS^- and OH^-. The following solubility reaction represents the pH dependency of the overall equilibrium.

$$MeS + H_2O \leftrightarrows Me^{2+} + HS^- + OH^-$$

Values of K^*_{sp} are sometimes reported in texts and online databases, and are about a factor of 10 smaller than the traditional $K_{sp} = [Me^{2+}][S^{2-}]$

A third version (K_{spa}) is called K_{spa}, or K_{sp} in acid. This is based on the work of Myers (1986) and is found in the *Handbook of Chemistry and Physics*. The solubility comes from equilibrium calculations of metal sulfides in acid solution using the reaction

$$MeS + 2H^+ \leftrightarrows Me^{2+} + H_2S$$

$$K_{spa} = \frac{[Me^{2+}][H_2S]}{[H^+]^2}$$

K_{spa} can also be derived from the second sulfide solubility product definition (K^*_{sp}) as follows:

1. Solve for $[HS^-]$ using the first ionization of H_2S.

$$H_2S \leftrightarrows HS^- + H^+ \quad \rightarrow \quad K_1 = \frac{[HS^-][H^+]}{H_2S} \quad \text{or} \quad [HS^-] = \frac{K_1[H_2S]}{[H^+]}$$

2. Solve for $[OH^-]$ from the ion product of water.

$$K_w = [H^+][OH^-] \rightarrow [OH^-] = K_w/H^+$$

3. Substitute these expressions for $[HS^-]$ and $[OH^-]$ into the definition of K_{sp}^* earlier to give

$$K_{sp}^* = [Me^{2+}][HS^-][OH^-]$$

$$= [Me^{2+}]\frac{K_1[H_2S]}{[H^+]}\frac{K_w}{[H^+]} = \frac{[Me^{2+}][H_2S]}{[H^+]^2}K_1K_w$$

4. Grouping the constants defines K_{spa}:

$$K_{spa} = \frac{K_{sp}^*}{K_1K_w} = \frac{[Me^{2+}][H_2S]}{[H^+]^2}$$

Using these different solubility product values makes very little difference. The K_{sp} values for metal sulfides can be misleading by suggesting that sulfide solubility is independent of pH and that free sulfide exists in aqueous solutions. K_{sp}^* and K_{spa} explicitly account for the variation of solubility with pH, and are based on the fact that the concentration of free sulfide ion is always negligible in aqueous solutions. The more important lesson to be learned is perhaps that this is the way science advances: What we believed and made students learn for nearly a century can be subject to revision by the careful work of a few individuals who are motivated to question "what everyone knows."

Example 8.1 Solubility of Calcium Fluoride

Find the equilibrium concentrations of Ca^{2+} and F^- dissolved in pure water when excess solid calcium fluoride (CaF_2) is added. Dissolution of one mole of crystalline calcium fluoride in water produces one mole of calcium and two moles of fluoride ions:

$$CaF_2 \leftrightarrows Ca^{2+} + 2\,F^- \quad K_{sp} = 2.7\times10^{-11}$$

When the liquid is in equilibrium with the solid phase, the general equilibrium relation gives

$$K_{sp} = \frac{[Ca^{2+}][F^-]^2}{[CaF_2]} = [Ca^{2+}][F^-]^2 = 2.7\times10^{-11}$$

The concentrations of Ca^{2+} and F^- are measured in mol/L, and they exist in a molar ratio of 2/1. Thus, if there is any amount of solid CaF_2 present, at equilibrium there will be X moles of calcium and $2X$ moles of fluoride in solution. Note that the amounts of Ca^{2+} and F^- in solution are independent of the amount of solid CaF_2 that is present. It is only important that there is some solid.

Substituting what is known gives

$$K_{sp} = X(2X)^2 = 4X^3 = 2.7\times10^{-11}$$

$$X^3 = 6.75\times10^{-12}$$

$$X = 0.00019 \text{ mol/L}$$

and

$$[Ca^{2+}] = 0.00019 \text{ mol/L}$$

$$[F^-] = 0.00038 \text{ mol/L}$$

The mass concentrations are as follows:

$$(0.00019 \text{ mol/L})(40 \text{ g } Ca^{2+}/\text{mol}) = 0.0076 \text{ g } Ca^{2+}/\text{L} = 7.6 \text{ mg } Ca^{2+}/\text{L}$$

$$(0.00038 \text{ mol/L})(19 \text{ g } F^{-}/\text{mol}) = 0.0072 \text{ g } F^{-}/\text{L} = 7.2 \text{ mg } F^{-}/\text{L}$$

Example 8.2 Copper Sulfide Solubility

Calculate the solubility of copper sulfide (CuS) in water using K_{sp} and K_{sp}^* solubility product values from Table 8.4.

1. K_{sp}, applies to the simple solubility reaction:

$$CuS \leftrightarrows Cu^{2+} + S^{2-}$$

$$K_{sp} = [Cu^{2+}][S^{2-}] = 6.3 \times 10^{-36}$$

At equilibrium, there will be X moles of copper and X moles of sulfide in solution.

$$K_{sp} = (X)(X) = X^2 = 6.3 \times 10^{-36}$$

$$X = 2.5 \times 10^{-18} \text{ M}$$

The equilibrium concentrations are

$$[Cu^{2+}] = [S^{2-}] = 2.5 \times 10^{-18} \text{ M}$$

It is not evident from this calculation, but the solubility depends on the pH. This can be shown using the definition of K_{sp}^*.

2. K_{sp}^* explicitly incorporates the effect of pH on the solubility of this metal sulfide. The reaction is

$$CuS + H_2O \leftrightarrows Cu^{2+} + HS^{-} + OH^{-}$$

with $$K_{sp}^* = [Cu^{2+}][HS^{-}][OH^{-}] = 6 \times 10^{-37}$$

Because HS^- (hydrosulfide) will hydrolyze to H_2S, the solubility, S^*, of CuS is

$$\text{Solubility} = S^* = [Cu^{2+}] = [HS^{-}] + [H_2S]$$

From the H_2S–HS^- equilibrium

$$H_2S \rightleftarrows HS^{-} + H^{+} \qquad K_1 = \frac{[HS^{-}][H^{+}]}{[H_2S]} = 8.9 \times 10^{-8}$$

Substituting $[H_2S] = [HS^-][H^+]/K_1$ into the solubility expression gives

$$S^* = [Cu^{2+}] = [HS^{-}] + \frac{[HS^{-}][H^{+}]}{K_1} = [HS^{-}]\left(1 + \frac{[H^{+}]}{K_1}\right)$$

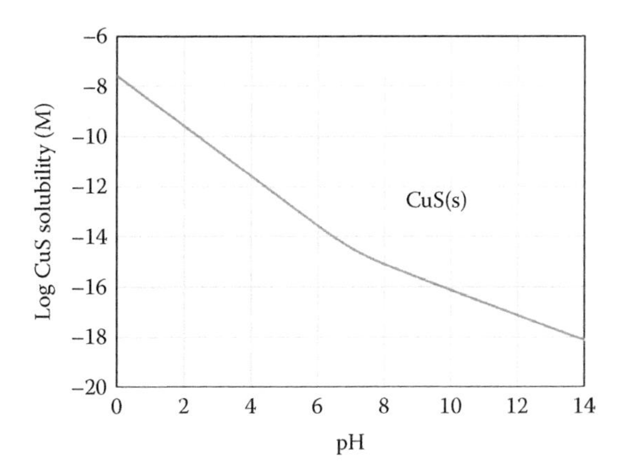

FIGURE 8.1 Solubility of CuS as a function of pH.

Multiplying by $[Cu^{2+}]$ gives

$$(S^*)^2 = [Cu^{2+}]^2 = [Cu^{2+}][HS^-]\left(1+\frac{[H^+]}{K_1}\right)$$

Noting that $[Cu^{2+}][HS^-] = K_{sp}^*/[OH^-]$ and $[OH^-] = K_w/[H^+]$, the expression can be rewritten as

$$(S^*)^2 = [Cu^{2+}]^2 = K_{sp}^*\frac{[H^+]}{K_w}\left(1+\frac{[H^+]}{K_1}\right)$$

The square root is the solubility of CuS as a function of pH.

$$S^* = \sqrt{K_{sp}^*\frac{[H^+]}{K_w}\left(1+\frac{[H^+]}{K_1}\right)}$$

At pH 5, the solubility is

$$S^* = \sqrt{K_{sp}^*\frac{[H^+]}{K_w}\left(1+\frac{[H^+]}{K_1}\right)} = \sqrt{6\times10^{-37}\frac{10^{-5}}{10^{-14}}\left(1+\frac{10^{-5}}{8.9\times10^{-8}}\right)} = 2.6\times10^{-13}\ \text{M}$$

At pH 9, the solubility is

$$S^* = \sqrt{K_{sp}^*\frac{[H^+]}{K_w}\left(1+\frac{[H^+]}{K_1}\right)} = \sqrt{6\times10^{-37}\frac{10^{-9}}{10^{-14}}\left(1+\frac{10^{-9}}{8.9\times10^{-8}}\right)} = 2.5\times10^{-16}\ \text{M}$$

Figure 8.1 shows the steadily decreasing solubility of CuS as a function of pH. The solubility calculated using K_{sp} occurs at a pH of about 13.

8.4 INVENTING A USEFUL PRECIPITATION PROCESS

How might we invent a useful precipitation process?

First, search tables of solubility products for compounds that have a "small" K_{sp}.

All insoluble materials have small solubility products. $K_{sp} = 10^{-10}$ is very small by normal standards. *Small* must be given a more precise operational meaning that relates to the precipitating metal and the effluent target concentration. We show this by example.

Example 8.3 Precipitation of a Hypothetical Compound of Mercury

The safe level of mercury in drinking water is 0.002 mg/L, which is only 1×10^{-8} mol/L. We can achieve this low level by precipitating a mercury compound of the form HgX only if the solubility product for HgX satisfies

$$K_{sp} \leq [Hg^{2+}][X^{2-}] = (1\times10^{-8})(1\times10^{-8}) = 1\times10^{-16}$$

In terms of a practical precipitation process to control toxic substances, a "small" value of K_{sp} is 10^{-16} or less. If we want to use precipitation to recover a material from solution and do not need an extremely low effluent concentration, a value of $K_{sp} = 10^{-9}$ might yield a feasible process.

Consider whether the reagent ion needed to shift the equilibrium is attractive in terms of cost, safety, ease of handling, etc. Compute an estimate of the theoretical solubility limit (at various pH levels if pH is important). The value you compute provides a useful lower bound. If the theoretical solubility limit is unacceptably high, the process concept fails; if it is below the target concentration, the process concept deserves more careful consideration. Table 8.5 lists some applications of precipitation reactions in pollution prevention and control.

TABLE 8.5
Useful Precipitation Reactions

Fluoride, from glass manufacture and cupola washwater (CaF_2 recovery)

$$2F^- + Ca(OH)_2 \rightarrow CaF_2(s) + 2OH^-$$

Aluminate from aluminum etching and anodizing

$$Na_2Al_2O_4 + H_2SO_4 + 2H_2O \rightarrow 2Al(OH)_3(s) + Na_2SO_4$$

Iron, from steel pickling liquor (recovery of iron and hydrochloric acid may be possible)

$$FeSO_4 + Ca(OH)_2 \rightarrow Fe(OH)_2(s) + CaSO_4\,(s)$$
$$FeCl_2 + Ca(OH)_2 \rightarrow Fe(OH)_2(s) + CaCl_2$$
$$6FeCl_2 + 6Ca(OH)_2 + O_2 \rightarrow 2Fe_3O_4(s) + 6CaCl_2 + 6H_2O$$

Sulfur dioxide from flue gas

$$SO_2 + CaCO_3 + 1/2O_2 \rightarrow CaSO_4(s) + CO_2$$

Phosphate, from municipal sewage and fertilizer plants

$$2PO_4^{3-} + Al_2(SO_4)_3 \rightarrow 2AlPO_4(s) + 3SO_4^{2-}$$
$$PO_4^{3-} + FeCl_3 \rightarrow FePO_4(s) + 3Cl^-$$
$$2PO_4^{3-} + 3FeSO_4 \rightarrow Fe_3(PO_4)_2(s) + 3SO_4^{2-}$$

Lime–soda ash water softening (removal of calcium and magnesium)

$$Ca(HCO_4)_2 + Ca(OH)_2 \rightarrow 2CaCO_3(s) + 2H_2O$$
$$Mg(HCO_4)_2 + Ca(OH)_2 \rightarrow Mg(OH)_2(s) + 2CaCO_3(s) + 2H_2O$$
$$CaSO_4 + Na_2CO_3 \rightarrow CaCO_3(s) + 2Na^+ + SO_4^{2-}$$

Alum flocculation for clarification of turbid water

$$Al_2(SO_4)_3 14H_2O + 3Ca(HCO_3)_2 \rightarrow 2Al(OH)_3(s) + 3CaSO_4 + 6CO_2 + 14H_2O$$

Heavy metals (Cd, Cu, Pb, Hg, Zn, etc.) removal from industrial wastewater

$$CdCl_2 + Ca(OH)_2 \rightarrow Cd(OH)_2(s) + CaCl_2$$
$$CuCl_2 + Ca(OH)_2 \rightarrow Cu(OH)_2(s) + CaCl_2$$
$$CdCl_2 + NaHS \rightarrow CdS(s) + NaCl + HCl$$
$$CuCl_2 + NaHS \rightarrow CuS(s) + NaCl + HCl$$
$$CdSO_4 + H_2S \rightarrow CdS(s) + H_2SO_4$$
$$CuSO_4 + H_2S \rightarrow CuS(s) + H_2SO_4$$

The effluent concentration of real processes will exceed the calculated equilibrium concentration. A real process is monitored in terms of

Total metal = Dissolved metal + Particulate metal

The dissolved metal concentration is likely to exceed the calculated theoretical value because (1) the process chemistry has been oversimplified and/or (2) equilibrium is not achieved. The removal of particulate metals will not be 100% efficient. The designer should make allowance for these inefficiencies. We have no definite rules for doing this, so jar tests and experience often come into the process design picture.

Example 8.4 Treatment of Lead- and Fluoride-Bearing Wastewater

The lead and fluoride wastes are produced separately and arrive for treatment in separate sewers. The first step toward evaluating the technical feasibility of removing lead or fluoride by precipitation is to look up solubility products in order to identify possible insoluble products. Some solubility products are:

Lead (Pb) Compounds		Fluoride (F) Compounds	
$PbCrO_4$	$K_{sp} = 2.8 \times 10^{-13}$	CaF_2	$K_{sp} = 2.7 \times 10^{-11}$
$PbCO_3$	$K_{sp} = 7.4 \times 10^{-14}$	CeF_2	$K_{sp} = 8.0 \times 10^{-16}$
$Pb(OH)_2$	$K_{sp} = 1.4 \times 10^{-15}$	CrF_3	$K_{sp} = 6.6 \times 10^{-11}$
$Pb_3(PO_4)_2$	$K_{sp} = 8.0 \times 10^{-43}$		
PbS	$K_{sp} = 8.0 \times 10^{-28}$		
$Pb_3(AsO_4)_2$	$K_{sp} = 4.0 \times 10^{-36}$		

The lead arrives in an acidic solution. A process based on $PbCO_3$ or $Pb(OH)_3$ will require a major adjustment of the pH since the minimum solubility is at high pH. PbS is very insoluble and it can be precipitated at pH 8. At a lower pH, PbS it is still very insoluble, but there would be problems with H_2S generation. By searching reference material, computing the equilibrium chemistry, or consulting with a chemist, we could discover that lead phosphate is very insoluble under acidic conditions. At pH 3.5 the solubility of lead is only 0.15 mg/L.

We have, at this point, identified some possibilities that could be checked in jar tests. A test of phosphate precipitation

$$3\,Pb^{2+} + 2\,Na_3PO_4 \rightarrow Pb_3(PO_4)_2(s) + 6\,Na^+$$

showed that 15 min of mixing followed by settling gave a supernatant with 0.2 mg/L Pb. It also showed that the required dosage of trisodium phosphate is greater than the stoichiometric amount. The excess phosphate will appear in the effluent. This might be a problem. Sometimes combinations of chemicals are feasible.

Chromium fluoride and copper fluoride are quite insoluble, but we should be cautious about adding one toxic substance (e.g., chromium or copper) to remove another. Calcium ion is provided as hydrated lime, which is cheap and relatively easy to handle and this suggests CaF_2 as a product. The reaction is

$$2\,HF + Ca(OH)_2 \rightarrow CaF_2(s) + 2\,H_2O$$

which is only efficient at high pH. Laboratory tests indicate that pH 12 is ideal. The theoretical solubility of CaF_2 gives $F^- = 7.8$ mg/L. This value also should be checked in jar tests.

Figure 8.2 is our preliminary idea for a treatment process.

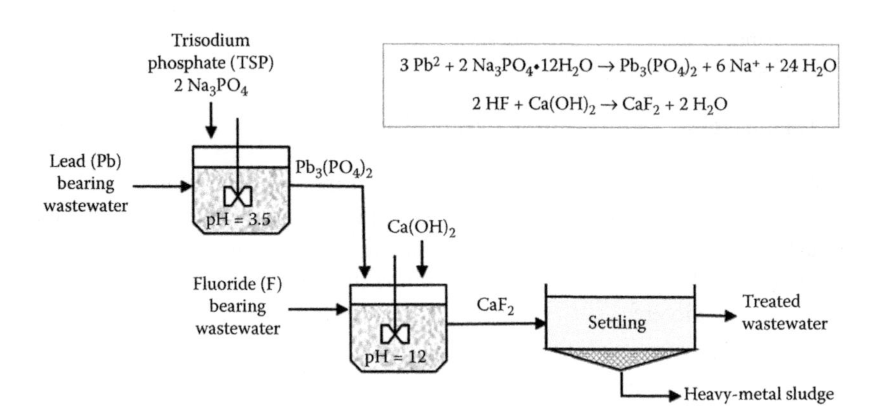

FIGURE 8.2 Possible integrated treatment process for combined lead and fluoride wastewaters.

Example 8.5 Separating Lead and Silver by Precipitation

The K_{sp} values for salts of silver and lead are listed in the following table:

Chloride Salts	K_{sp}	Sulfate Salts	K_{sp}
$AgCl$	1.8×10^{-10}	Ag_2SO_4	1.4×10^{-5}
$PbCl_2$	1.7×10^{-5}	$PbSO_4$	6.3×10^{-7}

Silver chloride is 100,000 times less soluble than lead chloride; there is a difference of 10^{-5} in the K_{sp} of AgCl and $PbCl_2$.

$$AgCl \quad K_{sp} = 1.8 \times 10^{-10}$$

$$PbCl_2 \quad K_{sp} = 1.7 \times 10^{-5}$$

The solubility ratio between Ag_2SO_4 and $PbSO_4$ is about 140.

Chloride, Cl^-, appears a good choice of negative ions for their separation.

$PbCl_2$ is rather soluble in warm water, but the solubility of AgCl is very small even at high temperatures. This suggests heating the solution 80°C, to keep Pb^{2+} ions in solution and precipitate AgCl as a solid.

Example 8.6 Differential Precipitation of Lead and Mercury

We have a solution of 0.010 M Hg^{2+} and 0.010 M Pb^{2+}. The goal is to achieve complete separation of these species by precipitation. *Complete separation* means that the concentration in solution of the analyte of interest must be <0.01% of its original value. That is, after treatment the mercury in solution cannot exceed

$$(0.010 \text{ M Hg}^{2+})(0.01/100) = 1.0 \times 10^{-6} \text{ M Hg}^{2+}$$

At this concentration, all the Pb^{2+} should still be in solution. We will use a solution of NaI, preferably not too concentrated, or it may precipitate too quickly and we will end up with PbI_2 in our HgI_2. The solubility products of HgI_2 and PbI_2 are 4.7×10^{-29} and 7.9×10^{-9}, respectively.

What does the remaining $[I^-]$ need to be to get 'all' the Hg^{2+} out as HgI_2?

$$K_{sp} = [Hg^{2+}][I^-]^2 = (1.0 \times 10^{-6})[I^-]^2 = 4.7 \times 10^{-29}$$

$$[I^-] = 6.8 \times 10^{-12} \text{ M}$$

Will any PbI_2 precipitate at this concentration?

The reaction of interest is

$$PbI_2 \rightleftarrows Pb^{2+} + 2\,I^-$$

The *reaction coefficient, Q,* is

$$Q = [Pb^{2+}][I^-]^2$$

Calculate the *reaction quotient, Q,* using the new solution concentrations.

$$Q = [Pb^{2+}][I^-]^2 = (0.010\ M)(6.8 \times 10^{-12}\ M)^2 = 4.6 \times 10^{-26}$$

If the solubility product is greater than the reaction quotient Q ($K_{sp} > Q$) the solubility has not been exceeded and a precipitate will not form.

$$K_{sp} \text{ for } PbI_2 = 7.9 \times 10^{-9} \gg Q = 4.6 \times 10^{-26}$$

PbI_2 precipitate will not form.

This selectivity is based on solubility equilibrium. It does not consider the kinetics of precipitation. If the rate of precipitation of HgI_2 is slow, there may be some precipitation of PbI_2, leading to an incomplete separation of Pb^{2+} and Hg^{2+}.

Example 8.7 Removal of Sulfur Dioxide from Stack Gas with Limestone

Sulfur dioxide emissions are a problem at any coal-fired power plant that burns high-sulfur coal. The approximate emissions from an uncontrolled 1000 MW coal-fired power plant are 1000 ton/day total ash, 700 ton/day fly ash, 1000 ton/day SO_2, 100 ton/day nitrogen oxides. In medieval times, the pungent odor of sulfur dioxide was a sign that the Devil was near (sulfur is the Biblical "brimstone"). This Devil is still with us and after a review of the more than 50 processes that have been suggested for removing SO_3 from flue gas, one starts to feel that exorcism might soon be added to the list.

One way to remove SO_2 is to convert it to a solid that is easily removed from the stack gas, either as a solid or a liquid waste stream. Limestone addition exploits simple chemistry

$$\underset{\text{limestone}}{CaCO_3} + \underset{\text{gas}}{SO_2} + \underset{\text{gas}}{0.5\,O_2} \rightarrow \underset{\text{solid}}{CaSO_4} + \underset{\text{gas}}{CO_2}$$

This chemistry can occur in the furnace (conventional or fluidized bed) if crushed limestone is fed along with the coal. The heat decomposes the limestone to calcium oxide and carbon dioxide, and the sulfur in the coal is converted to gaseous sulfur dioxide, which reacts with the calcium oxide and oxygen to form calcium sulfate. The calcium sulfate product will appear in the slag or as dust in the flue gas. It can be removed from the gas along with fly ash by filtration or electrostatic precipitation. This converts the air pollution problem into a solid waste disposal problem.

This chemistry can be accomplished outside the furnace by contacting the flue gas with water that is laden with dissolved limestone. The contactor in which this is done is called a wet scrubber. The scrubber water is a waste stream, so this converts an air pollution problem into a water pollution problem.

The stoichiometry predicts that about two kg of $CaSO_4$ will be produced for each kg of SO_2 captured. Experience confirms this. Only 70%–90% of SO_2 can be removed and excess limestone must be added to accomplish this. This unreacted limestone must be handled as a waste along with the coal ash.

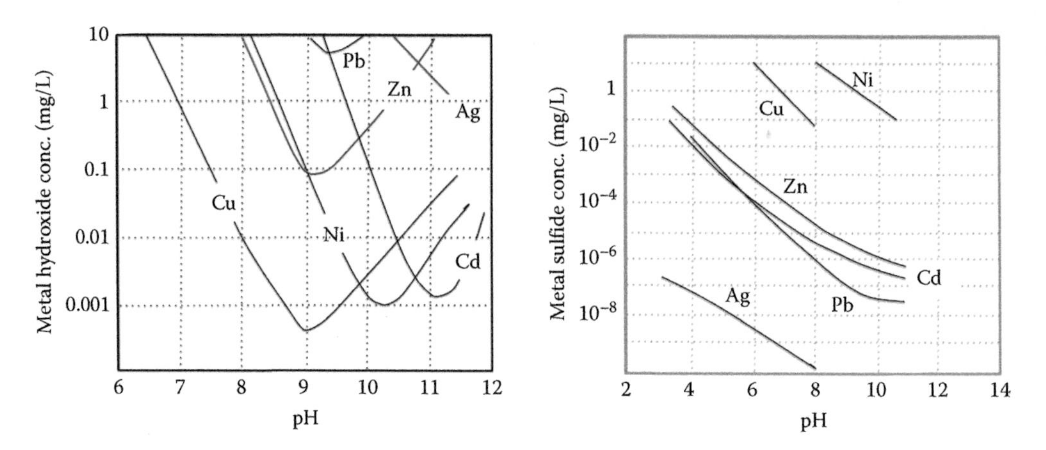

FIGURE 8.3 Comparison of the solubilities of metal hydroxides and sulfide (various EPA publications).

8.5 PRECIPITATING METALS AS HYDROXIDES

The simple precipitation of a dissolved metal (Me^{2+}) as a hydroxide occurs as

$$\underset{\text{soluble}}{Me^{2+}} + \underset{\text{soluble}}{2\,OH^-} \rightarrow \underset{\text{insoluble}}{M(OH)_2}$$

Figure 8.3 compares the theoretical solubility for the hydroxides and sulfides of cadmium, lead, copper, nickel, silver, and zinc. The solubility curves for the hydroxides have a minimum concentration—the solubility decreases and then increases once some critical pH is obtained. The minimums occur above pH 7.0.

The minimum solubility of the sulfides is lower than the hydroxides (note the difference in the vertical scale) and the sulfide precipitates form at much lower pH levels than the hydroxides.

Figure 8.4 shows the theoretical solubility of some heavy-metal hydroxides. The solubility curves were calculated by Lewis (2010) using Stream Analyzer (OLS Systems Inc. 2009). The curves differ from those in Figure 8.3 because Lewis included more ionic species in his model. The odd shape of the curve for lead is apparently due to the formation of multiple soluble lead species over the wide range of pH.

The hydroxides have the interesting property of having a minimum solubility at a particular pH. This occurs because these metals exist in different ionic forms, depending upon solution pH, and these forms have different solubility. Examples 8.8 and 8.9 show this behavior.

Example 8.8 Iron Precipitation (with a Warning)

One method of removing iron from water is to form a precipitate of ferric hydroxide $Fe(OH)_3$ that can be separated from the water by sedimentation or filtration. Assume that all the iron in solution is of the Fe^{3+} form. The solubility equilibrium condition is

$$Fe(OH)_3 \leftrightarrows Fe^{3+} + 3OH^-$$

with solubility product

$$K_{sp} = [Fe^{3+}][OH^-]^3 = 2.8 \times 10^{-39}$$

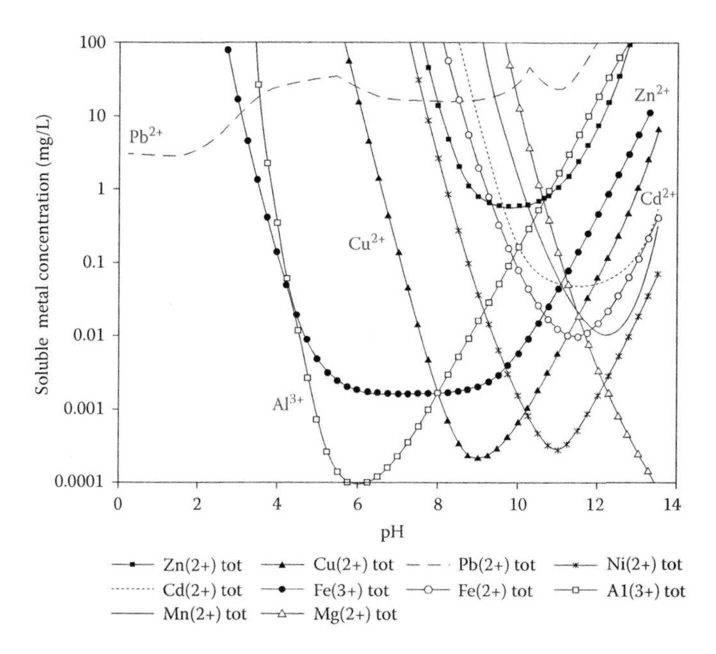

FIGURE 8.4 pH dependence of metal hydroxide solubility. Four of the curves have been labeled to clarify the symbols that identify the different metals. (Adapted from Lewis, A.E., 2010.)

Naive Calculation of Iron Solubility

This simple precipitation model assumes that $Fe(OH)_3$ is the only important insoluble species. The concentration of soluble ferric iron in solution at pH = 5 is calculated using $[OH^-] = 10^{-9}$.

$$Fe^{3+} = \frac{2.8\times10^{-39}}{[OH^-]^3} = \frac{2.8\times10^{-39}}{(10^{-9})^3} = \frac{2.8\times10^{-39}}{10^{-27}} = 2.8\times10^{-12}\ \text{mol/L}$$

The same calculation for pH = 10, $[OH^-] = 10^{-4}$, indicates ferric hydroxide is more than a trillion times less soluble at pH 10 than at pH 5.

$$Fe^{3+} = \frac{2.8\times10^{-39}}{[OH^-]^3} = \frac{2.8\times10^{-39}}{(10^{-4})^3} = \frac{2.8\times10^{-39}}{10^{-12}} = 2.8\times10^{-27}\ \text{mol/L}$$

Warning: This example calculation would be correct if the precipitation were controlled entirely by hydroxide species $Fe(OH)_3$. *It is not.* The iron chemistry becomes more complicated as the pH increases because the Fe^{3+} ion forms complexes with different numbers of hydroxyl ions.

Calculation Using a More Complex Model

The correct total concentration of ferric iron is the sum of several ionic species:

$$\text{Concentration of total iron} = [Fe^{3+}] + [FeOH^{2+}] + [Fe(OH)_2^+] + [Fe_2(OH)_2^{4+}] + [Fe(OH)_4^-]$$

A solubility equation for each of these compounds is needed. Thus, the correct calculation requires solving six simultaneous equations—five solubility equations plus the total iron equation. The solution is graphed in Figure 8.5. Not only is the simple calculation wrong in giving an iron concentration that is too low, it suggests that ferric iron becomes less soluble as pH increases, when in fact it becomes more soluble. It is least soluble at pH 8 and becomes more soluble at higher and lower pH.

Aluminum, zinc, and chromium (Cr^{3+}) are other metals that behave like ferric iron. Figure 8.6 shows the solubility of aluminum hydroxide as a function of pH. Unless one is an expert chemist, checking the empirical stoichiometry is essential when working with these metals.

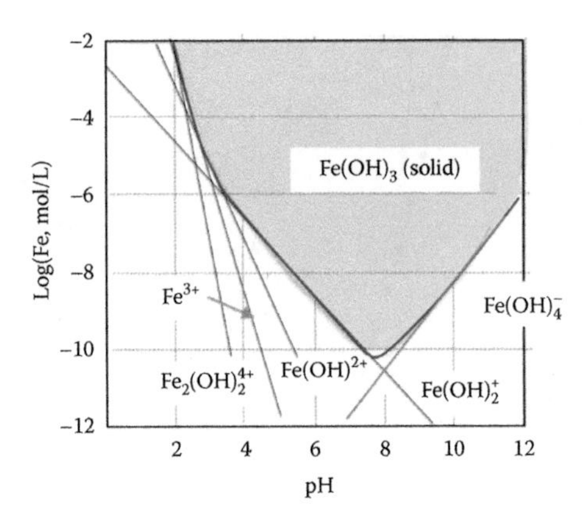

FIGURE 8.5 Solubility of iron (ferric hydroxide) is a function of several ionized species that are affected by pH.

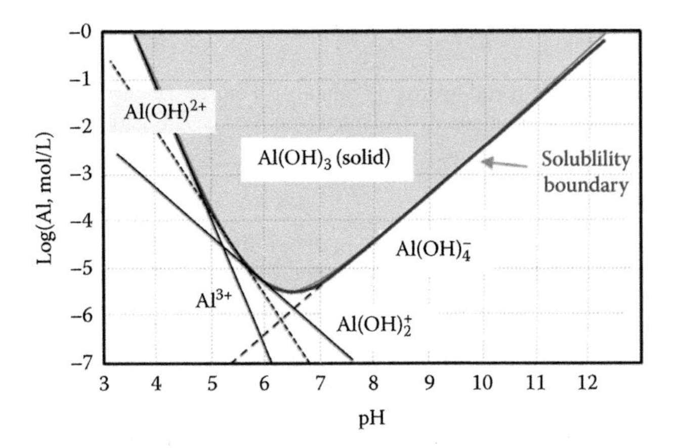

FIGURE 8.6 Solubility of aluminum (aluminum hydroxide) is a function of several ionized species that are affected by pH.

Example 8.9 Zinc Solubility

Prepare a solubility diagram (log C vs. pH) for water that is potentially at equilibrium with zinc hydroxide. Show all species, along with the total zinc (Zn_T) line. Indicate where precipitation will occur and the type of precipitate. Use the equation and K values in the following table:

Equilibrium Equation	log *K*
$ZnOH^+ \leftrightarrows Zn^{2+} + OH^-$	−5.04
$Zn(OH)_2^0 \leftrightarrows Zn(OH)^+ + OH^-$	−6.06
$Zn(OH)_3^{1-} \leftrightarrows Zn(OH)_2^0 + OH^-$	−2.50
$Zn(OH)_4^{2-} \leftrightarrows Zn(OH)_3^- + OH^-$	−1.20
$Zn(OH)_2(s) \leftrightarrows Zn^{2+} + 2OH^-$	−16.52

Equilibria for zinc hydroxide
Note: activity of $[Zn(OH)_2(s)] = 1$

$$K = \frac{[Zn^{2+}][OH^-]^2}{[Zn(OH)_2]} = [Zn^{2+}][OH^-]^2 = 10^{-16.52}$$

$$\log[Zn^{2+}] = -16.52 - 2\log[OH^-]$$

$$\log[Zn^{2+}] = -16.52 - 2[\log[H^+] - 14]$$

$$\log[Zn^{2+}] = 11.48 - 2pH$$

For the monohydroxide,

$$K = \frac{[Zn^{2+}][OH^-]}{[ZnOH^+]} = 10^{-5.04}$$

$$\log[ZnOH^+] = 5.04 + \log[Zn^{2+}] + \log[OH^-]$$

$$\log[ZnOH^+] = 5.04 + \log[Zn^{2+}] - 14 - \log[H^+]$$

$$\log[ZnOH^+] = -8.96 + \log[Zn^{2+}] + pH$$

$$\log[ZnOH^+] = -8.96 + (12.45 - 2pH) + pH$$

$$\log[ZnOH^+] = 3.49 - pH$$

For the dihydroxide,

$$K = \frac{[ZnOH^+][OH^-]}{[Zn(OH)_2^0]} = 10^{-6.06}$$

$$\log[Zn(OH)_2^0] = 6.06 + \log[ZnOH^+] + \log[OH^-]$$

$$\log[Zn(OH)_2^0] = 6.06 + \log[ZnOH^+] - 14 - \log[H^+]$$

$$\log[Zn(OH)_2^0] = -7.94 + \log[ZnOH^+] + pH$$

$$\log[Zn(OH)_2^0] = -7.94 + (3.49 - pH) + pH$$

$$\log[Zn(OH)_2^0] = -4.45$$

For the trihydroxide

$$K = \frac{[Zn(OH)_2^0][OH^-]}{[Zn(OH)_3^-]} = 10^{-2.5}$$

$$\log[Zn(OH)_3^-] = 2.50 + \log[Zn(OH)_2^0] + \log[OH^-]$$

$$\log[Zn(OH)_3^-] = 2.5 + \log[Zn(OH)_2^0] - 14 - \log[H^+]$$

$$\log[Zn(OH)_3^-] = -11.50 + \log[Zn(OH)_2^0] + pH$$

$$\log[Zn(OH)_3^-] = -11.50 + (-4.45) + pH$$

$$\log[Zn(OH)_3^-] = -15.95 + pH$$

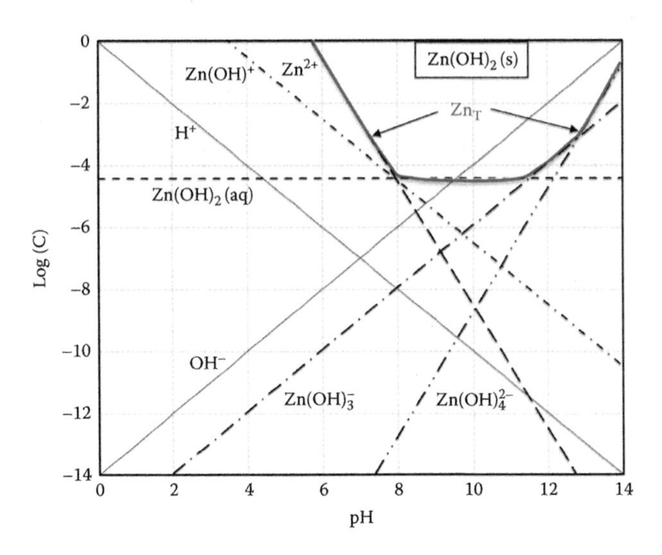

FIGURE 8.7 Zinc hydroxide solubility diagram.

For the tetrahydroxide,

$$K = \frac{[Zn(OH)_3^{-1}][OH^-]}{[Zn(OH)_4^{2-}]} = 10^{-1.20}$$

$$\log[Zn(OH)_4^{2-}] = 1.20 + \log[Zn(OH)_3^{-1}] + \log[OH^-]$$

$$\log[Zn(OH)_4^{2-}] = 1.20 + \log[Zn(OH)_3^{-1}] - 14 - \log[H^+]$$

$$\log[Zn(OH)_4^{2-}] = -12.80 + \log[Zn(OH)_3^{-1}] + pH$$

$$\log[Zn(OH)_4^{2-}] = -12.80 + (-15.95 + pH) + pH$$

$$\log[Zn(OH)_4^{2-}] = -28.75 + 2pH$$

The solubility diagram constructed from these equations is shown in Figure 8.7

The above examples are for the simple case of a solution that contains one metal species. If the water or wastewater contains chemicals that could form other metal complexes and precipitates, the equilibrium chemistry becomes more complicated.

One sees that calculating the equilibrium chemistry of actual water or wastewater may be very difficult to get right. The difficulty is not in making the calculations [there are computer programs to do this, e.g., MINTEQ, available from the U.S. Environmental Protection Agency (EPA)], but in getting an adequate model of the chemistry. In this context, *model* means including all the important chemical species and knowing the correct concentrations. Appendix C lists some software applications for making the calculations.

It is because the chemistry is difficult that the jar test (see Chapter 6) is so useful. Whenever possible, observe the empirical chemistry. Then, for a deeper understanding, you may wish to try to reproduce the jar test results with calculations. (Or do it the other way around. Just keep in mind that it may be a good idea to do both.)

Some limitations of metals precipitation processes with hydroxide are as follows:

- The theoretical minimum solubility for different metals occurs at different pH values.
- For a mixture of metal ions, it may not be possible for precipitation at a single pH to produce a sufficiently low effluent concentration of all metals. The question, then, is whether it is

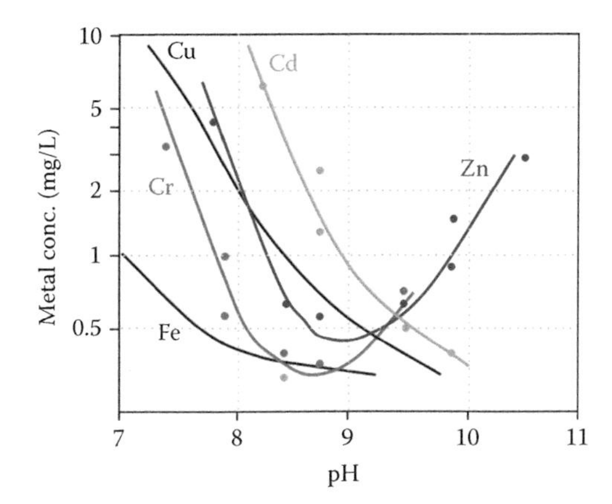

FIGURE 8.8 Metal concentrations observed in a jar test on an industrial wastewater that was treated with lime and settled for 30 min. (Data courtesy of Dan Conway.)

better to use two precipitations at different pH levels, to separate the metals at their sources and operate two separate treatment processes, or to use a different treatment method.

- Hydroxide precipitates tend to redissolve if the solution pH is increased or decreased from the minimum solubility point. Therefore, precise pH control is necessary.
- Complexing agents, such as ammonia, phosphate, and EDTA have an adverse effect on metal removal efficiency.

As a practical matter, finding the optimal hydroxide precipitation process requires bench-scale testing using samples of actual water or wastewater. Precipitation tests should be done at varying pH levels and with different chemicals to learn which produces the best result. Figure 8.8 shows jar test data for an industrial wastewater that contained five metals. Lime was added and the test samples were settled for 30 min before the metal concentrations were measured. The actual performance falls well short of the theoretical minimum concentrations in Figure 8.3.

The time required for mixing and settling can be investigated with jar tests, as can the need for polymers to enhance settling. The choice of reagent chemical should also consider the properties of the precipitate (sludge) that is produced. Zinc can be precipitated using NaOH or $Ca(OH)_2$. The pollutant removal efficiency may be the same, but the two chemicals will produce sludges that are substantially different in density and compaction properties. Lime ($Ca(OH)_2$) will produce more sludge than caustic soda (NaOH), but the sludge is usually easier to dewater.

Batch treatment may be economical for volumes up to 200 m^3. The batch reactor (a tank) is filled, a mixer is started to homogenize the contents, the reagent is added and mixed (using pH as an indication of sufficient dose), and 2–4 h settling is allowed. A sample could be taken and analyzed to determine that the effluent limit has been satisfied. For larger volumes, two batch reactors can be installed, one filling while the other functions as a reactor/settler.

When wastewater volumes are large, a continuous flow treatment system can be used. A blending tank may be included to dampen fluctuations in flow and concentration. The first treatment step is pH adjustment. Growing the precipitate requires reaction times that are typically 15–60 min. If the precipitates are small and slow in settling, polymer may be added to increase the particle size and aid settling. Detention times in settling basins usually range from 1–2 h. Sludge recirculation or the use of solids contact basins is beneficial.

A properly designed settling system will remove the bulk of the flocculated solids, but the effluent may still contain suspended solids that are laden with heavy metals. Post-settling filtration may be needed to meet a strict effluent limit. Gravity filters and pressure filters with a sand bed or multimedia bed are popular. Filters are cleaned intermittently by backwashing. Backwash water is collected and managed, perhaps by recirculation to the precipitation process so the solids can be disposed of with the settled sludge. One difficulty in doing this is that the backwash flow is intermittent and will put a heavy hydraulic pulse loading on the settling process. Therefore, storage and flow equalization may be needed.

The sludge collected from a settling process is frequently in the range of 1%–2% solids by weight. Hydroxide precipitation produces sludge that tends to be gelatinous and difficult to thicken or dewater. Lime, $Ca(OH)_2$, will produce more sludge than caustic soda, NaOH, but the lime sludge is usually easier to thicken and dewater. Sludge dewatering can be done by centrifuge, vacuum filter, pressure filter, and belt filter.

8.6 PRECIPITATING METALS AS SULFIDES

Most heavy metals will form stable metal sulfides. The exceptions are trivalent chromium and ferric iron. The precipitation of a dissolved metal as a sulfide (MeS) occurs as dissolved metal ions react directly with the soluble sulfide:

$$Me^{2+} + S^{2-} \rightarrow MeS$$

$$2Me^{3+} + 3S^{2-} \rightarrow Me_2S_3$$

If metal hydroxides are present they will redissolve and precipitate as sulfides:

$$Me(OH)_2(s) + S^{2-} \rightarrow MeS(s) + 2OH^-$$

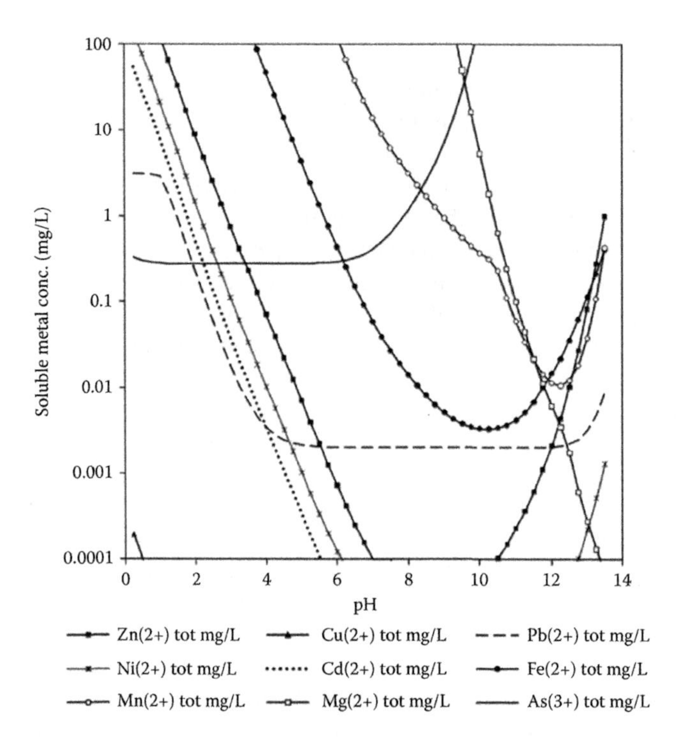

FIGURE 8.9 pH dependence of metal sulfide solubility. (Adapted from Lewis, A.E., 2010.)

NOTE: In a previous section, we said that explaining these reactions in terms of the S^{2-} ion, was not strictly correct because in aqueous solutions S^{2-} ion hydrolyzes to HS^-. The reaction with HS^- still results in the metal sulfide precipitate:

$$Me^{2+} + HS^- + H^+ \rightarrow MeS + 2H^+$$

$$2Me^{3+} + 3HS^- + 3H^+ \rightarrow Me_2S_3 + 6H^+$$

These reactions use a soluble form of sulfide, e.g., H_2S, Na_2S, or NaHS, under acid conditions. Writing a direct reaction between Me^{2+} and S^{2-} is simpler and more economical, and this is what we do in some of the examples that follow.

The sulfides are extremely insoluble, as shown in Figure 8.9.

Sulfide precipitation became more widely used when two problems were solved: (1) controlling the sulfide addition precisely enough to avoid accidental formation of H_2S and its odor and safety problems, and (2) removing the very fine precipitate solids.

Sulfide precipitation is practiced in two forms. The *soluble sulfide process (SSP)* uses a water-soluble sulfide compound like sodium hydrosulfide (NaHS). The *insoluble sulfide process (ISP)* uses a slurry of slightly soluble ferrous sulfide (FeS), which will dissociate to satisfy its solubility product to yield a dissolved sulfide concentration of approximately 0.02 parts per billion in the wastewater. The flowsheets of the sulfide processes are compared with hydroxide precipitation in Figure 8.10.

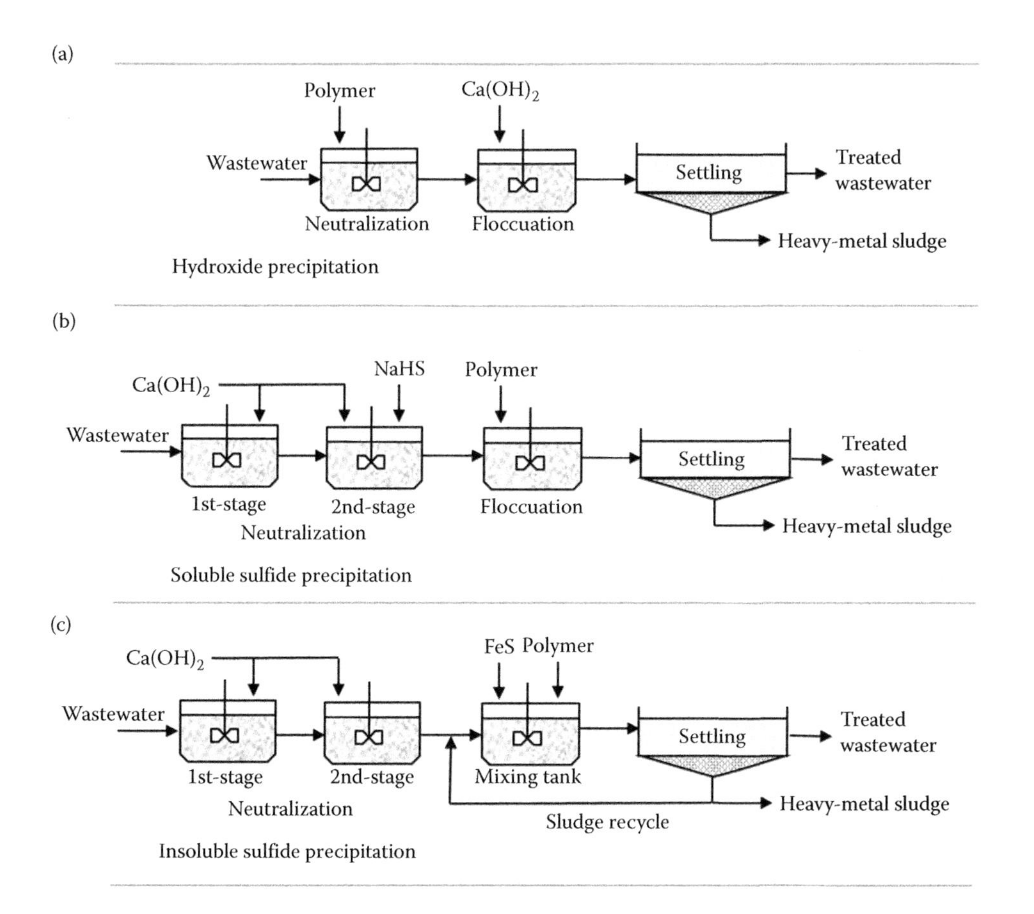

FIGURE 8.10 Hydroxide and sulfide precipitation processes. (a) Hydroxide precipitation, (b) soluble sulfide precipitation, and (c) insoluble sulfide precipitation.

The SSP chemistry is simple. A soluble form of sulfide is added to the wastewater and the sulfide reacts directly to form the metal sulfide precipitate. The sulfide reagent demand depends on the total metal concentration in the wastewater. If considerable hydroxides are present from a prior treatment step, or because the wastewater is naturally at high pH, it might be possible to save sulfide reagent by separating any hydroxide precipitate before adding the sulfide reagent.

Adding the soluble sulfide reagent provides a high concentration of S^{2-}. This promotes rapid precipitation, which often produces small particles and hydrated colloidal fines that are difficult to separate from the liquid. This problem can be solved by adding coagulants to form large, fast-settling particles.

Disadvantages of the SSP are safety and odor problems with hydrogen sulfide (H_2S). The odor detection level of H_2S is 0.1 to 1.0 ppm, and the Occupational Safety and Health Administration safety limitation is 10 parts per million (ppm). H_2S forms at acidic pH, so to avoid problems the process must be operated at pH 8.5 or above. In practice, considering the typical response lags of monitoring instruments and reagent control metering pumps and valves, control of H_2S odors is difficult, and it may be necessary to enclose and ventilate the process vessels.

Figure 8.11 is the flow diagram for the Boliden sulfide-lime process that is being used to remove copper, lead, zinc, arsenic, and mercury that come from process water sources and cleanup sources. Sodium sulfide (Na_2S) is used to precipitate the metals. The wastewater also contains fluoride, which is an anion and therefore does not react with sulfide. Sodium hydroxide (NaOH) is used to control the pH, so there are no safety or odor problems. The metal precipitates are removed by settling, with the aid of a polymer coagulant. Lime is added in a second stage of treatment to precipitate the fluoride as CaF_2. The metal sludge and the fluoride sludge are combined for dewatering.

The ISP adds the sulfide reagent as a slurry of solid ferrous sulfide (FeS). The solubility of FeS in water is about 0.02 μg/L. The FeS dissolves to replenish sulfide that has precipitated and to maintain a dissolved sulfide concentration of about 0.02 μg/L. One variation of the ISP mixes influent wastewater with recycled precipitate and FeS slurry, so precipitation occurs onto the mature and heavy particles.

The ISP will precipitate all metals that have a sulfide solubility less than that of FeS (the only heavy metal that is more soluble is MnS). Dissolved metals form sulfide precipitates directly, while metal hydroxides dissolve and then react with the sulfide ions. The conversion of metal hydroxides to metal sulfides is virtually complete. Ferrous iron will precipitate as a hydroxide. Maintaining low levels of ferrous ions in the effluent requires that the pH be controlled between 8.5 and 9.5. This has the disadvantage of producing considerably more sludge than a conventional hydroxide precipitation process.

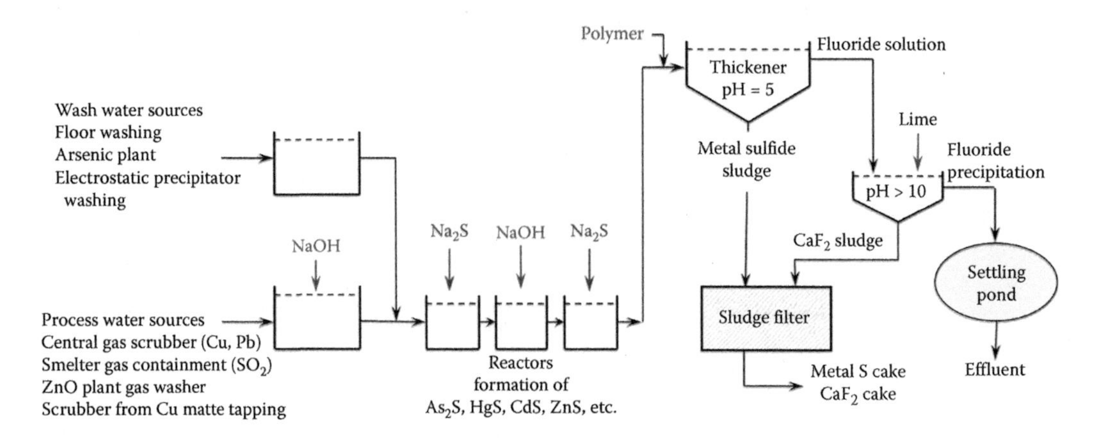

FIGURE 8.11 The Boliden sulfide–lime process to remove metals and fluoride.

The reactions are as follows:

Dissolution of FeS reagent	$FeS \rightarrow Fe^{2+} + S^{2-}$
Dissolution of metal hydroxides	$Me(OH)_2 \rightarrow Me^{2+} + 2OH^-$
Precipitation of metals	$Me^{2+} + S^{2-} \rightarrow MeS(s)$
Precipitation of ferrous iron	$Fe^{2+} + 2OH^- \rightarrow Fe(OH)_2(s)$

Neither hexavalent nor trivalent chromium is precipitated as a sulfide, but chromium will be removed as a hydroxide. The S^{2-} and Fe^{2+} will reduce hexavalent chromium (Cr^{6+}) to its trivalent state (Cr^{3+}), and under alkaline conditions the Cr^{3+} will precipitate as $Cr(OH)_3$. Thus, there is no need to segregate chromium wastes for separate treatment. (The SSP can also reduce Cr^{6+} and precipitate the Cr^{3+} if some ferrous iron is present.)

8.7 SOFTENING

Hard water contains calcium (Ca^{2+}) and magnesium (Mg^{2+}) ions. Hard water interferes with the cleansing action of soaps and detergents. This is a nuisance. Hard water also causes mineral deposits, mainly calcium carbonate ($CaCO_3$), in water heaters, kitchen appliances, boilers, cooling towers, pipes, and other equipment. This is an economic problem. Household problems caused by hard water are relieved when the hardness is reduced to about 60–80 mg/L $CaCO_3$. Solving industrial problems may require much lower levels.

Removing calcium and magnesium is called *softening*. Softening is an indirect pollution control process. It does not remove listed or regulated pollutants, but reducing scaling in pipes and boilers does save energy costs by lowering pumping and heating costs, and by reducing the need for anti-scaling chemical additives.

Total hardness is measured as the combined concentration of calcium and magnesium. It is expressed as the equivalent concentration of calcium carbonate; that is, as mg/L $CaCO_3$. For example, 40 mg/L Ca^{2+} is equivalent to 100 mg/L $CaCO_3$; 24.3 mg/L Mg^{2+} is equivalent to 100 mg/L $CaCO_3$. Water with 40 mg/L Ca^{2+} and 24.3 mg/L Mg^{2+} has a total hardness of 200 mg/L.

Table 8.6 gives the U.S. Geological Survey's classification of levels of hardness.

Water softening is accomplished either by chemical precipitation (lime–soda process) or ion exchange.

Some cities will soften the entire water supply using the lime–soda process, but it seems to be more common for individual households to deal with the problem by installing *ion exchange* water softeners. Ca^{2+} and Mg^{2+} ions are *exchanged* for sodium (Na^+) ions in a bed of *ion exchange resin particles*. This is a *cation exchange process*; the anions are not changed. Figure 8.12 shows the balanced exchange.

TABLE 8.6
Levels of Hardness

Level of Hardness	Concentration (mg/L $CaCO_3$)	Concentration (mmol/L)
Soft	<75	<0.75
Moderately hard	75–150	0.75–1.50
Hard	150–300	1.50–3.0
Very hard	>300	>3.0

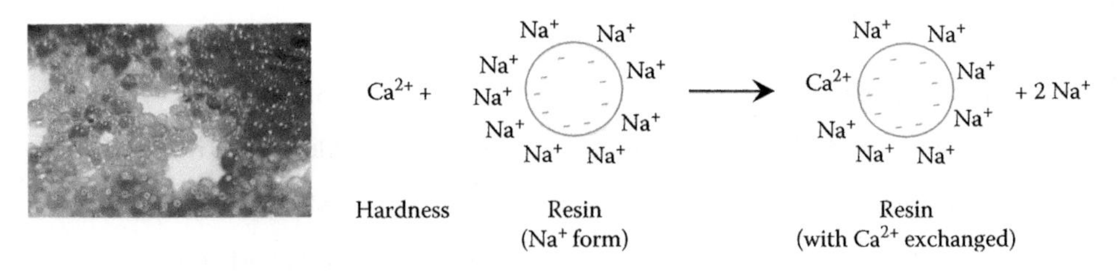

FIGURE 8.12 Ion exchange: Resin photo (left) and exchange reaction (right).

The exchange must maintain the balance of ionic charges (electroneutrality), so two ions of Na^+ are released from the resin when one ion of Ca^{2+} is captured. The same is true for Mg^{2+}. The mass of 2 Na^+ is greater than the mass of 1 Ca^{2+} or 1 Mg^{2+}, so ion exchange softening increases the total dissolved solids concentration of the water.

Ion exchange will remove virtually all the calcium and magnesium. In some applications, this is undesirable because the water will be corrosive. Untreated water can be blended with the softened water to give a mixture that is *soft* but stable. Often only the hot water supply is softened since this is where the calcium carbonate scaling problems occur (the solubility of calcium, unlike most ions, decreases with higher temperatures).

There are ion exchange *demineralization* processes that combine a cation exchange resin that exchanges H^+ ions for all the cations in water with an *anion exchange resin* that exchanges OH^- ions for all the anions. In this way, all the minerals are removed and replaced with water (H_2O). This method is used by industries that require high-purity water.

Precipitation softening processes include cold lime softening (15°C–60°C), warm lime softening (60°C–85°C), and hot lime softening (90°C–105°C). The residual hardness of these processes is 80–100 mg/L (cold), 30–50 mg/L (warm), and 15–25 mg/L (hot), all measured as $CaCO_3$. Warm softening and hot softening are popular for industrial treatment of boiler makeup water.

The cold lime–soda process has historically been used in drinking water treatment. Hard water, as it arrives for treatment, typically has a pH of 7–8 and contains alkalinity in the form of bicarbonate (HCO_3^-). This alkalinity is converted to carbonate (CO_3^{2-}) by adding lime ($Ca(OH)_2$) to raise the pH according to the equilibrium

$$HCO_3^- + OH^- \rightleftarrows CO_3^{2-} + H_2O$$

The addition of lime will precipitate calcium and magnesium as $CaCO_3$ and $Mg(OH)_2$. The process is effective because it uses the naturally occurring bicarbonate alkalinity (HCO_3^-) as the source of the carbonate for precipitating the hardness. Sometimes soda ash, Na_2CO_3, must be added if the alkalinity is insufficient to precipitate all the hardness. Lime addition also converts any carbon dioxide (CO_2) in the water to carbonate (CO_3^{2-}).

At first glance, it seems strange to remove calcium by adding more calcium, but the chemistry is straightforward.

1. Lime addition to neutralize CO_2 and raise the pH

$$CO_2 + Ca(OH)_2 \rightarrow CaCO_3 + H_2O$$

2. Lime addition to remove calcium hardness by reaction with natural alkalinity

$$Ca^{2+} + 2HCO_3^- + Ca(OH)_2 \rightarrow 2CaCO_3 + 2H_2O$$

3. Lime addition to remove magnesium hardness by reaction with natural alkalinity

$$Mg^{2+} + 2HCO_3^- + 2Ca(OH)_2 \rightarrow Mg(OH)_2 + 2CaCO_3 + 2H_2O$$

4. Soda ash addition to remove calcium hardness in excess of alkalinity

$$Ca^{2+} + Na_2CO_3 \rightarrow CaCO_3 + 2Na^+$$

5. Removal of magnesium hardness in excess of alkalinity
   a. Lime addition for magnesium removal

$$Mg^{2+} + Ca(OH)_2 \rightarrow Mg(OH)_2 + Ca^{2+}$$

   b. Soda ash addition for removing calcium from added lime

$$Ca^{2+} + Na_2CO_3 \rightarrow CaCO_3 + 2Na^+$$

The stoichiometric chemical requirements for these reactions are as follows:

Lime: $Ca(OH)_2 = CO_2 + \text{Alkalinity} + Mg^{2+}$

Soda ash: $Na_2CO_3 = Ca^{2+} + Mg^{2+} - \text{Alkalinity}$

These doses will reduce the original calcium and magnesium, and the calcium that was added by the lime down to the solubility limit. With proper pH control (typically pH = 10–11) the residual calcium hardness will be 35–50 mg/L $CaCO_3$. The theoretical solubility is less than this, but the solubility does not approach the theoretical within the detention times that are used in practice.

Example 8.10 Solubility of Calcium Carbonate

The solubility of salts of weak acids is strongly pH dependent. Calcium carbonate (solubility product $K_{sp} = [Ca^{2+}][CO_3^{2-}] = 3.3 \times 10^{-9}$) is an important example. When calcium carbonate dissolves, one mole of $CaCO_3$ yields one mole of Ca^{2+} and one mole of CO_3^{2-}.

$$CaCO_3 \rightleftarrows Ca^{2+} + CO_3^{2-}$$

Some of the carbonate hydrolyzes to bicarbonate

$$CO_3^{2-} + H_2O \rightleftarrows HCO_3^- + OH^-$$

This drives the solubility equilibrium to the right and increases the solubility of $CaCO_3$.

If we neglect the effect of the carbonate hydrolysis, the solubility, S^*, of $CaCO_3$ is

$$S^* = [Ca^{2+}] = [CO_3^{2-}] = \sqrt{K_{sp}} = \sqrt{3.3 \times 10^{-9}} = 5.7 \times 10^{-5}\ \text{M} = 5.7\ \text{mg/L as } CaCO_3$$

If we include the effect of carbonate hydrolysis, the solubility of $CaCO_3$ is

$$S^* = [Ca^{2+}] = [CO_3^{2-}] + [HCO_3^-]$$

From the carbonate–bicarbonate equilibrium,

$$HCO_3^- \rightleftarrows CO_3^{2-} + H^+ \qquad K_2 = \frac{[CO_3^{2-}][H^+]}{[HCO_3^-]} = 4.7 \times 10^{-11}$$

Solving for $[HCO_3^-]$ and substituting into the solubility expression gives

$$S^* = [Ca^{2+}] = [CO_3^{2-}] + \frac{[CO_3^{2-}][H^+]}{K_2} = [CO_3^{2-}]\left(1 + \frac{[H^+]}{K_2}\right)$$

Multiplying by $[Ca^{2+}]$ gives

$$(S^*)^2 = [Ca^{2+}]^2 = [Ca^{2+}]\left([CO_3^{2-}] + \frac{[CO_3^{2-}][H^+]}{K_2}\right) = [Ca^{2+}][CO_3^{2-}]\left(1 + \frac{[H^+]}{K_2}\right)$$

Substituting K_{sp} for $[Ca^{2+}][CO_3^{2-}]$ and taking the square root gives the solubility of $CaCO_3$ as a function of pH.

$$S^* = \sqrt{K_{sp}\left(1 + \frac{[H^+]}{K_2}\right)}$$

At pH 7, the solubility is

$$S^* = \sqrt{K_{sp}\left(1 + \frac{[H^+]}{K_2}\right)} = \sqrt{3.3\times10^{-9}\left(1 + \frac{[10^{-7}]}{4.7\times10^{-11}}\right)} = 2.6\times10^{-3}\ \text{M} = 260\ \text{mg/L as}\ CaCO_3$$

At pH 9, the solubility is

$$S^* = \sqrt{K_{sp}\left(1 + \frac{[H^+]}{K_2}\right)} = \sqrt{3.3\times10^{-9}\left(1 + \frac{[10^{-9}]}{4.7\times10^{-11}}\right)} = 2.7\times10^{-4}\ \text{M} = 27\ \text{mg/L as}\ CaCO_3$$

This analysis does not account for bicarbonate hydrolysis to carbonic acid.

$$H_2CO_3 \rightleftarrows HCO_3^- + H^+ \qquad K_1 = \frac{[HCO_3^-][H^+]}{[H_2CO_3]} = 4.5\times10^{-7}$$

Including that effect gives the exact solution for the solubility of $CaCO_3$ as a function of pH to

$$S^* = \sqrt{K_{sp}\left(1 + \frac{[H^+]}{K_2} + \frac{[H^+]}{K_2}\frac{[H^+]}{K_1}\right)}$$

As a comparison, at pH 7 the exact solubility $CaCO_3$ is

$$S^* = \sqrt{K_{sp}\left(1 + \frac{[H^+]}{K_2} + \frac{[H^+]}{K_2}\frac{[H^+]}{K_1}\right)} = \sqrt{3.3\times10^{-9}\left(1 + \frac{[10^{-7}]}{4.7\times10^{-11}} + \left(\frac{[10^{-7}]}{4.7\times10^{-11}}\right)\frac{[10^{-7}]}{4.5\times10^{-7}}\right)}$$

$$= 2.9\times10^{-3}\ \text{M} = 290\ \text{mg/L as}\ CaCO_3$$

This is about 10% higher than the 260 mg/L as $CaCO_3$ we computed when the bicarbonate hydrolysis was neglected. At pH 9, the exact solubility is 27.1 mg/L as $CaCO_3$, which is about 1% higher. Clearly, the solubility of $CaCO_3$ decreases as a function of pH, as shown in Figure 8.13.

Figure 8.14 shows a system that might be used to soften water when turbidity removal is not needed. Aeration reduces the CO_2 concentration, which raises the pH slightly and lowers the requirement for lime. Lime, CaO, and soda ash, Na_2CO_3, (if necessary) are added to form the $CaCO_3$ and $Mg(OH)_2$

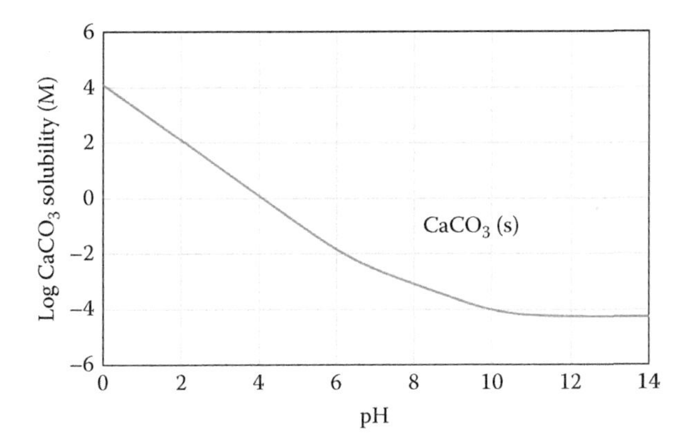

FIGURE 8.13 Solubility of $CaCO_3$ as a function of pH (exact solution).

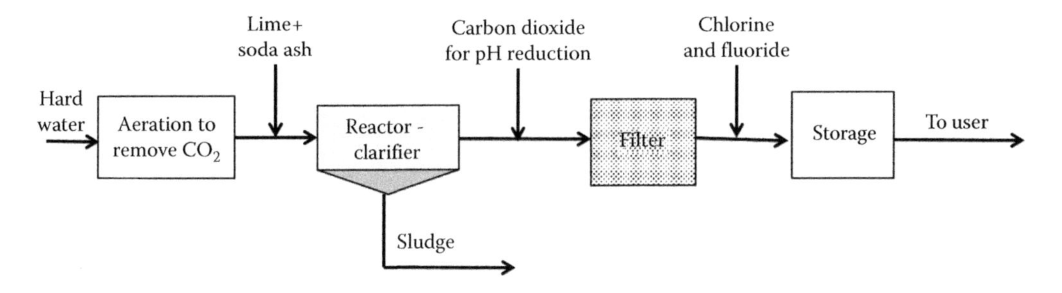

FIGURE 8.14 Lime–soda softening process for a drinking water system, assuming the incoming water does not need turbidity removal.

precipitates. This process operates at pH 10–11, so the pH must be reduced to avoid precipitation in the filter and the water distribution system. A common way to do this is to add CO_2. Filtration is needed because the settling process will not remove 100% of the precipitate solids. Assuming this is a drinking-water supply, chlorine is added for disinfection and fluoride is added to prevent tooth decay. This would not be needed if the water is going to an industrial process.

Example 8.11 Lime Softening Stoichiometry

Water that has 150 mg/L HCO_3^- as $CaCO_3$, 90 mg/L Ca^{2+}, 10 mg/L Mg^{2+}, 5 mg/L CO_2, pH 7, and a temperature of 10°C will be softened by lime precipitation.

Convert all concentrations to equivalent $CaCO_3$.

Ca^{2+}	(90 mg/L)(100/40) = 225 mg/L $CaCO_3$
Mg^{2+}	(10 mg/L)(100/24.3) = 41 mg/L $CaCO_3$
HCO_3^-	150 mg/L $CaCO_3$
CO_2	(5 mg/L)(100/44) = 11 mg/L $CaCO_3$

Chemical requirements are as follows:

$$\text{Lime} = CO_2 + \text{Alk} + Mg^{2+} = 11 + 150 + 41 = 202 \text{ mg/L as } CaCO_3$$

$$\text{Soda ash} = Ca^{2+} + Mg^{2+} - \text{Alk} = 225 + 41 - 150 = 116 \text{ mg/L as } CaCO_3$$

Typical drinking-water softening practice is to adjust the stoichiometric amounts. Excess lime (about 20 mg/L as $CaCO_3$) is added to raise the pH to >11 so $Mg(OH)_2$ will precipitate. Most applications strive for a total hardness between 75 and 120 mg/L as $CaCO_3$. Magnesium concentrations greater than 40 mg/L as $CaCO_3$ cause scale, thus common practice is to remove Mg only in excess of 40 mg/L.

8.8 CHEMICAL PHOSPHORUS REMOVAL

Phosphorus can exist in three ionic forms, depending on the pH. The reactions and the equilibrium expressions are as follows:

$$\text{Low pH:} \quad H_3PO_4 \leftrightarrows H^+ + H_2PO_4^-$$

$$H_2PO_4^- \leftrightarrows H^+ + HPO_4^{2-}$$

$$\text{High pH:} \quad HPO_4^{2-} \leftrightarrows H^+ + PO_4^{3-}$$

$$K_1 = \frac{[H^+][H_2PO_4^-]}{[H_3PO_4]} = 6.9 \times 10^{-3}$$

$$K_2 = \frac{[H^+][HPO_4^{2-}]}{[H_2PO_4^-]} = 6.2 \times 10^{-8}$$

$$K_3 = \frac{[H^+][PO_4^{3-}]}{[HPO_4^{2-}]} = 4.8 \times 10^{-13}$$

At neutral pH, pH = 7.0,

$$[H^+] = 10^{-pH} = 10^{-7}$$

Substituting into the equilibrium equations gives

$$K_1 = \frac{[10^{-7}][H_2PO_4^-]}{[H_3PO_4]} = 6.9 \times 10^{-3} \quad \Rightarrow \quad \frac{[H_2PO_4^-]}{[H_3PO_4]} = 6.9 \times 10^{4}$$

$$K_2 = \frac{[10^{-7}][HPO_4^{2-}]}{[H_2PO_4^-]} = 6.2 \times 10^{-8} \quad \Rightarrow \quad \frac{[HPO_4^{2-}]}{[H_2PO_4^-]} = 0.62$$

$$K_3 = \frac{[10^{-7}][PO_4^{3-}]}{[HPO_4^{2-}]} = 4.8 \times 10^{-13} \quad \Rightarrow \quad \frac{[PO_4^{3-}]}{[HPO_4^{2-}]} = 4.8 \times 10^{-6}$$

At pH 7, only $H_2PO_4^-$ and HPO_4^{2-} are present in significant amounts. The proportions are 62% HPO_4^{2-} and 38% $H_2PO_4^-$.

Figure 8.15 shows the molar percentages of the four phosphoric acid species as a function of pH. Figure 8.16 is a log concentration-pH diagram for 10^{-3} M phosphoric acid that shows the concentrations of the four ionic species as a function of pH.

Chemical phosphorus removal can be done with iron salts or aluminum salts.

Iron salts. Iron is commercially available in four forms: ferric chloride ($FeCl_3$), ferrous chloride ($FeCl_2$), ferric sulfate ($Fe_2(SO_4)_3$), and ferrous sulfate ($FeSO_4$). All are corrosive and must be carefully handled. The ferric salts, Fe(III), are commonly used for chemical phosphorus removal. They react with the dominant forms of phosphorus, $H_2PO_4^-$ and HPO_4^{2-} (indicated as $H_nPO_4^{3-n}$, where n is typically 1 or 2) as follows:

$$\text{Ferric chloride } (FeCl_3) \quad FeCl_3 + H_nPO_4^{3-n} \rightarrow FePO_4 + nH^+ + 3Cl^-$$

$$\text{Ferric sulfate } (Fe_2(SO_4)_3) \quad Fe_2(SO_4)_3 + 2H_nPO_4^{3-n} \rightarrow 2FePO_4 + 2nH^+ + 3SO_4^{2-}$$

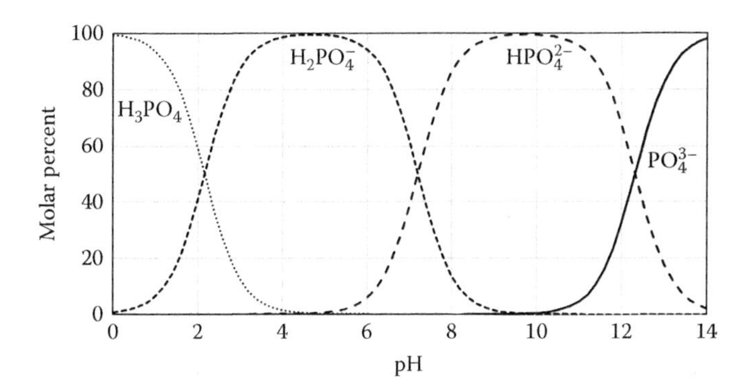

FIGURE 8.15 Distribution of phosphoric acid species as a function of pH.

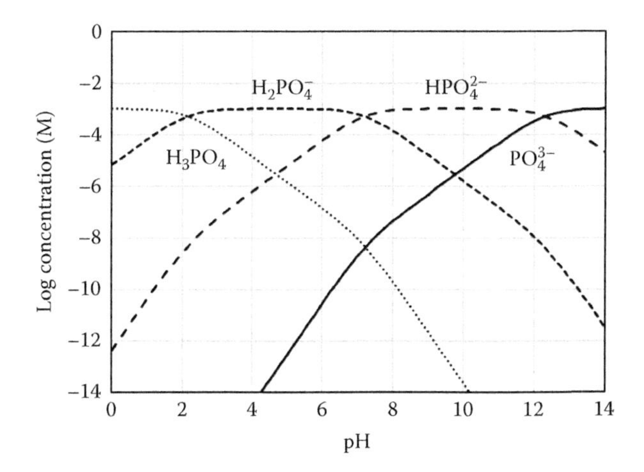

FIGURE 8.16 Log concentration-pH diagram for 10^{-3} M phosphoric acid.

Theoretically, 1.8 kg of ferric iron is required to remove 1 kg of phosphorus (as P). However, much more is required to achieve low phosphorus concentrations. The excess ferric iron generates H^+ that causes the wastewater pH to drop approximately 0.1 pH unit per 10 mg/L of iron added. Approximately 1 mg/L of alkalinity is consumed for each mg/L of iron added. The theoretical mass of sludge produced is 4.87 mg $FePO_4$/mg P (at Fe/P = 1/1).

If ferrous salts, Fe[II], present in some waste products like pickle liquor, are used, the phosphorus precipitation reactions are:
Ferrous chloride ($FeCl_2$)

$$3FeCl_2 + 2H_nPO_4^{3-n} \rightarrow Fe_3(PO_4)_2 + 2nH^+ + 6Cl^-$$

Ferrous sulfate ($FeSO_4$)

$$3FeSO_4 + 2H_nPO_4^{3-n} \rightarrow Fe_3(PO_4)_2 + 2nH^+ + 3SO_4^{2-}$$

Theoretically, 2.7 kg of ferrous iron is required to remove 1 kg of phosphorus (as P). The theoretical mass of sludge produced is 5.76 mg $Fe_3(PO_4)_2$/mg P (at Fe/P = 3/2).

Ferrous iron salts can be used if the ferrous iron is oxidized to ferric form. The phosphorus precipitation reactions are:
Ferrous chloride ($FeCl_2$)

$$FeCl_2 + H_nPO_4^{3-n} + 0.25O_2 + 0.5H_2O \rightarrow FePO_4 + nH^+ + OH^- + 3Cl^-$$

Ferrous sulfate ($FeSO_4$)

$$2FeSO_4 + 2H_nPO_4^{3-n} + 0.5O_2 + H_2O \rightarrow 2FePO_4 + 2nH^+ + 2OH^- + 2SO_4^{2-}$$

The typical dissolved oxygen concentration in an activated sludge aeration basin is 1–2 mg/L, and this is enough to oxidize the ferrous salts.

Iron works over a wide pH range. Iron salt solutions contain up to 75–100 mg/L trace metals, depending on the product.

Aluminum salts. Aluminum is commercially available in five forms: aluminum sulfate (alum, $Al_2(SO_4)_3$), poly-aluminum chloride (PACl), aluminum chloride ($AlCl_3$), aluminum chlorohydrate, and sodium aluminate ($Na_2Al_2O_4$). All dissolve to release Al^{3+} ions and the theoretical chemistry is

$$\text{Aluminum salts} \qquad Al^{3+} + H_nPO_4^{3-n} \rightarrow AlPO_4 + nH^+$$

Theoretically one mole of Al will precipitate one mole of P.

9.6 mg alum will remove 1 mg P.
0.5 mg alkalinity will be consumed per mg of alum added.

In practice, it takes 2–15 times the theoretical dose because some of the metal forms a hydroxide precipitate:

$$Fe^{3+} + 3OH^- \rightarrow Fe(OH)_3$$

$$Al^{3+} + 3OH^- \rightarrow Al(OH)_3$$

The pH must be reduced to reach a concentration of 0.1–0.5 mg/P/L. The dosing chemicals (Table 8.7) are Al and Fe salts. This is illustrated by Figures 8.17 and 8.18. The left panel

TABLE 8.7
Chemicals Used for Phosphorus Removal

Chemical	Formula	Removal Mechanism	Effect on pH
Aluminum sulfate (Alum)	$Al_2(SO_4)_3 \cdot 14.3(H_2O)$ MM = 599.4	Metal hydroxides	Removes alkalinity
Ferric chloride	$FeCl_3$ MM = 162.3	Metal hydroxides	Removes alkalinity
Poly-aluminum chloride	$Al_nCl_{(3n-m)}(OH)_m$ $Al_{12}Cl_{12}(OH)_{24}$	Metal hydroxides	None
Ferrous sulfate (pickle liquor)	$FeSO_4$	Metal hydroxides	Removes alkalinity
Lime	CaO, $Ca(OH)_2$	Insoluble precipitate	Raises pH to above 10

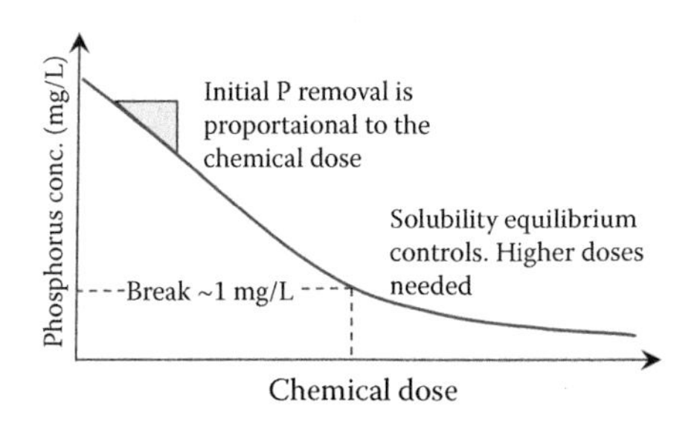

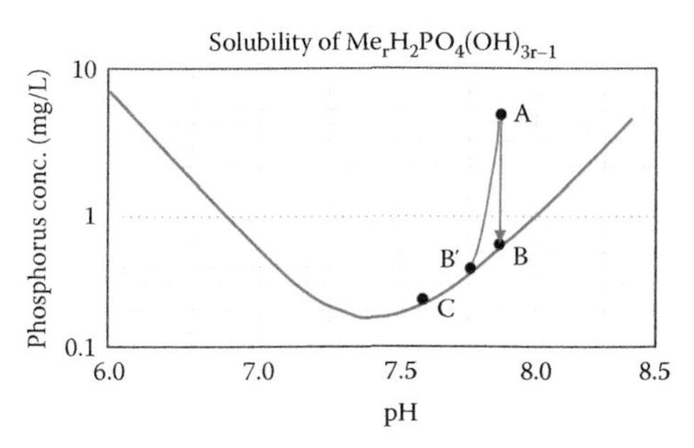

FIGURE 8.17 Phosphorus removal is proportional to chemical dose down to about 1 mg/L P.

of Figure 8.17 shows how the phosphorus concentration decreases as Al or Fe dosage increases. The removal is stoichiometric until about 1 mg P/L. To accomplish more reduction, an excess of reagent is needed. The right panel shows on the pH-solubility curve. The initial concentration is A. Stoichiometric removal brings the concentration from point A to point B. Excess chemical will reduce the pH (the reagents act as acids) and bring the final P concentration to point B or point C.

The amount of alkalinity consumed depends on the form of aluminum used. Alum uses approximately 0.5 mg/L of alkalinity for each mg/L of alum added. Aluminum chloride uses 1 mg/L. PACl uses almost no alkalinity. The optimum pH is 6.5.

Iron and aluminum salts consume alkalinity in the wastewater. This is helpful if the effluent phosphorus limits are less than 1 mg/L because that concentration cannot be achieved without reducing the pH. If reducing the alkalinity and the pH is a problem, liquid sodium aluminate provides aluminum in an alkaline solution, and contains more aluminum that its acidic counterparts. It will provide 2.5–3 times the amount of aluminum as liquid alum.

A kilogram of aluminum and a kilogram of iron are not equally effective for precipitating phosphorus. An aluminum molecule has a mass less than half an iron molecule. This means that 1 kg of aluminum has twice as many molecules as 1 kg of iron and can combine with two times as much phosphorus. Chemically removing 1 kg of phosphorus requires 0.87 kg of aluminum or 1.8 kg of iron.

Tests show that there is a fraction of soluble P that is nonreactive; it cannot be precipitated for removal. The amount is typically 0.01–0.02 mg/L, but some wastewaters have values ranging from 0.05–0.07 mg/L.

Example 8.12 Alum Precipitation of Phosphorus

Aluminum sulfate (alum) will be added to reduce P from 3 to 0.5 mg/L. The flow rate is 10,000 m^3/day (2.6 mg day). From experiments (Figure 8.18) it is estimated that this requires a dose of 1.7 mol of Al per mole P. Assume that 1 mole of Al will react to precipitate phosphorus, and 0.7 mol will form aluminum hydroxide. Calculate the mass of alum needed, and estimate the (dry) mass of sludge produced.

The mass of phosphorus removed is

$$(0.0025\ \text{kg/m}^3)(10{,}000\ \text{m}^3/\text{day}) = 25\ \text{kg/day}$$

The molar mass of P = 30.97 kg/kg-mol
The moles of P removed = (25 kg P/day)/(31 kg P/kg-mol) = 0.806 kg-mol P/day
The production of aluminum phosphate sludge, assuming one mole of Al per mole P

$$Al^{3+} + PO_4^{3-} \rightarrow AlPO_4$$

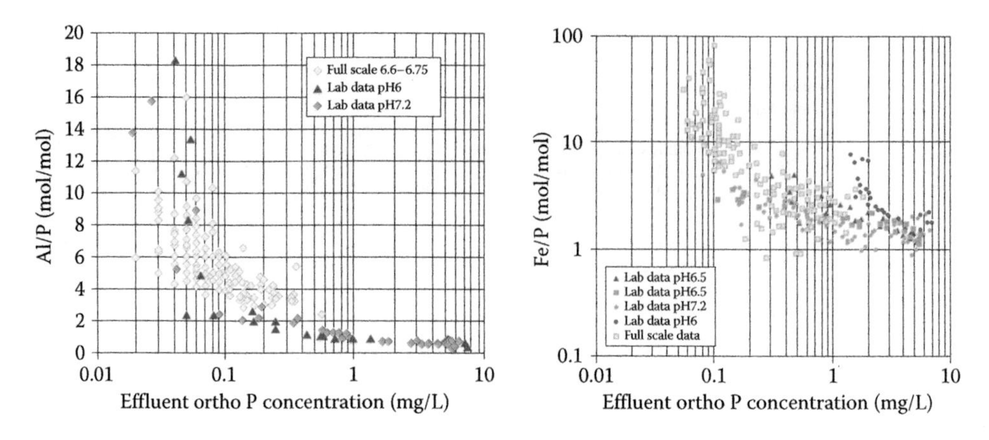

FIGURE 8.18 Molar dose ratios from full-scale and laboratory tests. (Hermanowicz, S.W., 2006.)

Adding one mole of Al^{3+} yields one mole of $AlPO_4$.
The molar mass of $AlPO_4$ = 27 + 32+ 4(6) =122 kg/kg-mol
Dry mass $AlPO_4$ = (0.806 kg-mol $AlPO_4$/day)(122 kg $AlPO_4$/kg-mol) = 98.3 kg/day

The production of aluminum hydroxide sludge, assuming 0.7 mol Al

$$Al^{3+} + 3OH^{-} \rightarrow Al(OH)_3$$

Adding 0.7 mol Al^{3+} yields 0.7 mol of $Al(OH)_3$
The molar mass of $Al(OH)_3$ = 27 + 3(17) = 78 kg/kg-mol
Dry mass $Al(OH)_3$ = (0.7 kg-mol $Al(OH)_3$/day)(78 kg $AlPO_4$/kg-mol) = 54.6 kg/day

Total chemical sludge (dry mass) production = 98.3 + 54.6 = 152.9 kg/day

Figure 8.18 shows that direct reaction of ferric ions with phosphate:

$$Fe^{3+} + PO_4^{3-} \rightarrow FePO_4(s)$$

is only close to being correct when the final phosphorus concentration is 1 mg/L or more. For many years the effluent requirement for the relatively few treatment plants that had to remove phosphorus was 1 mg/L, and this simple chemistry was useful to designers. As discharge permits have started to require lower effluent concentrations, the chemistry of interest is no longer the 1:1 ratio of metal to PO_4.

At ortho-P (PO_4^{3-}) concentrations lower than 1 mg/L, and correspondingly higher metal ion doses, the usual chemical structure of the precipitates, $FePO_4$ or $AlPO_4$, no longer applies. Empirical tests indicate that an empirical formula of this general form will be better (Sedlak 1991):

$$rMe^{3+} + H_2PO_4^{-} + (3r - 1)OH^{-} \rightleftarrows Me_rH_2PO_4(OH)_{3r-1}$$

where Me_r is either Fe or Al. According to Jenkins & Hermanowicz (1991) the value for a stoichiometric coefficient differs depending on the metal salt. For aluminum salts $r_{Al} = 0.8$ while $r_{Fe} = 1.6$. This gives

$$0.8Al^{3+} + H_2PO_4^{-} + 1.4OH \rightleftarrows Al_{0.8}H_2PO_4(OH)_{1.4.}$$

and

$$1.6Al^{3+} + H_2PO_4^{-} + 3.8OH \rightleftarrows Fe_{1.6}H_2PO_4(OH)_{3.8}$$

8.9 STRUVITE PRECIPITATION AND NUTRIENT RECOVERY

Struvite is crystalline magnesium ammonium phosphate (MAG), $MgNH_4PO_4$. The theoretical solubility stoichiometry is

$$Mg^{2+} + NH_4^+ + PO_4^{3-} + 6H_2O \leftrightarrows MgNH_4PO_4 \cdot 6H_2O \quad K_{sp} = 2.5 \times 10^{-13}$$

The molecular mass of struvite is 245.1 g/mol. It is 9.9% magnesium, 12.6% phosphorus, and 5.7% nitrogen, by mass.

Anaerobic sludge digestion systems have high ammonia and phosphorus concentrations, and in locations that have hard water the magnesium concentration is also high. This creates the tendency for struvite to precipitate on pipe walls, pump impellors, and heat exchanger surfaces, as shown in Figure 8.19. The scale is difficult to remove and it is a serious maintenance problem.

Nucleation for struvite precipitation is strongly dependent on the super saturation level, S. This is measured by

$$S = IAP / K_{Sp} = [Mg^{2+}][NH_4^+][PO_4^{3-}] / K_{sp}$$

where IAP = ion activity potential, and the quantities in brackets are the activities of the ions shown. Values of S in excess of 1.5 are usually required to initiate precipitation.

The solubility of struvite is strongly pH dependent (like calcium carbonate in softening), because it is the salt of a weak acid (phosphoric) and a weak base (ammonium). Figure 8.20 shows that struvite is least soluble at pH about 10.

The usual approach to preventing scaling is to reduce the concentration of one or more of the component minerals so super saturation does not occur. The addition of iron salts (usually $FeCl_3$) to precipitate the phosphorus as $Fe_3(PO_4)_2 \cdot 8H_2O$ (vivianite) is sometimes helpful, but not trouble free. The iron salts are acidic and they destroy alkalinity and reduce the pH in the digester, which is undesirable, and they increase the chloride content of the effluent. Also, vivianite itself can also cause scale.

Struvite precipitation is also a problem in cattle and swine manure storage and digestion systems because the wastes are very high in ammonia. These systems have become popular because usable methane gas is produced and the application of the wastes to farmland is more controlled than when manure is applied.

On the positive side, struvite can be precipitated and recovered under controlled conditions to form spherical particles that make an excellent slow-release fertilizer. Slow-release means that the nitrogen and phosphorus become available to plants over time, so the material can be applied at high rates without damage to plant roots. This also reduces nutrient losses by runoff to streams and infiltration into groundwater. Nutrient recovery by controlled struvite precipitation is also possible for some industrial wastes and from urine (Mohan et al., 2001; Zhang, 2010).

FIGURE 8.19 Struvite scaling in a pipe.

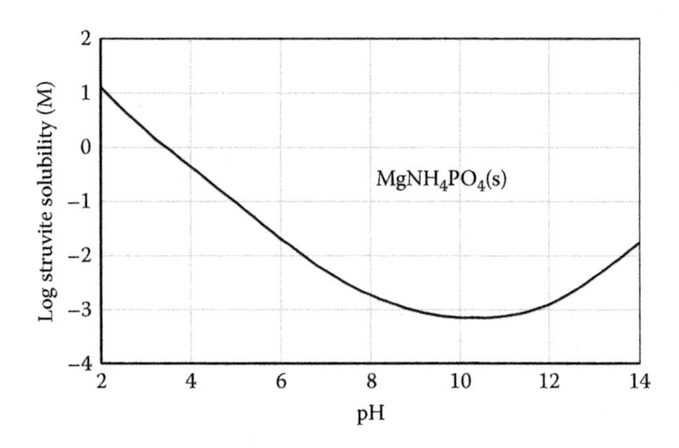

FIGURE 8.20 Solubility of struvite as a function of pH.

In the municipal sludge digestion situation, the goal of struvite recovery is to remove the phosphorus. In the agricultural situation, the goal is more often to recover the ammonia as a struvite fertilizer. Phosphorus is also recovered, but the amounts are much less than ammonia.

Example 8.13 Struvite Precipitation at the Nine Springs Wastewater Treatment Plant, Madison, Wisconsin

The Nine Springs Wastewater Treatment Plant, Figure 8.21, has an average flow of 159,000 m³/day (42 million gallons per day). A Modified Cape Town biological nutrient removal process is used to remove biochemical oxygen demand (BOD), phosphorus, and nitrogen. The process has three biological reactors. An anaerobic zone followed by an anoxic zone facilitates the biological release

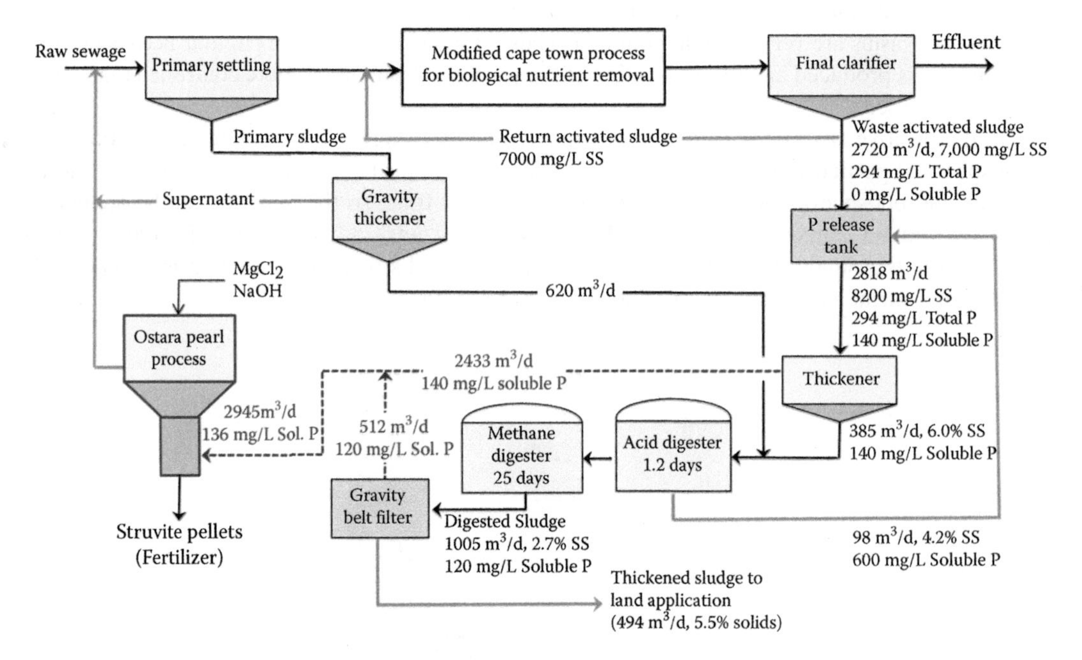

FIGURE 8.21 Sludge processing system with struvite precipitation at the Nine Springs Wastewater Treatment Plant, Madison, WI. (Courtesy of Steve Reusser, Madison Metropolitan Sewerage District.)

FIGURE 8.22 The Ostara Pearl 500 equipment and the struvite pellet product. (Courtesy of Steve Reusser, Madison Metropolitan Sewerage District.)

of soluble phosphorus. These are followed by aerobic-activated sludge treatment, which removes carbonaceous BOD, nitrifies ammonia to nitrate, and produces enhanced removal of phosphorus.

The mixed liquor and waste activated sludge (WAS) suspended solids is 5.5% phosphorus on a mass basis. This high concentration of phosphorus has been a problem in the past because it is solubilized in the anaerobic sludge digesters and it then precipitates as struvite and vivianite in pipes and heat exchangers.

To reduce the inconvenient precipitation of struvite in the equipment, the Ostara Pearl fluidized bed process was installed to remove struvite under controlled conditions and to recover phosphorus and ammonia as pellets that are sold as fertilizer (Ohlinger 1999 and Ohlinger et al. 1998).

The WAS contains 292 mg/L total phosphorus and almost no soluble phosphorus. The feed to the Ostara process must contain soluble phosphorus. The WAS flows into a phosphorus release tank, which is simply an anaerobic holding tank where most of the total P is converted to soluble P. The WAS is then thickened, and the thickener supernatant, which contains all the soluble P, is sent to the Ostara process. The thickened sludge is combined with thickened primary sludge for digestion.

The Nine Springs sludge treatment system is unusual because it uses a two-stage anaerobic digestion process. The first stage solubilizes the volatile solids and starts the digestion by making volatile fatty acids (mainly acetic acid and propionic acid). The second digestion stage converts the fatty acids to methane gas and carbon dioxide.

The digestion also produces ammonia and releases more soluble P. The digested sludge is thickened from 5.5% solids to 27% solids with a gravity belt thickener. The filtrate from this process is high in ammonia and soluble P.

The digested sludge filtrate is combined with the WAS thickener supernatant to form the feed to the struvite precipitation process. This is a flow of 2886 m^3/day that contains 135 mg/L soluble P. Magnesium chloride ($MgCl_2$) and sodium hydroxide (NaOH) are added to the process. The Mg provides the stoichiometric requirement to form the struvite, $MgNH_4PO_4$. The NaOH is needed to control the pH.

Eighty percent of the soluble P is recovered in the struvite. About 60% of the total P is recovered.

The chemistry of the Ostara process has the stoichiometry described earlier. The recovery is not 100% because the process is a combined precipitation and solids separation system. The flow of the high phosphorus stream is upward through a fluidized bed of the struvite pellets. Some of the phosphorus forms colloidal particles that are flushed out of the reactor. This recycles through the primary settling process and the biological treatment process (Figure 8.22).

Example 8.14 Ammonia Recovery from Swine Waste

A swine manure waste has 3200 mg/L COD, 1400 mg/L NH_4^+, 24 mg/L PO_4^{3+}, 22 mg/L Mg^{2+}, 21 mg/L Ca^{2+}, and 2150 mg/L K^+. The objective is to recover the ammonia in the form of struvite particles. The volume of liquid manure to be processed each day is 100 m^3, which gives the mass

TABLE 8.8
Material Balance for the Basic Struvite Precipitation Process Shown in Figure 8.23

	Manure Slurry			Reagents Added		Recovered		Lost in Centrate	
Constituent	**(mg/L)**	**(kg/d)**	**(mol/d)**	**(kg/d)**	**(mol/d)**	**(kg/d)**	**(mol/d)**	**(kg/d)**	**(mol/d)**
NH_4^+	1,400	140	7.8	0	0	139.5	7.75	0.9	0.05
PO_4^{3-}	24	2.9	0.03	807.5	8.5	736.2	7.75	74.1	0.78
Mg^{2+}	22	2.2	0.09	206.6	8.5	188.3	7.75	20.4	0.84
Struvite						1064.0	7.75		

quantities in Table 8.8. The slurry is decanted from the digester prior to processing, but it is not free of suspended solids.

It will be necessary to add magnesium and phosphate. Tests showed that a slight excess above the stoichiometric amounts was required. Also, caustic (NaOH) must be added to raise the pH to 10 for the best performance. The material balance data are given in Table 8.8.

Figure 8.23 is the process flow diagram. The "lost" chemicals are in the centrate.

The struvite crystals that leave the centrifuge are contaminated with any suspended solids that were carried into the reactor with the manure slurry. These solids have not been shown in the diagram. The material is dark and storage will be a problem because of odors and biological degradation. A purification step will yield a clean, white crystal. The modified process is shown in Figure 8.24.

The modified process recovers approximately 84% of the ammonia in the influent.

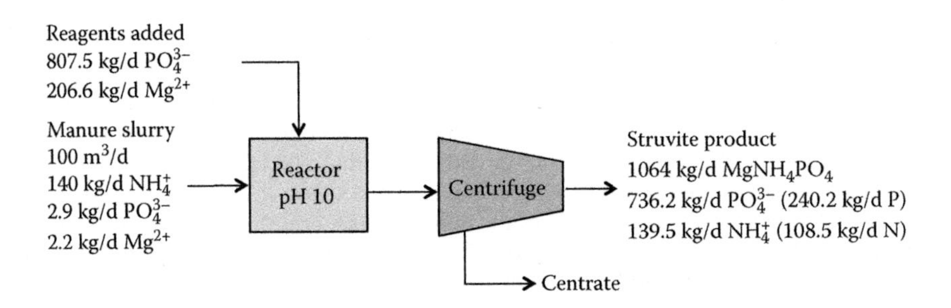

FIGURE 8.23 Process flow diagram for ammonia recovery from swine waste.

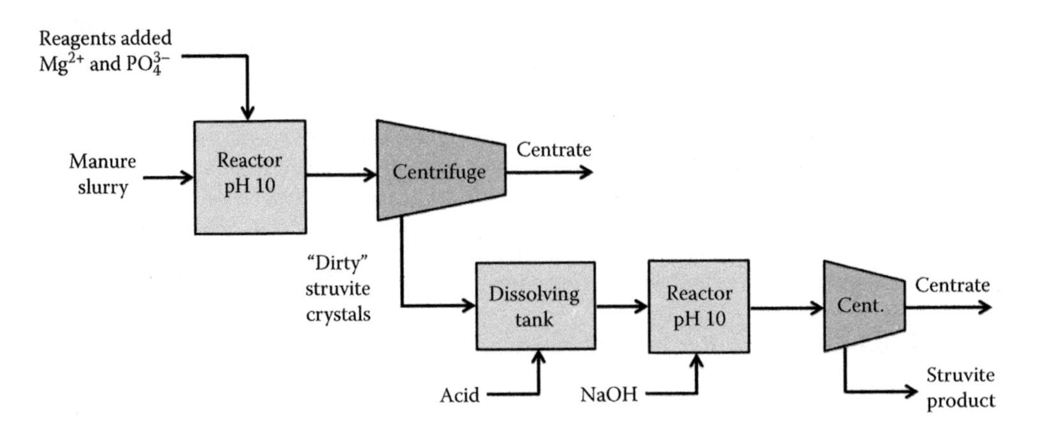

FIGURE 8.24 Struvite precipitation with a purification step to produce white spherical crystals.

TABLE 8.9
Comparison of Leachability of Hydroxide and Sulfide Sludges

	Metal Hydroxide Sludge			Metal Sulfide Sludge		
Metal	Sludge Cake Composition (mg/kg Dry Solids)	Leachate Composition (mg/kg Sludge)	% of Metal Leached from Sludge Cake	Sludge Cake Composition (mg/kg Dry Solids)	Leachate Composition (mg/kg Sludge)	% of Metal Leached from Sludge Cake
Pb	21,000	91	0.43	25,000	<27	<0.11
Cr	15,000	390	0.26	17,000	<27	<0.06
Zn	10,000	39	0.39	12,000	25	0.21
Cu	73,000	440	0.60	79,000	<27	<0.34
Ni	57,000	280	0.49	52,000	2,100	4.0

8.10 LEACHING METALS FROM SLUDGE

Metals leaching in sludge should remain fixed in the sludge solids so they do not migrate into the soil or groundwater. The sludge pH should remain in the range of 7.5–9 to fix the metals.

In another situation, you may want to release the metals for recovery. To release the metals, the sludge would be acidified. Selective release is possible, just as selective precipitation is possible.

Sludge that contains incidental amounts of metals, such as most municipal sewage sludge, can be applied to land, provided certain restrictions are met (Berthouex and Brown, 2013). Sludge from a dedicated metals precipitation process will be classified as a hazardous material and special handling will be required. One issue will be whether the metals in the sludge will dissolve and become mobile after placement in landfill or other disposal site.

A waste material has the characteristic of *toxicity* (under U.S. EPA regulations) if it will leach specific heavy metals, organic compounds, or pesticides into the soil or groundwater under landfill conditions. The *toxicity characteristic leaching procedure* (TCLP) *test* is used to measure this potential. The test procedure is complex and the details are omitted; they can be found in readily available references. The test conditions simulate the worst case for co-disposal of wastes in a landfill environment. The waste solids are subjected to leaching for 18 h in a buffered extraction fluid that has pH 4.9 (a mixture of glacial acetic acid and sodium hydroxide).

Table 8.9 shows that sulfide precipitates are more stable than hydroxide precipitates, which is not surprising since most of these metals exist in the earth as sulfide ores. These are not TCLP test results. Details of the sludge characteristics and the leaching procedure are omitted. The only purpose of this table is to illustrate that metal sulfide sludge is more stable than hydroxide sludge.

8.11 CONCLUSION

The brief treatment of solubility in this book (and other introductory texts) usually makes the chemistry too simple. The value of the fundamentals presented is to (1) help us understand that the solubility of a substance depends on the concentrations in solution, (2) show how solubility often depends on solution pH (perhaps in a complicated way), and (3) give a suggestion of how solubility is calculated.

Precipitation is used to convert dissolved pollutants into solid particles that can be removed by a physical separation method like settling or filtration. Precipitation is commonly used to remove metals from industrial wastewater, remove phosphorus from wastewater, and to soften drinking water and boiler feed water.

The solubility of inorganic compounds is defined by the solubility product, K_{sp}. The smaller the value of K_{sp}, the less soluble the compound. One step in inventing a precipitation process is to examine the solubility product of the various precipitates that might be formed when different precipitating agents are used. For example, precipitation with either iron or aluminum can be used to precipitate phosphorus from wastewater. The choice will depend on cost, which includes not only the reagent cost but also sludge disposal.

Precipitation processes are always supported by some kind of a separation process. Effluent limits for metals are given as total metal: *total* = *dissolved* + *particulate*. The ultimate effluent quality will depend on the efficiency of process that is used to remove the precipitate from the solution. Separation systems are a vital element in all treatment systems.

Two widely used options for removing heavy metals are hydroxide precipitation and sulfide precipitation. The sulfides are less soluble than the hydroxides, but the sulfide processes require more precise control over reagent addition and they tend to yield smaller particles that are more difficult to remove.

Controlled struvite precipitation is a rather new process that is used to recover phosphorus, in the form of a pelletized fertilizer, at wastewater treatment plants. It is also used to recover ammonia and phosphorus at animal waste treatment plants.

9 Oxidation–Reduction Reactions

9.1 THE DESIGN PROBLEM

Oxidation causes an increase in ionic charge, and reduction causes a decrease. *Oxidation* of a chemical (or element) involves the loss of electrons. *Reduction* is the reverse, a gain in electrons. The oxidized substance is the electron donor and the reduced substance is the electron acceptor.

When one substance is oxidized, another must be reduced. The name for these paired reactions is *oxidation–reduction* reactions, or *redox* reactions.

In this chapter, we are interested in how to use redox reactions to

- Change the ionic state of inorganic chemicals so they are easier to remove.
- Destroy or detoxify organic compounds.

Oxidation–reduction reactions are used to promote the precipitation of contaminants (especially in the case of inorganic pollutants such as heavy metals). Ferric iron (Fe^{3+}) can easily be removed by precipitation; ferrous iron (Fe^{2+}) cannot. Oxidation from Fe^{2+} to Fe^{3+} might be useful in an iron removal process. Trivalent chromium (Cr^{3+}) will precipitate, but hexavalent chromium (Cr^{6+}) will not. Reduction could be one step in a chromium removal process.

Organic compounds can be oxidized to carbon dioxide and water. Reduction processes may be used to convert toxic organics to a much less toxic form, or to an easily biodegraded form. All biological reactions are paired oxidation–reduction reactions.

Phenol, a toxic organic chemical, can be oxidized to carbon dioxide and water with hydrogen peroxide and an iron catalyst (Fenton's reaction). Cyanide can be oxidized with chlorine or ozone. Odorous sulfur compounds can be destroyed by oxidation. Disinfection of drinking water is an oxidation process that involves chlorine or ozone.

9.2 OXIDATION NUMBERS

The balanced equation for a redox reaction does not show the electrons that are being transferred. In order to tell whether a redox reaction has occurred or not, we need a way to keep track of electrons. The way to do so is by assigning oxidation numbers to the atoms or ions involved in a chemical reaction.

Oxidation numbers are hypothetical numbers assigned to an individual atom or ion and to groups of atoms or ions. *Oxidation numbers* (*oxidation states*) can be positive, negative, or zero.

Tables 9.1 and 9.2 give the oxidation number(s) for some common elements and compounds. Many of these exist in only one oxidation state, at least in the most common pollution control problems. Pure metal is in the completely reduced state, but it usually exists as an oxidized form in other compounds and in solution.

The oxidation number is +1 for H, Na, K, and Ag. The oxidation number is +2 for Mg, Ca, Ni, Zn, and Cd. It is –2 for O, and –1 for F, and +3 for Al. Iron (Fe) is either +2 or +3. These fixed values are used to work out the oxidation state of elements they combine with to form compounds.

Carbon, nitrogen, sulfur, chlorine, chromium, manganese, and a few others have multiple oxidation states.

TABLE 9.1
Atomic Number, Atomic Mass, and Oxidation Number for Some Elements That Are Common in Pollution Control Problems

Atomic Number	Element	Atomic Mass	Oxidation Numbers	Atomic Number	Element	Atomic Mass	Oxidation Numbers
1	H	1.008	1	24	Cr	51.996	6, 3, 2
6	C	12.011	4, 2, –4	25	Mn	54.938	7, 4, 3, 2
7	N	14.007	5, 4, 3, 2, 1, –3	26	Fe	55.845	3, 2
8	O	15.999	–2	27	Co	59.993	3, 2
9	F	18.998	–1	28	Ni	58.693	2
11	Na	22.980	1	29	Cu	63.546	2, 1
12	Mg	24.305	2	30	Zn	65.38	2
13	Al	26.982	3	33	As	74.922	5, 3, –3
14	Si	28.086	4	47	Ag	107.868	1
15	P	30.974	5, 3	48	Cd	112,41	2
16	S	32.065	6, 4, –2	50	Sn	118.710	4, 2
17	Cl	35.453	7, 5, 4, 3, 1, –1	79	Au	196.967	3, 1
19	K	39.098	1	80	Hg	200.59	2, 1
20	Ca	40.078	2	82	Pb	207.2	4, 2

TABLE 9.2
Oxidation Numbers for Some Nitrogen, Sulfur, and Phosphorus Compounds

Oxidation Number	Increasing Level of Oxidation									
	–3	–2	–1	0	+1	+2	+3	+4	+5	+6
					Nitrogen					
Aqueous solution and salts	NH_4^+						NO_2^-		NO_3^-	
	NH_3						HNO_2		HNO_3	
Gas phase	NH_3			N_2	N_2O	NO		NO_2		
					Sulfur					
Aqueous solution and salts		H_2S						H_2SO_3		H_2SO_4
		HS^-						HSO_3^-		HSO_4^-
										SO_4^{2-}
		S^{2-}	S_2^{2-}					SO_2		SO_3
Gas phase		H_2S						SO_2		SO_3
					Phosphorus					
Aqueous solution and salts							H_3PO_3		H_3PO_4	
							HPO_2		Na_3PO_4	
									PO_4^{3-}	

The following rules are used to assign oxidation numbers:

- The oxidation number for an atom in its elemental form is always zero. A substance is elemental if both of the following are true: (1) only one kind of atom is present and (2) charge = 0
  - Oxidation number of S = 0
  - Oxidation number of Fe = 0
- The oxidation number of a monoatomic ion = charge of the monatomic ion.
  - Oxidation number of S^{2-} is –2
  - Oxidation number of Al^{3+} is +3.
- The oxidation number of Na and K = +1 (unless elemental).
- The oxidation number of Mg and Ca = +2 (unless elemental).
- The oxidation number of H = +1 (unless it is bonded to a metal).
- Oxygen (O) has two possible oxidation numbers:
  - –1 in peroxides, H_2O_2
  - –2 in all other compounds (most common).
- The oxidation number of fluoride (F^-) = –1.
- The sum of the oxidation numbers of all atoms (or ions) in a neutral compound = 0.
- The sum of the oxidation numbers of all atoms in a polyatomic ion = charge on the polyatomic ion.

Example 9.1 Oxidation Number of Each Element in Na_2SO_4

There are three elements present, Na, S, and O. The substance is ionic.

Known oxidation states: Na = +1; O = –2.

S can exist in multiple oxidation states.

The sum of the oxidation numbers in a neutral compound must = 0.

Sodium: 2(+1) = +2

Oxygen: 4(–2) = –8

Sodium + Oxygen = +2 – 8 = –6

To balance, S = +6.

Example 9.2 Oxidation Numbers

Determine the oxidation number of the elements in the compounds listed in Table 9.3.

TABLE 9.3
Oxidation Numbers

Compound	Oxidation Numbers		
H_2CO_3	H = +1	O = –2	C = +4
N_2	N = 0		
$Zn(OH)_4^{2-}$	Zn = +2	H = +1	O = –2
$KMnO_4$	Mn = +7	K = +1	O = –2
K_2CrO_4	Cr = +6	K = +1	O = –2

TABLE 9.4
Sulfur Oxidation Numbers

Compound	Hydrogen	Oxygen	Relation	Sulfur
SO_2		$2(-2) = -4$	$S - 4 = 0$	$S = +4$
H_2S	$2(+1)$		$S + 2 = 0$	$S = -2$
S			Elemental S	0
SO_4^{2-}		$4(-2) = -8$	$-8 + S = -2$	$S = +6$
H_2SO_4	$2(+1)$	$4(-2) = -8$	$2 - 8 + S = 0$	$S = +6$
SO_3^{2-}		$3(-2) = -6$	$-6 + S = -2$	$S = +4$

TABLE 9.5
Phosphorus Oxidation Numbers

Compound	Hydrogen	Sodium	Magnesium	Oxygen	Relation	Phosphorus
Na_3PO_3		$3(+1) = +3$		$3(-2) = -6$	$+3 - 6 + P = 0$	$P = +3$
H_3PO_4	$3(+1) = +3$			$4(-2) = -8$	$+3 - 8 + P = 0$	$P = +5$
HPO_3^{2-}	$1(+1) = +1$			$3(-2) = -6$	$+1 - 6 + P = -2$	$P = +3$
$Mg_2P_2O_7$			$2(+2) = +4$	$7(-2) = -14$	$+4 - 14 + 2P = 0$	$P = +5$
NaH_2PO_4	$2(+1) = +2$	$1(+1) = +1$		$4(-2) = -8$	$+2 + 1 - 8 + P = 0$	$P = +5$

Example 9.3 Oxidation Numbers for Sulfur

Determine the oxidation number for sulfur (S) in the compounds shown in Table 9.4.

Example 9.4 Oxidation Numbers for Phosphorus

Determine the oxidation number for phosphorus (P) in the compounds shown in Table 9.5.

9.3 OXIDATION–REDUCTION REACTIONS

An example of reduction for metal reclamation is found in the copper industry. Diluted leach waters from copper ore extraction contain copper ions (Cu^{2+}) in a dilute solution. The copper can be recovered as a pure metal by passing the leach water over scrap iron. Solid iron is oxidized and goes into solution, and Cu^{2+} is reduced and appears as solid elemental copper. Fine elemental copper is settled, filtered, and recovered as cake for reclamation. The reaction is

$$\underset{\text{oxidized copper}}{Cu^{2+}} + \underset{\text{metal iron}}{Fe} \rightarrow \underset{\text{elemental copper}}{Cu} + \underset{\text{ferrous iron}}{Fe^{2+}}$$

The stoichiometric equation shows the amounts (moles) of each chemical species involved in the reaction. Mass quantities are calculated, as illustrated earlier (the atomic mass of Cu and Cu^{2+} are the same). In addition, the oxidation–reduction reaction equation shows the oxidation state of the chemical species and how many electrons are transferred.

The number of electrons exchanged during a redox reaction must balance. The net charge must be the same on both sides of the equation in order to account for the electrons that are transferred. The number of electrons gained by the molecules containing the oxidizing species (which is reduced) must equal the number of electrons lost by the molecules containing the reducing species (which is oxidized).

TABLE 9.6
Oxidizing and Reducing Agents

Oxidizing Agents		Reducing Agents	
Oxygen (or air)	O_2	Ferrous chloride (Fe^{2+})	$FeCl_2$
Ozone	O_3	Ferrous sulfate (Fe^{2+})	$FeSO_4$
Chlorine	Cl_2	Sulfur dioxide	SO_2
Chlorine dioxide	ClO_2	Hydrogen sulfide	H_2S
Calcium hypochlorite	$Ca(OCl)_2$	Sodium bisulfite	$NaHSO_3$
Sodium hypochlorite	$NaOCl$	Sodium metabisulfite	$Na_2S_2O_5$
Hydrogen peroxide	H_2O_2	Sodium borohydride	$NaBH_4$
Potassium permanganate	$KMnO_4$		
Ferric chloride (Fe^{3+})	$FeCl_3$		

The mnemonic for this is

LEO = Loss of Electrons = Oxidation
GER = Gain of Electrons = Reduction

Table 9.6 lists some common oxidizing and reducing agents. Air, chlorine, and ferric iron (Fe^{3+}) have been the most widely used oxidizing agents, but there is strong interest in advanced oxidation processes that use ozone and hydrogen peroxide. Ferrous iron (Fe^{2+}) and sulfur dioxide (SO_2) have been the most often-used reducing agents.

The oxidizing agents are frequently used to remove sulfur dioxide and hydrogen sulfide from air and water. They are also used to convert ferrous iron, which is soluble in water at ambient conditions, to ferric iron, which will precipitate and can be easily removed by settling and filtration.

Example 9.5 Oxidations and Reductions I

Identify the species being oxidized and reduced for the reactions shown in Table 9.7.

Example 9.6 Oxidations and Reductions II

Indicate whether an oxidizing agent or reducing agent is needed for the reactions shown in Table 9.8 to occur.

TABLE 9.7
Oxidation–Reduction Reactions

Reaction	Oxidized	Reduced
$Cr^{+} + Sn^{4+} \rightarrow Cr^{3+} + Sn^{2+}$	$Cr^{+} \rightarrow Cr^{+3} + 2e^{-}$	$Sn^{4+} + 2e^{-} \rightarrow Sn^{2+}$
$3Hg^{2+} + 2Fe(s) \rightarrow 3Hg + 2Fe^{3+}$	$Fe(s) \rightarrow Fe^{3+} + 3e^{-}$	$Hg^{2+} + 2e^{-} \rightarrow Hg$
$2As(s) + 3Cl_2(g) \rightarrow 2AsCl_3$	$As(s) \rightarrow As^{3+} + 3e^{-}$	$Cl(g) + e^{-} \rightarrow Cl^{-}$
$4Fe + 3O_2 \rightarrow 2Fe_2O_3$	$Fe \rightarrow Fe^{3+} + 3e^{-}$	$O + 2e^{-} \rightarrow O^{2-}$
$3CuO + 2NH_3 \rightarrow 3Cu + N_2 + 3H_2O$	$N^{3-} \rightarrow N + 3e^{-}$	$Cu^{2+} + 2e^{-} \rightarrow Cu$

TABLE 9.8
Oxidation and Reducing Agents

Reaction	Agents Required
$ClO_3^- \rightarrow ClO_2$	Reducing agent
$SO_4^{2-} \rightarrow S^{2-}$	Reducing agent
$Mn^{2+} \rightarrow MnO_2$	Oxidizing agent
$Zn \rightarrow ZnCl_2$	Oxidizing agent

Example 9.7 Balanced Oxidation and Reduction Reaction

In this oxidation–reduction reaction,

$$aH_2S + bHNO_3 \rightarrow cH_2SO_4 + dNO_2 + eH_2O$$

S has changed oxidation state from –2 to +6, losing 8 electrons.
N has changed oxidation state from +5 to +4, gaining 1 electron.
The eight electrons lost by the sulfur are accepted by the nitrogen, giving $a/b = 1/8$.
The balanced reaction is

$$H_2S + 8HNO_3 \rightarrow H_2SO_4 + 8NO_2 + 4H_2O$$

9.4 USEFUL OXIDATION–REDUCTION REACTIONS

Table 9.9 lists some specific redox reactions that are useful in pollution prevention and control. The list is not complete because a reaction that can be accomplished with the reagent shown can probably be accomplished with other reagents. Cyanide, for example, is shown being oxidized by ozone, oxygen, oxygen with a copper catalyst, hydrogen peroxide, and chlorine.

Chlorine is the most widely used and important oxidizing agent. In addition to its importance in water disinfection, it is used for odor control (to destroy hydrogen sulfide) and to destroy toxic chemicals (e.g., cyanide). These are shown in Examples 9.8 and 9.9.

Chlorine as an oxidation agent—Chlorine gas dissolves in water, where it hydrolyzes according to the reaction:

$$Cl_2 + H_2O \leftrightarrows HOCl + H^+ + Cl^-$$

Hypochlorous acid (HOCl) is a weak acid, which dissociates to form hypochlorite:

$$HOCl \leftrightarrows H^+ + OCl^-$$

Sodium and calcium hypochlorite dissolve to release hypochlorite which hydrolyses to hypochlorous acid:

$$NaOCl \rightarrow Na^+ + OCl^-$$

$$Ca(OCl)_2 \rightarrow Ca^{2+} + 2OCl^-$$

$$H^+ + OCl \leftrightarrows HOCl$$

The sum of the OCl^- and HOCl concentrations is called the *free available chlorine*. The distribution of the ionic species in equilibrium is a strong function of pH.

Ozone is a strong oxidizing agent that is very effective as a decolorizing agent and as an oxidant of organic material. It is a gas at normal pressure and temperature. Its solubility in water is a

TABLE 9.9
Some Widely Used Oxidation and Reduction Reactions

Contaminant	Reagent	Reaction
		Oxidations
Cyanide	Ozone	$NaCN + O_3 \rightarrow NaCNO + O_2$
Cyanide	Oxygen with Cu catalyst	$2CN^- + O_2 \rightarrow 2CNO^-$
Cyanate	Oxygen with Cu catalyst	$CNO^- + 2H_2O + 2H^+ \rightarrow CO_2 + NH_4^+ + H_2O$
Cyanide	Hydrogen peroxide	$NaCN + H_2O_2 \rightarrow NaCNO + H_2O$
Cyanide	Chlorine	$NaCN + Cl_2 \rightarrow CNCl + NaCl$
		$CNCl + 2NaOH \rightarrow NaCNO + NaCl + H_2O$
Cyanate	Chlorine	$2NaCNO + 3Cl_2 + 4NaOH \rightarrow N_2 + 2CO_2 + 6NaCl + 2H_2O$
Ferrous iron	Chlorine	$2Fe^{2+} + HOCl + 5H_2O \rightarrow 2Fe(OH)_3 + Cl^- + 5H^+$
Sulfide	Hydrogen peroxide	$H_2O_2 + H_2S \rightarrow 2H_2O + S$
		$4H_2O_2 + S^{2-} \rightarrow SO_4^{2-} + 4H_2O$
Sulfide	Chlorine	$H_2S + Cl_2 \rightarrow S + 2HCl$
Sulfide	Potassium permanganate	$4KMnO_4 + 3H_2S \rightarrow 2K_2SO_4 + S + 3MnO + MnO_2 + 3H_2O$
Organics	Ozone/UV	$CH_3CHO + O_3 \rightarrow CH_3COOH + O_2$
Organics	Peroxide/UV	$CH_2Cl_2 + 2H_2O_2 \rightarrow CO_2 + 2H_2O + 2HCl$
		Reductions
Chromium (+6)	Sulfur dioxide	$3SO_2 + 3H_2O \rightarrow 3H_2SO_3$
		$2CrO_3 + 3H_2SO_3 \rightarrow Cr_2(SO_4)_3 + 3H_2O$
Chromium (+6)	Ferrous sulfate	$2CrO_3 + 6FeSO_4 + 6H_2SO_4 \rightarrow Cr_2(SO_4)_3 + 3Fe_2(SO_4)_3 + 6H_2O$
Copper	Sodium borohydride	$NaBH_4 + 8Cu^+ + 2H_2O \rightarrow 8Cu + NaBO_2 + 8H^+$

Note: The reagent is the oxidizing agent or reducing agent.

function of its partial pressure and temperature. It is generated by high-voltage discharge in air or oxygen. It is unstable and tends to revert to oxygen:

$$O_3 + 2H^+ + 2e^- \rightarrow O_2 + H_2O$$

Hydrogen peroxide (H_2O_2) is a strong oxidant that is used for the treatment of cyanides and wastewaters containing organic materials. It is used as a 30%–70% solution. In the presence of a catalyst (such as iron) the reaction is

$$H_2O_2 + 2H^+ + 2e^- \rightarrow 2H_2O$$

Wet oxidation is a process in which oxygen dissolved in the wastewater is used as an oxidizing agent at high temperature and pressure. Typical temperatures and pressures are 150°C–250°C and 2000–20,000 kPa (gauge pressure), respectively. The process is extremely effective in oxidizing organic materials, organic sulfur, cyanides, pesticides, and other toxic compounds, with removal efficiencies typically exceeding 99%.

This oxidation–reduction reaction is used to remove iron from water:

$$O_2 + 4Fe^{2+} + 4H^+ \rightarrow 4Fe^{3+} + 2H_2O$$

Oxygen is the oxidizing agent, and the ferrous iron (Fe^{2+}) is oxidized to ferric iron (Fe^{3+}). Ferric iron is easily precipitated as ferric hydroxide, $Fe(OH)_3$, which is removed from the water by settling and filtration. Any other oxidizing agent listed in Table 9.6 could be used to accomplish this oxidation of iron.

Example 9.8 Hydrogen Sulfide Control

Hydrogen sulfide, H_2S, a corrosive, odorous, and toxic compound, can be oxidized to elemental sulfur, S, or to sulfur dioxide, SO_2, with chlorine.

$$H_2S + Cl_2 \rightarrow S + 2HCl$$

$$H_2S + 2H_2O + 3Cl_2 \rightarrow SO_2 + 6HCl$$

Example 9.9 Destruction of Cyanide by Oxidation

Oxidation of cyanide with chlorine is the classic method of destroying this toxic compound. Partial destruction results from oxidizing cyanide to cyanate. This takes about 30 min at pH above 8.5. The reaction is

$$\underset{\text{cyanide}}{CN^-} + \underset{\text{chlorine}}{Cl_2} + H_2O \rightarrow \underset{\text{cyanate}}{CNO^-} + 2HCl$$

Complete destruction requires first converting cyanide to cyanate, and then converting cyanate to carbon dioxide and nitrogen.

$$2CNO^- + 3Cl_2 + 4NaOH \rightarrow N_2 + 2CO_2 + 2Cl^- + 4NaCl + 2H_2O$$

The two reactions will occur in the same reactor in about 40 min at pH 8.5. The overall reaction is

$$2CN^- + 5Cl_2 + 4NaOH \rightarrow N_2 + 2CO_2 + 4HCl + 4NaCl + 2Cl^-$$

Sometimes the two steps are done in separate reactors, as shown in Figure 9.1.

Table 9.9 shows some other reactions for oxidizing cyanide.

Example 9.10 Removal of Iron and Manganese from Wastewater

Soluble ferrous (Fe^{2+}) and manganous (Mn^{2+}) ions are removed by precipitation as $Fe(OH)_3$ and MnO_2, respectively, via oxidation. The reaction rate is a function of pH, alkalinity, and impurities that can act as catalysts. Oxygen (O_2), chlorine (Cl_2), or permanganate (MnO_4^-) are popular oxidation agents.

Reactions with oxygen (10–20 min)

$$2Fe^{2+} + 1/2O_2 + 5H_2O \rightarrow 2Fe(OH)_3(s) + 4H^+ \quad pH = 7$$

$$Mn^{2+} + 1/2O_2 + H_2O \rightarrow MnO_2(s) + 2H^+ \quad pH = 10$$

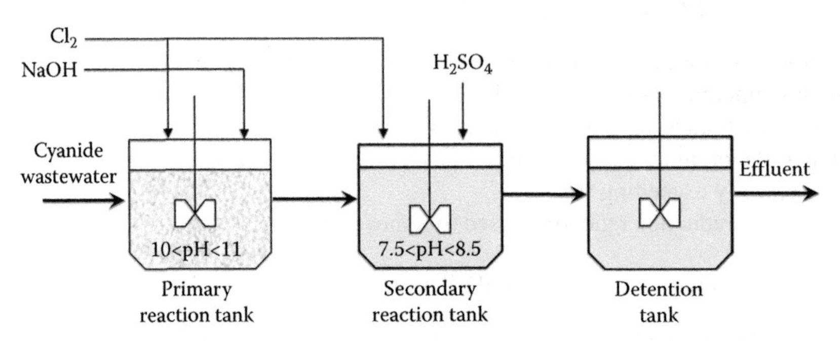

FIGURE 9.1 Two-stage cyanide oxidation process.

Reactions with Cl_2 (fast reaction)

$$Fe^{2+} + 1/2Cl_2 + 3H_2O \rightarrow Fe(OH)_3(s) + Cl^- + 3H^+$$

$$Mn^{2+} + Cl_2 + 2H_2O \rightarrow MnO_2(s) + 2Cl^- + 4H^+$$

Reactions with $KMnO_4$ (pH = 6–9; very fast reaction)

$$3Fe^{2+} + MnO_4^- + 7H_2O \rightarrow 3Fe(OH)_3(s) + MnO_2(s) + 5H^+$$

$$3Mn^{2+} + 2MnO_4^- + 2H_2O \rightarrow 5MnO_2(s) + 4H^+$$

Example 9.11 Reduction of Chrome Toxicity

Dilute rinse water from a metal plating works contains hexavalent chromium in the form of dichromate $Cr_2O_7^{2-}$. The goal is to remove the toxic hexavalent chromium from the rinse water. The solubility products (Chapter 8) show that all forms of Cr^{6+} are soluble, but the following forms of Cr^{3+} are insoluble: CrF_3, $Cr(OH)_3$, Cr_2O_3, Cr_2S_3. Also, CrO and CrS are two insoluble forms of Cr^{2+}.

A solution begins to suggest itself. Reduce the Cr^{6+} to Cr^{3+} and then use precipitation. Intuitively, the reduction from valence 6+ to 3+ will be easier to accomplish than a reduction from 6+ to 2+. This reduction is accomplished by adding a reducing agent such as ferrous iron, sulfur dioxide, or sodium metabisulfite. The oxidations and reductions that occur using ferrous iron are as follows:

$$\text{Oxidation of ferrous iron:} \quad 6Fe^{2+} \rightarrow 6Fe^{3+} + 6e^-$$

$$\text{Reduction of hexavalent chromium:} \quad Cr_2O_7^{2-} + 14H^+ + 6e^- \rightarrow 2Cr^{3+} + 7H_2O$$

$$\text{Overall oxidation-reduction reaction:} \quad Cr_2O_7^{2-} + 14H^+ + 6Fe^{2+} \rightarrow 2Cr^{3+} + 7H_2O + 6Fe^{3+}$$

The $14H^+$ on the left-hand side indicates that acidic conditions facilitate this reaction. At pH 2 the reaction is essentially instantaneous. At pH 3 it reaches completion after about 3 min. The reduction with ferrous iron can occur at higher pH, however. (Also, sodium hydrosulfite can reduce chromium under alkaline conditions).

The reduced chromium can be precipitated with lime (or NaOH):

$$2Cr^{3+} + 3Ca(OH)_2 \rightarrow 2Cr(OH)_3(s) + 3Ca^{2+}$$

Providing an excess of hydroxyl ions shifts the reaction to produce the hydroxide, which favors lower chromium concentrations. The optimal pH is in the comfortable range of pH 7.5–8.5.

Metal plating industries often have cyanide waste and chromium waste. The treatment can be integrated, as shown in Figure 9.2. Cyanide oxidation with chlorine is done best at alkaline conditions. The reduction of Cr^{6+} to Cr^{3+} must be done in acidic conditions. After the chrome has been reduced and the cyanide has been oxidized, the alkaline and the acidic effluents can be mixed. Reagents are added as needed for neutralization. Other nontoxic wastes may be blended in. The chrome precipitate, and any others that may have formed during neutralization, can be coagulated and removed by settling. The advantage of integrated treatment is a savings in neutralization chemicals and a smoothing of variations by blending different waste streams. A disadvantage is the chrome contamination of the sludge.

9.5 FENTON'S CHEMISTRY

Fenton's chemistry, Fenton's reaction, and Fenton's reagent all refer to the same sequence of chemical reactions. Fenton chemistry has been used to attack a variety of toxic organic chemicals, including

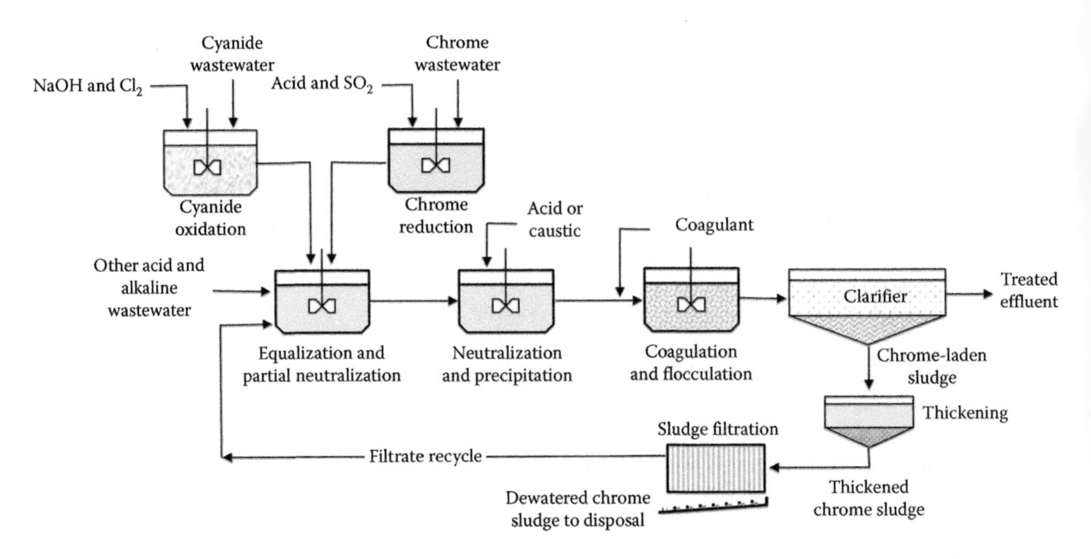

FIGURE 9.2 Integrated treatment of cyanide wastewater and chromium wastewater.

cyanide, phenol, analine dyes, formaldehyde, chlorophenols, dinitrotoluene, and pesticides. It has been used for odor control and to treat specific industrial wastes such as tannery waste, textile waste, and pharmaceutical wastes.

It can be used to completely mineralize many organic compounds and to alter molecules to make them more biodegradable. Biochemical oxygen demand (BOD) and carbon oxygen demand (COD) can be removed by Fenton oxidation.

Fenton's reagent, a solution of hydrogen peroxide and an iron catalyst, is one of the most effective methods for the oxidation of organic pollutants.

Iron(II) sulfate is typically used as the iron catalyst in Fenton's reagent and can be used to destroy organic compounds such as trichloroethylene (TCE) and tetrachloroethylene (PCE). The exact mechanisms of the redox cycle are uncertain, and non-hydroxyl radical ($OH^•$) oxidizing mechanisms of organic compounds have been suggested. Therefore, it may be appropriate to broadly discuss Fenton's chemistry rather than a specific Fenton reaction.

Iron(II) is oxidized by hydrogen peroxide to iron(III), forming a hydroxyl radical ($HO^•$) and a hydroxide ion in the process. Iron(III) is then reduced back to iron(II) by another molecule of hydrogen peroxide, forming a hydroperoxyl radical ($HOO^•$) and a proton. The net effect is hydrogen peroxide creating two different oxygen-radical species, with water ($H^+ + OH^-$) as a by-product.

$$Fe^{2+} + H_2O_2 \rightarrow Fe^{3+} + HO^\bullet + OH^-$$

$$Fe^{3+} + H_2O_2 \rightarrow Fe^{2+} + HOO^\bullet + H^+$$

The free radicals ($HO^•$ and $HOO^•$) engage in secondary reactions. For example, the hydroxyl radical is a powerful, nonselective oxidant. Oxidation of an organic compound by Fenton's reagent is rapid and results in the oxidation of contaminants to primarily carbon dioxide and water.

In spite of many applications of Fenton's chemistry, the specific chemistry is not well understood. We still rely on experiments to determine the best ratio of chemical reagents and to assess the need and benefit of catalysts, which may include UV radiation (photo-Fenton reactions).

Example 9.12 Fenton's Oxidation of Nitrobenzene

A large number of experiments were run by Ghosh et al. (2012) to study the oxidation of nitrobenzene [NB]. (Details are omitted.) The results for three experiments at three nitrobenzene

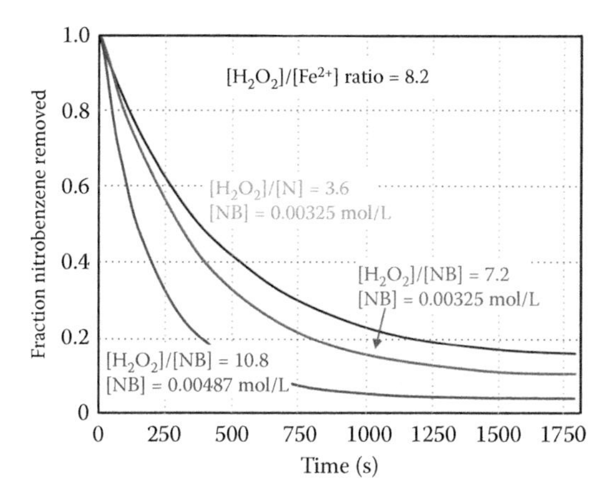

FIGURE 9.3 Oxidation of nitrobenzene with Fenton chemistry. (From Ghosh, P. et al., 2012.)

concentrations are shown in Figure 9.3. The $[H_2O_2]/[Fe^{2+}]$ ratio was 8.2 and the $[H_2O_2]/[NB]$ ratio is varied accordingly. Increasing the $[H_2O_2]/[NB]$ ratio increased the removal rate and reduced the final nitrobenzene concentration. We draw no general conclusions from this one example. Its sole purpose is to introduce Fenton's chemistry and to suggest the nature of the experiments that were done to understand it.

9.6 CASE STUDY: REHABILITATION OF WASTE PICKLE LIQUOR

Iron oxides on the surface of steel parts are removed by an acid washing process called *pickling*. Any strong acid will serve the purpose, but hydrochloric acid (HCl) is commonly used. After a period of use, the acid bath accumulates iron and other impurities and will be replaced with fresh solution. The discarded solution is called *waste pickle liquor*. The pickle liquor could be neutralized and discarded, but it would be preferable to rehabilitate and reuse it. Hydrochloric acid is easier to recover than sulfuric acid, and HCl is used for this example.

The pickle liquor contains ferrous chloride ($FeCl_2$). Under the proper conditions, heat and oxygen will decompose the ferrous chloride, liberate the chloride from solution as hydrogen chloride gas (HCl), and oxidize the iron to form solid ferric oxide (Fe_2O_3). The reaction is

$$2FeCl_2 + 2H_2O + 0.5O_2 \rightarrow Fe_2O_3(s) + 4HCl(g)$$

The hydrogen chloride gas is absorbed into water to form hydrochloric acid. Acid recovery can be 99.5% complete, and the regenerated acid is virtually iron free. Figure 9.4 shows the Lurgi process that exploits this chemistry. The cyclone, scrubber, separator, and absorber are separation processes that support the chemistry.

How do we know that the oxidized form of iron will be Fe_2O_3 and not $Fe(OH)_2$ or $Fe(OH)_3$? The answer lies in the field of chemical thermodynamics. An industrial chemist and a chemical engineer are valuable members of the pollution prevention team.

9.7 CASE STUDY: AN INTEGRATED ACID RECOVERY PROCESS

The pickling process, introduced in Section 9.6, is used to eliminate rust from steel plates. Rust is a mixture of iron(II) and iron (III) oxides, FeO and Fe_2O_3, and represented as Fe_3O_4. The pickling reaction dissolves iron oxide scale from steel:

$$\underset{\text{iron oxide rust}}{Fe_3O_4} + \underset{\text{acid}}{8HCl} + \underset{\text{iron}}{Fe} + (\text{Excess acid}) \rightarrow \underset{\text{ferrous chloride}}{4FeCl_2} + 4H_2O + (\text{Excess acid})$$

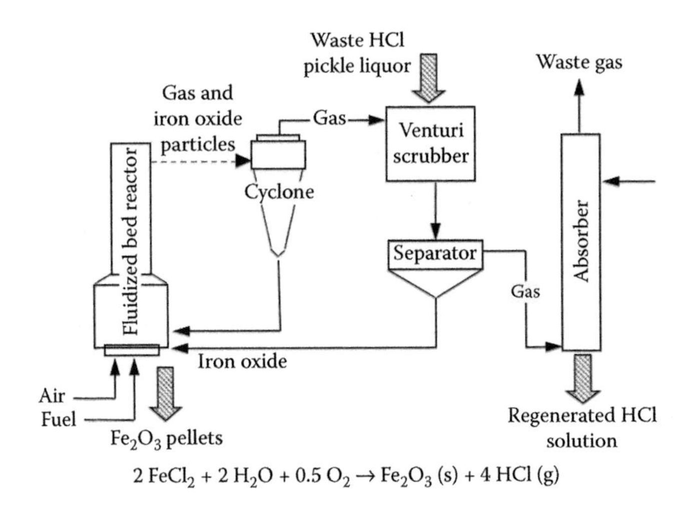

FIGURE 9.4 Lurgi process applied to recover iron oxide and hydrochloric acid from waste pickle liquor.

Hydrochloric acid and iron can be recovered from the waste aqueous solution by integrating the use of neutralization, oxidation, and precipitation. The process development strategy might go something like this.

Process 1: Neutralization

Neutralize the excess acid with lime:

$$2HCl + Ca(OH)_2 \rightarrow CaCl_2 + 2H_2O$$

This solves the acid problem, but it is not a good total solution because a simultaneous reaction converts the dissolved iron ($FeCl_2$) to a ferrous hydroxide solid ($Fe(OH)_2$) and a strong calcium chloride solution.

$$FeCl_2 + Ca(OH)_2 \rightarrow Fe(OH)_2 + CaCl_2$$

Ferrous hydroxide sludge is difficult to dewater by physical means, and it may not dry out for years if it is put into a storage lagoon. Also, the acid is destroyed and the iron is lost with the useless sludge. This means that neutralization alone is not attractive. Can it be incorporated into a broader scheme that would recover iron or acid, or both?

Process 2: Iron Oxide Recovery

We might oxidize the ferrous iron to form ferric oxide instead of ferrous hydroxide. This can be accomplished in the neutralized calcium chloride solution if the reaction pH and temperature are controlled at proper levels. The oxygen can be supplied by aeration.

Neutralization $$2HCl + Ca(OH)_2 \rightarrow CaCl_2 + 2H_2O$$

$$FeCl_2 + Ca(OH)_2 \rightarrow Fe(OH)_2 + CaCl_2$$

Oxidation $$4Fe(OH)_2 + O_2 \rightarrow 2Fe_2O_3 + 4H_2O$$

This yields iron oxide in a calcium chloride solution. The ferric oxide is a solid that can be recovered by filtration. The acid is destroyed.

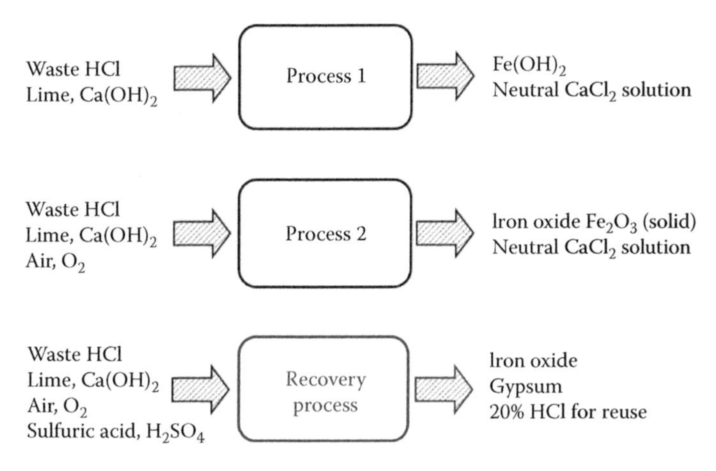

FIGURE 9.5 Summary of process development using waste pickle liquor for an integrated acid recovery system.

Process 3: An Integrated Acid Recovery Process

Process 2 recovered the iron oxide and destroyed the acid. Can the calcium chloride be converted back to hydrochloric acid? This requires adding an acid in a way that will remove the calcium. Adding sulfuric acid will drive the following reaction to produce gypsum and hydrochloric aid:

$$\underset{\text{sulfuric acid}}{H_2SO_4} + CaCl_2 \rightarrow \underset{\text{gypsum}}{CaSO_4} + \underset{\text{acid}}{2HCl}$$

Both reaction products have some value. Calcium sulfate (gypsum) will be insoluble if the concentration of Ca^{2+} or SO_4^{2-} is high. It is counterproductive to buy Ca^{2+} or SO_4^{2-} for this purpose because we can increase the concentrations by evaporating water. This simultaneously concentrates HCl, which is beneficial because we would like to have 20% HCl (by weight) to recycle.

Figure 9.5 shows the process that converts a waste HCl into three usable products (iron oxide, gypsum, and hydrochloric acid) that are easily separated for reuse. Figure 9.6 is the integrated acid recovery system for wastewater flow of 12,000 gallon per day and concentrations of 12% $FeCl_2$ and 1% HCl.

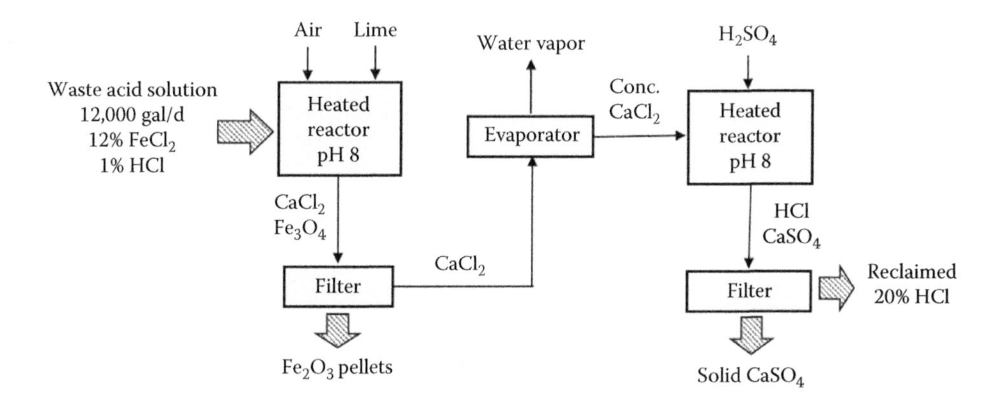

FIGURE 9.6 Integrated recovery of iron oxide, gypsum, and hydrochloric acid from waste pickle liquor.

9.8 CONCLUSION

Oxidation of a chemical (or element) involves the loss of electrons. *Reduction* is the reverse, a gain in electrons. An oxidation always accompanies a reduction. One substance is oxidized and another substance must be reduced. The electrons donated from the oxidized substance are accepted by the reduced substance.

The balanced equation for a redox reaction does not show the electrons that are being transferred. In order to tell whether a redox reaction has occurred or not, we need a way to keep track of electrons. The best way to do so is by assigning oxidation numbers to the atoms or ions involved in a chemical reaction. *Oxidation numbers* are assigned to an individual atom or ion present in a substance using a set of rules.

Many elements and compounds exist in only one oxidation state, at least in the most common pollution control problems. The oxidation number is +1 for H and Na, +2 for Mg and Ca. It is −2 for O and +3 for Al. Iron (Fe) is either +2 or +3. Carbon, nitrogen, chlorine, and sulfur have multiple oxidation states.

Oxidation is used in pollution control to destroy organic chemicals and as a means of converting chemicals so they will precipitate. The oxidation of cyanide and the oxidation of iron and manganese are examples.

An example of reduction is the conversion of hexavalent chromium (Cr^{6+}) to trivalent chromium (Cr^{3+}). Trivalent chromium can be precipitated but hexavalent chromium cannot.

Integrated processes for recovering valuable materials from waste streams may use oxidation–reduction reactions in combination with other reactions that will yield gases or solids to facilitate separation of products and wastes.

10 Green Chemistry

10.1 THE DESIGN PROBLEM

Green chemistry, clean chemistry, clean manufacturing, and pollution prevention embody the same important idea: *Try to do no harm*!

Manufacturing has the potential to damage human health and environmental quality, but current practice demonstrates that it does not want this to happen. Its preference is to eliminate or minimize the generation of hazardous substances. In addition, it wants to protect against whatever cannot be eliminated by minimizing exposure.

Pollution prevention is a relatively broad concept because it applies to every human activity, and not just to chemical manufacturing. Pollution prevention is engineering with the goal of eliminating pollution at the source and reducing risk to human health and to the environment. It is sometimes called *green engineering* or *clean engineering*. Earlier chapters, as well as Berthouex and Brown (2014), include a variety of examples.

The early practice of pollution prevention was directed toward housekeeping solutions such as the following:

- Reducing leaks in piping systems
- Reducing evaporation of volatile chemicals from vats and vessels
- Reducing the loss of material through overspray coating applications
- Water conservation by using high efficiency sprays and counter-current

These practices picked the "low-hanging fruit." They are implemented by making reasonable changes in an existing physical manufacturing system. *Green engineering* deals with many non-chemical issues that are linked to a chemical process.

Green chemistry is about the chemical process more than the physical process. It adds the concepts of *benign by design* and *life cycle sustainability* (Anastas and Zimmerman, 2003; Jiminez-Gonzalez and Constable, 2011; Marteel-Parrish and Abraham, 2014). The goals are to:

- Reduce the use of hazardous materials in favor of materials that are environmentally benign. For example, use water as a solvent rather than a chlorinated organic chemical.
- Minimize the creation of waste materials in order to minimize the need for waste collection, treatment, and disposal.
- Design for efficient separation of the product from the bulk flow of material.
- Design for energy efficiency and energy recovery. Operate at ambient conditions whenever possible.
- Design products that can be easily disassembled, so materials can be recovered and recycled. Minimize the number of different materials in the final product.
- Use renewable materials.

The chemical process industry has special pollution prevention problems, and green chemistry is a wonderful option for that industry.

Ideal chemical reactions would be simple, safe, high yield, and energy efficient, and use renewable and recyclable reagents and raw materials. Not all these goals can be satisfied simultaneously, but chemists and engineers can search for ways to maximize the balance of desirable attributes. The idea is to eliminate steps in the chemical reaction sequence that use toxic chemicals or create pollutants. This is green chemistry.

The previous chapters dealt with using industrial chemistry to convert undesirable chemicals into harmless or even useful products. The focus was on material recovery or waste treatment. This chapter will develop and illustrate the basic ideas of pollution prevention as they specifically apply to chemistry of manufacturing. After a general introduction, three case studies will show how process chemistry can be changed.

The first is the manufacture of soda ash [sodium carbonate (Na_2CO_3)], which was originally made by the highly polluting Le Blanc process. The Le Blanc process was replaced by the less-polluting Solvay process, which has a waste stream consisting mainly of nontoxic calcium chloride salts. Still, the chloride salt load was more than most streams could safely accept, so alternates for the Solvay process have been developed.

The second example is about handling red mud from bauxite processing for aluminum manufacture. This is presented in the form of a plausible (but fictitious) discussion between a process engineer and an industrial chemist as they try to find a less polluting process.

The third example is the evolution of chemistry to make adipic acid, which is used in the manufacture of nylon 66.

Many excellent examples can be found in industrial practice, but they involve organic chemistry that is beyond our scope. Catalysis is another critical part of green chemistry; it is another specialty area that we reluctantly choose to leave for self-study.

10.2 THE PRINCIPLES OF GREEN CHEMISTRY

The chemical industry's earliest and maybe its best opportunity to eliminate pollution is when the chemistry is designed. The following questions should be answered as the chemistry is designed:

What hazardous wastes will be generated?
What toxic substances will need to be handled by workers making the product?
What toxic contaminants might be present in the product?
What regulatory compliance issues are associated with making this product?
What liability concerns are there with the manufacture of this product?
What waste treatment and disposal costs will be incurred?

The reaction stoichiometry given by the laboratory chemist is almost never the stoichiometry of the actual production process. The chemist tells us that A can be reacted with B to form the product P.

$$\text{Ideal reaction:} \quad A+B \rightarrow P$$

What actually happens is likely to be one of the following:

$$\text{Excess raw material used:} \quad (1+n)A+B \rightarrow P+nA$$

$$\text{By-product X is formed:} \quad A+B \rightarrow P+X$$

$$\text{Raw material contains impurity Y:} \quad (A+Y)+B \rightarrow P+Y$$

$$\text{All of the above occur:} \quad (1+n)(A+Y)+B \rightarrow P+X+(1+n)Y+nA$$

All of these are undesirable because it will be necessary to separate the product from the excess raw material, the by-product, or the impurities in the raw material. Additional chemicals and energy are consumed in doing this and it may be difficult and expensive. The separated materials may be voluminous, difficult to handle, and perhaps even toxic.

Possible solutions are to

- Purchase raw materials that do not contain the undesirable impurity.
- Adjust the operating conditions (temperature, pressure, etc.) so excess raw material is not needed.
- Adjust operating conditions to reduce the production of by-product.

These goals may be incompatible. The process conditions that minimize by-product may be those that require an excess of raw material.

Another solution is to find a different chemical route to the product. This might use different raw materials, such as a feedstock that is renewable rather than a petroleum-based chemical.

Benign by design is a challenge to design products and manufacturing routes that use and generate fewer hazardous substances. "This type of prophylactic chemistry is as important to avoiding environmental problems as preventive medicine is to avoiding medical problems" (Anastas and Farris, 1994).

Catalysis is an important part of benign by design. A catalyst creates an alternative pathway for making and breaking chemical bonds. Catalysts make it possible to lower the pressures and temperatures in reactors, and sometimes to achieve reactions that simply were not technically feasible without catalysis. They promote the synthesis of chemicals by new routes that are less polluting than the old ways. This provides the opportunity to make a quantum jump in pollution prevention by entirely replacing polluting plants with cleaner ones.

A majority of chemical manufacturing processes use catalysts, so this is a field rich with green engineering opportunities. Discovering or inventing a more specific or more effective catalyst is a great way to reduce pollution in manufacturing.

> While significant emission reductions can be achieved through the direct application of well known engineering principles to existing process plants, major emission reductions from the chemical industry will be the result of business decisions which involve the closing of heavily polluting obsolescent manufacturing sites. ... the closing of a polluting plant implies the availability of a superior less-polluting technology to take its place in the economy. ... the major challenge ... is the development of radically new chemical technology to replace antiquated processes.
>
> *(Cortright and Crittenden, 1998)*

Table 10.1 lists the 12 principles of *Green Chemistry* that were proposed by Anastas and Warner (1998). Since then, chemical manufacturing, environmental regulations, and consumer expectations have evolved. Green chemistry is an established industrial practice that can bestow environmental and economic benefits.

10.3 TYPES OF CHEMICAL REACTIONS

Let us consider the general types of chemical reactions in terms of how effectively the atoms in the reactants are converted to products.

A *rearrangement reaction* is a broad class of organic reactions where the carbon skeleton of a molecule is rearranged to give a structural isomer of the original molecule. No atoms are added or removed. No intrinsic waste is generated. In the example below, the reactive group, R, simply moves from carbon atom 1 to carbon atom 2.

$$CH_2R\text{-}CH_2\text{-}CH_3 \rightarrow CH_3\text{-}CHR\text{-}CH_3$$

An *addition reaction* adds the elements of the reactant to the elements of the reagent. There is an equimolar quantity of reacting species and all reagents are consumed in the reaction. No additional by-products are formed. An example is the chlorination of ethylene.

TABLE 10.1
Twelve Principles of Green Chemistry

1	Atom economy	Maximize the incorporation of material from the starting reagents into the final product
2	Prevention	It is better to prevent waste than to clean up waste after it is formed
3	Less hazardous material synthesis	Synthetic materials should be designed to maximize the incorporation in the final product of all materials used in the process
4	Design safer chemicals	Wherever practicable, synthetic methodologies should be designed to use and generate substances that possess little or no toxicity to human health and the environment
5	Safer solvents and auxiliaries	The use of auxiliary substances (e.g., solvents, separation agents, etc.) should be made unnecessary wherever possible and innocuous when used
6	Energy efficiency	Energy requirements should be minimized. Synthetic methods should be conducted at ambient temperature and pressure
7	Use renewable feedstocks	A raw material feedstock should be renewable rather than depleting, wherever technically and economically possible
8	Reduce molecular manipulation	Minimize manipulations and rearrangements of molecules, because each manipulation usually requires additional reagents and solvents and generates waste
9	Catalysis	Catalytic reagents (as selective as possible) are superior to stoichiometric reagents
10	Design for degradation	Chemical products should be designed so that at the end of their function they break down into innocuous degradation products and do not persist in the environment
11	Real-time analysis for pollution prevention	Analytical methodologies need to be developed to allow for real-time in-process monitoring and control prior to formation of hazardous substances
12	Safer chemicals for accident prevention	Substances should be chosen so as to minimize the potential for chemical accidents, including releases, explosions, and fires

Source: Anastas, P.T. and Warner, J.C., 1998.

$$\underset{\text{ethylene}}{C_2H_4} + \underset{\text{chlorine}}{Cl_2} \rightarrow \underset{\text{dichloroethane}}{C_2H_4Cl_2}$$

In *substitution reactions,* the substituting species displaces a leaving species. These reactions often need a catalyst. In general

$$AB + CD \rightarrow AC + BD$$

C is substituted for B to create the desired product AC. BD is a by-product or a waste. Containment, capture, destruction, or recycling of the by-product or waste becomes part of the environmental protection plan. An example is making toluene from benzene.

$$\underset{\text{benzene}}{C_6H_5H} + \underset{\text{methyl chloride}}{CH_3Cl} \rightarrow \underset{\text{toluene}}{C_6H_5CH_3} + \underset{\text{hydrochloric acid}}{HCl}$$

Elimination reactions are the reverse of addition reactions. Reagents do not become part of the final product and eliminated atoms are lost as waste. An example is the dehydration of an alcohol to give an olefin (an olefin is a hydrocarbon with one or more carbon double bonds).

$$\underset{\text{ethyl alcohol}}{H_3CCH_2OH} \rightarrow \underset{\text{ethylene}}{H_2C{=}CH_2} + \underset{\text{water}}{H_2O}$$

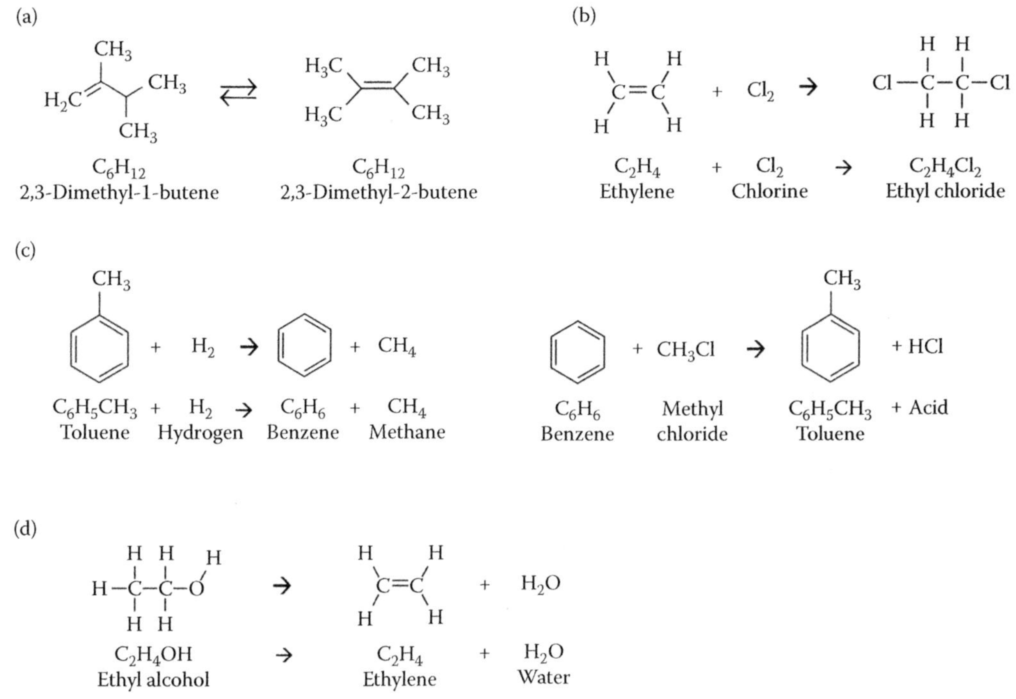

(a)

C_6H_{12} 2,3-Dimethyl-1-butene ⇄ C_6H_{12} 2,3-Dimethyl-2-butene

(b)

C_2H_4 Ethylene + Cl_2 Chlorine → $C_2H_4Cl_2$ Ethyl chloride

(c)

$C_6H_5CH_3$ Toluene + H_2 Hydrogen → C_6H_6 Benzene + CH_4 Methane

C_6H_6 Benzene + CH_3Cl Methyl chloride → $C_6H_5CH_3$ Toluene + HCl Acid

(d)

C_2H_4OH Ethyl alcohol → C_2H_4 Ethylene + H_2O Water

FIGURE 10.1 Examples of (a) rearrangement, (b) addition, (c) substitution, and (d) elimination reactions.

Water is removed from ethyl alcohol to form ethylene. In this case, the by-product is water, which is innocuous. In a more general case, the leaving group (by-product) may need to be separated from the product and managed as a waste.

Figure 10.1 shows examples of rearrangement, addition, substitution, and elimination reactions.

Oxidation–reduction (redox) reactions are used to control the oxidation state of a molecular species. The oxidation of methane to methanol, methanol to formaldehyde, formaldehyde to formic acid, and formic acid to carbon dioxide (and water) is an oxidation sequence. The sequence can be reversed by carrying out reduction reactions. The oxidation or reduction is accomplished chemically or electrochemically by a "courier" molecule that shuttles the accepted or extracted electron from the molecule being oxidized or reduced. Electrochemical reactions use electrical current to directly remove or add electrons to or from the target molecule. Chemical redox agents may be toxic (benign redox agents are not widely available) and their use must be carefully controlled.

Rearrangement and addition reactions are the most "green" because they have no waste products, assuming they run at a high degree of conversion. If the reactants are only partially converted into product there will be waste, some of which can be handled by efficient separation of the product and recycle of unreacted feedstock. Substitution and elimination reactions always have by-products and often these are waste products. Green chemistry has the goal of finding new reaction pathways to eliminate unwanted by-products and to find better catalysts so process yields are high.

Example 10.1 A Rearrangement Reaction

A rearrangement reaction has an atom or bond that moves or migrates from one site in a reactant molecule to a different site in a product molecule. This example is the Beckman rearrangement. There are two rearrangements. The ethyl group ($-CH_2CH_3$) migrates from the carbon atom to the nitrogen atom. The oxygen atom moves from the hydroxyl group ($-OH$) on the nitrogen atom to the carbon atom.

$$CH_3(CH_3CH_2)C{=}N{-}OH \xrightarrow[H_2SO_4]{\text{Catalyst}} CH_3C({=}O){-}NH{-}CH_2CH_3$$

Example 10.2 An Addition Reaction

When bromine is added to an olefin (any hydrocarbon with one or more carbon double bonds), both reactants are fully incorporated into the product molecule. The reaction is 100% atom efficient.

$$Br_2 + \underset{\text{Olefin}}{C_6H_{10}} \rightarrow C_6H_{10}Br_2$$

Example 10.3 A Substitution Reaction

Potassium iodide (KI) is used to remove a methyl group ($-CH_3$) from a carboxylic acid compound [a carboxylic acid group has the structure R–C(=O)–OH]. The resulting methyl iodide (ICH_3) is a waste compound that must be separated from the product carboxylic acid salt.

$$\underset{\text{Carboxylic acid methyl ester}}{R-C(=O)-O-CH_3} + \underset{\text{Potassium iodide}}{KI} \longrightarrow \underset{\text{Carboxylic acid salt}}{R-C(=O)-OK} + \underset{\text{Methyl iodide}}{ICH_3}$$

Example 10.4 An Elimination Reaction

Water is removed from propyl alcohol to form propene (an olefin). The by-product is water, which is innocuous. In a more general case the leaving group may need to be separated from the product and managed as a waste.

$$\underset{\text{Propyl alcohol}}{H_3C\text{-}CH_2\text{-}CH_2\text{-}OH} \rightarrow \underset{\text{Propene}}{H_3C\text{-}CH{=}CH_2} + H_2O$$

10.4 MEASURES OF REACTION EFFICIENCY

The most memorable definition of an ideal process was given by Sir John Cornforth (Ander 2000):

> The ideal chemical process is that which a one-armed operator can perform by pouring the reactants into a bath tub and collecting pure product from the drain hole.

Lacking such an ideal process, we need some measures of success in designing a clean chemical synthesis. The following examples will define and illustrate the *atom economy* (AE), the *process yield*, the *environmental factor yield,* the *environmental factor* (E-Factor), and the *effective mass yield* (EMY). Constable et al. (2002) is a good reference.

AE is the conversion efficiency of a chemical process in terms of all atoms involved in an ideal process. In an ideal process, the mass of starting materials equals the mass of products. The AE is 100% if all the reactants are converted into product.

$$\text{Atom economy} = \frac{\text{Molar mass of desired products}}{\text{Molar mass of all reactants}} \times 100\%$$

For A + B → C or for A + B → C + D, where C is the desired product

$$\text{Atom economy} = \frac{\text{MM of C}}{\text{MM of A} + \text{MM of B}} \times 100\%$$

For a multistage reaction where G is the desired product

$$A + B \rightarrow C \quad \text{then} \quad C + D \rightarrow E \quad \text{followed by} \quad E + F \rightarrow G$$

$$\text{Atom economy} = \frac{\text{MM of G}}{\text{MM (A + B + D + F)}} \times 100\%$$

The AE of rearrangement and addition reactions, as shown in Examples 10.1 and 10.2, is 100%.

Example 10.5 AE for Two Routes of Maleic Anhydride Manufacture

Maleic anhydride is produced in large amounts for use in coatings and polymers. The traditional chemical route was from benzene. Today most maleic anhydride is produced from butene. Both reactions have innocuous by-products (water and carbon dioxide) but they have different atom economies. The stoichiometric reactions are (Figure 10.2)

$$\text{Molar masses: Benzene(Bz)} = 78, \text{Butene(Bu)} = 56, O_2 = 32, \text{Maleic anhydride} = 98$$

Benzene route: $$\text{Atom economy} = \frac{98 \text{ kg } C_4H_2O_3}{78 \text{ kg Bz} + 144 \text{ kg } O_2} \times 100\% = \frac{98 \text{ kg}}{222 \text{ kg}} \times 100\% = 44.1\%$$

Butene route: $$\text{Atom economy} = \frac{98 \text{ kg } C_4H_2O_3}{56 \text{ kg Bu} + 96 \text{ kg } O_2} \times 100\% = \frac{98 \text{ kg}}{152 \text{ kg}} \times 100\% = 64.5\%$$

The butene route has an atom efficiency nearly 1.5 times that of the benzene route. A second attractive feature of this reaction is the elimination of CO_2, a greenhouse gas, as a product.

Yield is the amount of product obtained in a chemical reaction. The yield of a chemical process is different. A process can have 100% reaction yield and still generate more waste than product because of inefficient use of reagents, residues from separation processes, or the formation of undesirable by-products.

This reaction has 100% yield of product C with respect to reactant B, but there is waste because of the need for excess A, impurities in the raw materials, and the formation of by-products and waste.

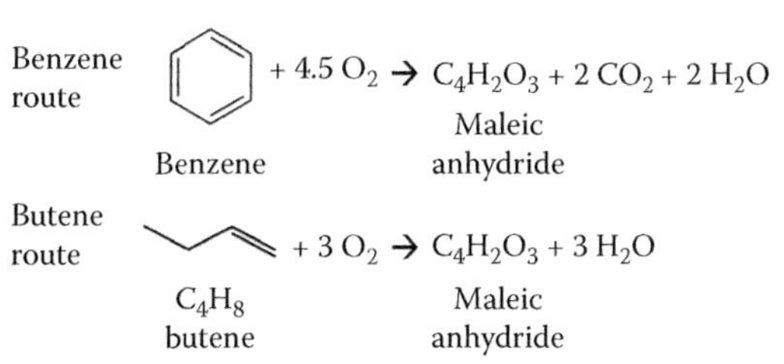

Benzene route: Benzene $+ 4.5\ O_2 \rightarrow C_4H_2O_3 + 2\ CO_2 + 2\ H_2O$ (Maleic anhydride)

Butene route: C_4H_8 butene $+ 3\ O_2 \rightarrow C_4H_2O_3 + 3\ H_2O$ (Maleic anhydride)

FIGURE 10.2 Maleic acid production from benzene and butene.

$$A + B + \text{excess } A + \text{impurities} \rightarrow C + \text{unreacted } A + \text{by-products} + \text{waste}$$

Process yield is based on the theoretical product yield and the actual product yield. The actual can be much less than the theoretical.

$$\text{Process yield} = \frac{\text{Actual quantity of desired products}}{\text{Theoretical quantity of product achievable}} \times 100\%$$

Example 10.6 Yield of Maleic Acid Production

Both reactions in Example 10.5 operate in a real reactor with a catalyst at 400°C. The benzene route has an actual yield of 65 kg of maleic anhydride from 100 kg of benzene consumed in the reaction. The alternate route produces 55 kg of maleic anhydride from 100 kg of butene.

$$\text{Yield from benzene} = (65\text{ kg}/100\text{ kg})100\% = 65\%$$
$$\text{Yield from butene} = (55\text{ kg}/100\text{ kg})100\% = 55\%$$

Example 10.7 Yield on a Molar Basis: Benzene from Toluene

Benzene is to be produced from toluene by the reaction:

$$\underset{\text{Toluene}}{C_6H_5CH_3} + H_2 \rightarrow \underset{\text{Benzene}}{C_6H_6} + CH_4$$

One mole of toluene is required to produce one mole of benzene. Some of the benzene formed enters a secondary reaction and becomes diphenyl (a double ring compound):

$$\underset{\text{Benzene}}{2C_6H_6} \rightarrow \underset{\text{Diphenyl}}{C_{12}H_{10}} + H_2$$

The amount, in mol/h, of each component in the reactor feed and output are as follows:

Component	Feed Rate (mol/h)	Output Rate (mol/h)
Hydrogen (H_2)	1858	1583
Methane (CH_4)	804	1083
Benzene (C_6H_6)	11	282
Toluene ($C_6H_5CH_3$)	372	93
Diphenyl ($C_{12}H_{10}$)	0	4

$$\text{Toluene conversion} = \frac{\text{Toluene consumed in the reactor}}{\text{Toluene fed to the reactor}} = \frac{372-93}{372} = 0.75$$

$$\text{Benzene yield from toluene} = \left(\frac{\text{Benzene produced in the reactor}}{\text{Toluene consumed in the reactor}}\right) = \left(\frac{282-11}{372-93}\right) = 0.97$$

$$\text{Reactor yield from toluene} = \left(\frac{\text{Benzene produced in the reactor}}{\text{Toluene fed to the reactor}}\right) = \left(\frac{282-11}{372}\right) = 0.73$$

The *environmental factor*, also known as the *E-Factor*. E = everything but the desired product. This is one of the earliest and most useful green chemistry metrics. The definition is

$$\text{E-Factor} = \frac{\text{Total mass of waste}}{\text{Mass of product}}$$

A high value indicates a relatively inefficient process.

The *effective mass yield*, EMY, incorporates information about the mass of waste produced and the mass of environmentally undesirable material put into the process. The definition is the percentage of desired product mass relative to the mass of all non-benign materials used in the product synthesis.

$$\text{EMY} = 100 \times \frac{\text{Mass of product}}{\text{Mass of non-benign raw material used}}$$

One problem with the EMY is that there is no general agreement on what constitutes an environmentally benign material. Some of the most troublesome wastes to treat are large volumes of water that contain trace amounts of toxic chemicals. The water may not be benign, but to include it will make the EMY calculation useless, so water and inorganic salts are ignored. Therefore, the EMY will not be able to discriminate alternate processes on the basis of the substances that make the water polluted.

Example 10.8 E-Factor and EMY

Butyl acetate ($C_6H_{12}O_2$) is formed by mixing 37 g butanol (C_4H_9OH) with 60 g glacial acetic acid ($C_2H_4O_2$) and a small amount of sulfuric acid (H_2SO_4) catalyst. Following the reaction, the mixture is added to 250 g water. The crude butyl acetate is further washed with water (100 g), and then with 25 g saturated sodium bicarbonate ($NaHCO_3$) solution, and with 25 g water. Finally, it is dried over 5 g of anhydrous sodium sulfate (Na_2SO_4). The crude ester is distilled to give 40 g of product—butyl acetate ($C_6H_{12}O_2$). This is a yield of 69%. Calculate the E-Factor and EMY for this process.

To summarize the inputs,

$$37\text{ g butanol} + 60\text{ g acetic acid} + 25\text{ g } NaHCO_3 + 375\text{ g } H_2O + 5\text{ g } Na_2SO_4 = 502\text{ g raw materials}$$

The mass of acid catalyst is ignored.

E-Factor:

An amount of 40 g of product has been made from a total input of 502 g.
The waste produced is 502 g input - 40 g product = 462 g waste.
The E-Factor is

$$E = (462\text{ g waste})/(40\text{ g product}) = 11.6$$

This high value indicates a relatively inefficient process.

EMY:

The EMY is based on the nonbenign materials used. The benign materials (water, inorganic salts, and acetic acid) are ignored. The nonbenign material is butanol.

$$\text{EMY} = 100\left(\frac{\text{Mass of product}}{\text{Mass of non-benign raw material used}}\right)$$

$$\text{EMY} = 100\left(\frac{40\text{ g product}}{37\text{ g butanol}}\right) = 108\%$$

The *economic potential* (*EP*) compares the value of the products and the cost of the raw materials. It is not an estimate of revenue or profit because the raw material purchase price and process operation cost are not included. It is a useful number because if the material input costs more than the selling price of the products, the reaction sequence must be abandoned or made more efficient. Usually the selling price is fixed (within limits), so the *EP* must be increased by changing to less costly raw materials, or by increasing the conversion into product.

$$\text{EP} = (\text{Value of products}) - (\text{Cost of raw materials})$$

Example 10.9 EP of Vinyl Chloride from Acetylene

One synthesis route to vinyl chloride (C_2H_3Cl) is via chlorination of acetylene (C_2H_2):

$$\underset{\text{Acetylene}}{C_2H_2} + \underset{\text{Acid}}{HCl} \rightarrow \underset{\text{Vinylchloride}}{C_2H_3Cl}$$

Molar mass (kg) 26 36 62

Calculate the EP of this reaction pathway if acetylene costs \$1.00/kg, hydrogen chloride costs \$0.35/kg, and vinyl chloride sells for \$0.45/kg.

The ideal stoichiometry says that the materials combine as follows:

26 kg acetylene + 36 kg hydrogen chloride yields 62 kg of vinyl chloride.

$$\text{EP} = (\$0.45/\text{kg})(62\text{ kg}) - (\$1.00/\text{kg})(26\text{ kg}) - (\$0.35/\text{kg})(36\text{ kg})$$

$$= -\$10.70 / \text{mol vinyl chloride product}$$

The cost of the reactants to produce vinyl chloride is greater than the price for which it can be sold, thus making vinyl chloride via this synthesis route economically unattractive.

10.5 SOLVENTS

Solvents are widely used and widely problematic. Many are listed air pollutants and regulated water pollutants. Many are toxic and carcinogenic. Some examples are described in Table 10.2.

There are three goals for solvent control:

- Reduce solvent use.
- Substitute more benign solvents for toxic ones.
- Recover and reuse solvents.

Industry has had great success with all three practices.

Some solvents are so undesirable they have been abandoned or are being replaced as quickly as possible. Table 10.3 lists these and suggests suitable replacements. Those in the *usable* column can

TABLE 10.2
Examples of Toxic Solvents

Solvent	Disadvantage of Solvent	Substitute	Characteristics of Substitute
Benzene	Toxic, causing blood disorders, suspected of causing leukemia, metabolized to toxic phenol	Toluene	Much less toxic than benzene because of the presence of a metabolically oxidizable methyl substituent group; produces hippuric acid metabolite
n-Hexane	Neurotoxic causing peripheral neuropathy manifested by mobility loss, reduced sensations in extremities	2,5-Dimethylhexane	Lacks toxicity characteristics of n-hexane; significantly higher boiling point may be a disadvantage
Glycol ethers	Ethylene glycol monomethyl ether and ethylene glycol monoethyl ether have adverse reproductive and developmental effects in animals	1-methoxy-2-propanol	Less toxic than glycol ethers but still effective as a solvent
Various organic solvents	Flammability, toxicity, poor biodegradability, tendency to contribute to photochemical smog	Supercritical fluid carbon dioxide	Widely available, good solvent for organic solutes, readily removed by evaporation, nonpolluting, except as a greenhouse gas if allowed to escape

TABLE 10.3
Solvent Selection Table

Preferred	Usable	Undesirable
Water	Cyclohexane	Pentane
Acetone	Heptane	Hexane(s)
Ethanol	Toluene	Di-isopropyl ether
2-Propanol	Methylcyclohexane	Diethyl ether
1-Propanol	Tertiarybutylmethylether (TBME)	Dichloromethane
Ethyl acetate	Iso-octane	Dichloroethane
Isopropyl acetate	Acetonitrile	Chloroform
Methanol	2-Methyl tetrahydrofuran	*N*-Methyl-2-pyrrolidone (NMP)
Methylethylketone (MEK)	Tetrahydrofuran (THF)	Dimethylformamide (DMF)
1-Butanol	Xylenes	Pyridine
t-Butanol	Dimethyl sulfoxide (DMSO)	Dimethylacetamide (DMAc)
	Acetic acid	Dioxane
	Ethylene glycol	Dimethoxyethane
		Benzene
		Carbon tetrachloride

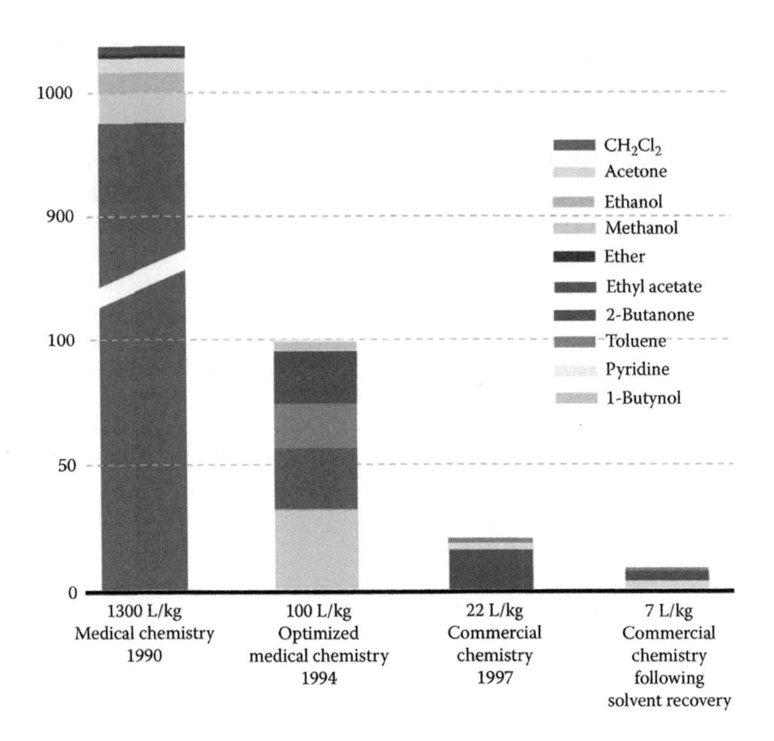

FIGURE 10.3 Reduction in the use of solvents in the development and manufacture of Sildenafil. (Adapted from Dunn, P.J., et al. 2004.)

be managed. The preferred solvents are *green solvents*. Green solvents have low toxicity, are easy to recycle (no disposal), are easy to remove from the product, and have low reactivity.

The pharmaceutical industry is a heavy user of solvents. The active pharmaceutical ingredient may be only a small percentage of the material coming from the process. The product must be recovered from this mass of material and purified. Solvents are essential for this. Because of this, solvent management is a great opportunity for pollution prevention.

A spectacular pollution prevention example is Pfizer's manufacture of Sildenafil, as shown Figure 10.3. Solvent use in the laboratory (the medical phase) may appear to be extravagant but this is acceptable because solvent cost is unimportant and solvent vapors are safely controlled. Once the chemical reaction sequence has been worked out, the focus shifts to developing *green chemistry* for full-scale manufacturing. Solvents that are acceptable in the laboratory may be too expensive for larger-scale production, or pollution control problems may dictate a change. For a variety of such reasons, solvent use is reduced and process improvement becomes an ongoing part of the manufacturing process. Figure 10.3 shows two simultaneous changes: Different solvents were used, and the quantity used was reduced. The commercial chemistry in 1997 used 22 L/kg of product. Solvent recovery was implemented, and the total solvent use decreased to 7 L/kg. The industry has a goal of reducing this to 4 L/kg.

10.6 CATALYSIS

About 90% of chemical production involves a catalyst. A catalyst will accelerate a reaction or activate it with less energy. It does not alter the equilibrium position or change the stoichiometry. It does change the conditions (temperature and pressure) under which a reaction will occur. As an example, ammonia can be made from nitrogen and hydrogen ($N_2 + 3H_2 \rightarrow 2NH_3$) using a certain catalyst at 1000°C. An improved catalyst makes it possible to run the reaction at 500°C. This is a considerable reduction in cost and CO_2 emissions.

The catalyst itself is not changed or consumed by the reaction. It can be reused while the reaction products move downstream for further processing or purification.

Example 10.10 Catalytic Hydrogenation of Lactic Acid to 1, 2 Propanediol

The manufacture of chemicals from renewable resources, such as corn, is a popular endeavor. Corn can be converted to corn starch and subsequently to glucose. Glucose has many possible uses. One is to make lactic acid, which is a widely used feedstock in chemical manufacturing. These steps typically would be done in different facilities.

$$\text{Corn} \rightarrow \text{Corn starch} \rightarrow \text{Glucose} \rightarrow \text{Lactic acid}$$

Lactic acid can be used to make 1,2 propanediol (propylene glycol) or other polymers, and esters. One use of 1,2 propanediol is to make fibers for carpeting. Figure 10.4a shows one possible scheme. The difficulty with this scheme is that lactic acid can react to form propionic acid as well as propylene glycol. Forming propylene glycol requires a hydrogenation (adding two hydrogen atoms). The predominant reaction, however, is to add only one hydrogen, forming propionic acid. Without the proper catalysts, the conversion to propionic acid is very fast and the conversion to propylene glycol is very slow. The result is a poor yield of the desired product.

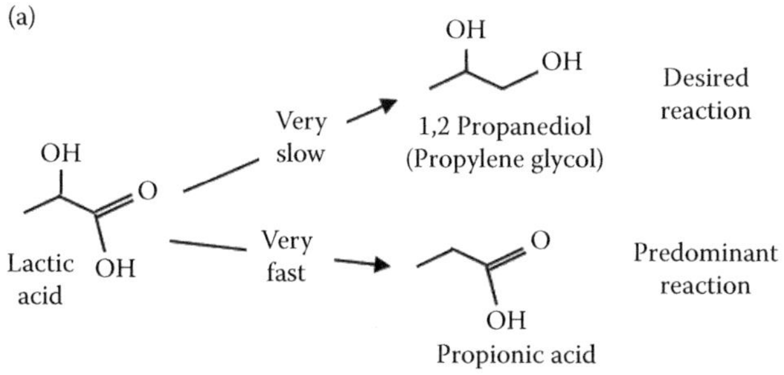

Inefficient reaction has a high yield of propionic acid and a low yield of propylene glycol.

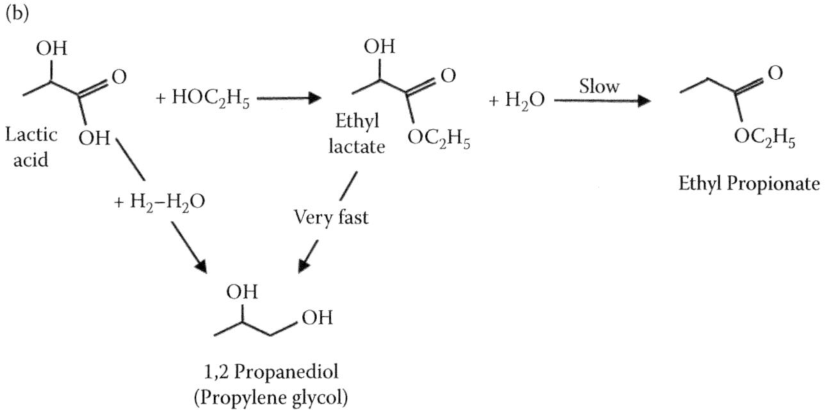

Reaction sequence with a copper chromite catalysts (200°C and 1 atm) has a high yield of propylene glycol and a low yield of ethyl propionate

FIGURE 10.4 The use of a suitable catalyst can replace the inefficient production of propylene glycol (a) with the very efficient production route shown in (b). (From Cameron, D.C. et al., 1998.)

Figure 10.4b shows how the reaction sequence is altered to obtain propylene glycol as the predominant product. It involves a reaction between lactic acid and ethyl acetate, catalyzed with copper chromite at atmospheric pressure and 200°C. The more efficiently catalyzed reaction forms propylene glycol by two routes, first directly by hydrogenation and second by the formation and subsequent transformation of ethyl lactate. A small quantity of the ethyl lactate is converted to ethyl propionate, but the predominant product is propylene glycol.

Catalysis is not trouble free. The catalysts are not consumed in the reaction, but fouling with organic or inorganic materials may require replacement or regeneration. Emissions or effluents are generated with catalyst activation or regeneration. Emissions and effluents associated with catalyst handling and regeneration can be reduced by extending the catalyst life. In situ regeneration eliminates unloading/loading emissions and effluents compared to offsite regeneration or disposal. Catalysts composed of noble metals, because of their cost, are generally recycled. Catalyst attrition and carryover into product requires de-ashing facilities that are a likely source of wastewater and solid waste.

10.7 CASE STUDY: SODA ASH PRODUCTION

Soda ash is the common name for sodium carbonate, which is used in the manufacture of glass, paper, soap, detergents, aluminum, and other consumer products. In 2010, the U.S. production of soda ash was over 9 million T/year, or roughly 24,500 T/day (54,000 U.S. tons/day). Global production is and over 40 million T/year.

About one-third of the U.S. production come from mining natural deposits in Wyoming and California, and the remaining two-thirds is manufactured synthetically from limestone and salt by the Solvay process, which replaced the LeBlanc process.

The LeBlanc process was a terrible polluter. Unfortunately, the Solvay process (see Figure 10.5) is also a polluter of significant magnitude, particularly by the release of wastewater containing large amounts of calcium chloride, sodium chloride, and other salts.

To realize the extent of calcium chloride waste disposal problems caused by using the Solvay process, we need only examine the net reaction, which indicates that for each molecule of sodium carbonate manufactured, one molecule of waste calcium chloride is formed. The net chemical reaction for the Solvay process is

$$2NaCl + CaCO_3 \rightarrow Na_2CO_3 + CaCl_2$$

To put this into perspective, a soda ash factory (closed in 1972) produced 2,000,000 lb/day (1000 tons/day) of soda ash per day and discharged 950 tons/day of calcium chloride and 575 ton/day of excess sodium chloride (1.5 tons of dissolved salts for each ton of product).

Table 10.4 gives a typical wastewater analysis (note the concentrations are in g/L, not our usual mg/L). These wastes are not amendable to treatment except at great cost, and no treatment was provided. The river downstream from the factory's waste discharge had nearly 5000 mg/L chloride concentration, while the upstream concentration rarely exceeded 230 mg/L. Hardness increased from 130 to 1500 mg/L due to calcium and magnesium in the waste discharges. Pressure to abate the pollution was one factor that finally led to the plant closing down. How important this was relative to other factors we do not know, but we are certain that such an operation would not be permitted under today's pollution control laws.

What alternative is there to closing the plant or treating the salty wastewater?

A superficial understanding of the process is not sufficient to develop a solution to the problem. The overall reaction in the Solvay process is accomplished by using the six separate chemical reactions described below (Berthouex and Rudd, 1977).

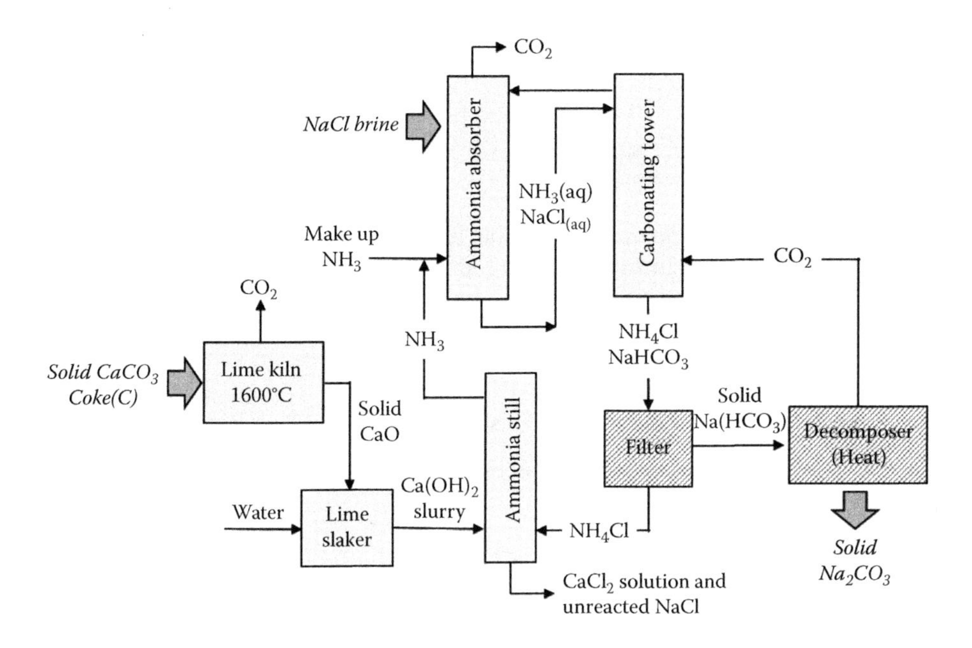

FIGURE 10.5 Solvay soda ash (Na_2CO_3) process.

TABLE 10.4
Typical Solvay Soda Ash Wastewater Analysis

Soluble	(g/L)	Relatively Insoluble	(g/L)
$CaCl_2$	85–95	$CaCO_3$	6–15
NaCl	45–50	$CaSO_4$	3–5
CaO	2–4	$Mg(OH)_2$	3–10
NH_3	0.009	Fe_2O_3 and Al_2O_3	1–3
SiO_2	1–4		

Reaction 1—Limestone calcination. At 1600°C limestone decomposes into calcium oxide (quicklime) and carbon dioxide.

$$\underset{\text{limestone}}{CaCO_3} \rightarrow \underset{\text{quicklime}}{CaO} + CO_2$$

Reaction 2—Lime slaking. Solid lime is slaked with water to produce a slurry of hydrated lime.

$$\underset{\text{quicklime}}{CaO} + H_2O \rightarrow \underset{\text{hydrated lime}}{Ca(OH)_2}$$

Reaction 3—Ammonia production. Adding the lime slurry to an ammonium chloride solution releases ammonia and generates the waste calcium chloride.

$$\underset{\text{lime}}{Ca(OH)_2} + \underset{\text{ammonium chloride}}{2NH_4Cl} \rightarrow \underset{\text{ammonia}}{2NH_3} + \underset{\text{calcium chloride}}{CaCl_2} + 2H_2O$$

Reaction 4—Ammonia absorption. Ammonia gas is absorbed into water to produce an ammonium hydroxide solution.

$$\underset{\text{ammonia}}{NH_3} + H_2O \rightarrow \underset{\text{ammonium hydroxide}}{NH_4OH}$$

Reaction 5—Bicarbonate production. In solution, ammonium hydroxide, carbon dioxide, and salt are reacted to produce ammonium chloride and sodium bicarbonate.

$$\underset{\text{ammonium hydroxide}}{NH_4OH} + \underset{\text{carbon dioxide}}{CO_2} + \underset{\text{sodium chloride}}{NaCl} \rightarrow \underset{\text{ammonium chloride}}{NH_4Cl} + \underset{\text{sodium bicarbonate}}{NaHCO_3}$$

Excess sodium chloride is used in this step. This appears in the waste stream.

Reaction 6—Bicarbonate decomposition. Calcination of the bicarbonate produces the sodium carbonate product and carbon dioxide.

$$\underset{\text{sodium bicarbonate}}{2NaHCO_3} \rightarrow \underset{\text{sodium carbonate}}{Na_2CO_3} + \underset{\text{carbon dioxide}}{CO_2} + \underset{\text{water}}{H_2O}$$

The direct reaction between salt (NaCl) and limestone ($CaCO_3$) does not occur in this six-step sequence.

The objective of any process alteration is to maintain the production of soda ash while converting the reaction products into forms that are more easily disposed of than the calcium chloride (which is formed in reaction 3). The obvious step of evaporating the wastewater to obtain dry calcium chloride and salt is impractical because the calcium chloride recovered from only one soda ash factory would satisfy the nation's annual demand for this chemical.

Now consider alternate chemistry that will be cleaner. Ammonium chloride decomposes when heated to form ammonia and hydrogen chloride. Both gases can be separated to sufficiently pure form for use.

New Reaction—Decomposition of ammonium chloride

$$\underset{\text{ammonium chloride}}{NH_4Cl} \rightarrow \underset{\text{ammonia}}{NH_3} + \underset{\text{hydrogen chloride}}{HCl}$$

Can this reaction be used to modify the Solvay process so that chlorine atoms are released as hydrogen chloride for sale rather than as calcium chloride for waste?

Reaction 3 in the Solvay process is the one that generates the ammonia required for reaction 4. It also generates the waste calcium chloride. Reaction 2 is used to generate the hydrated lime needed by reaction 3. The net effect of using the new reaction is to replace reactions 2 and 3 and, in doing so, to create a process effluent of HCl and to eliminate the need for the CaO generated by reaction 1. The reaction sequence now becomes:

Reaction 1—Limestone calcination (for CO_2 in Reaction 5)

$$CaCO_3 \rightarrow CaO + CO_2$$

New *Reactions 2 and 3—Ammonium chloride decomposition*

$$NH_4Cl \rightarrow NH_3 + HCl$$

Reaction 4—Ammonia absorption

$$NH_3 + H_2O \rightarrow NH_4OH$$

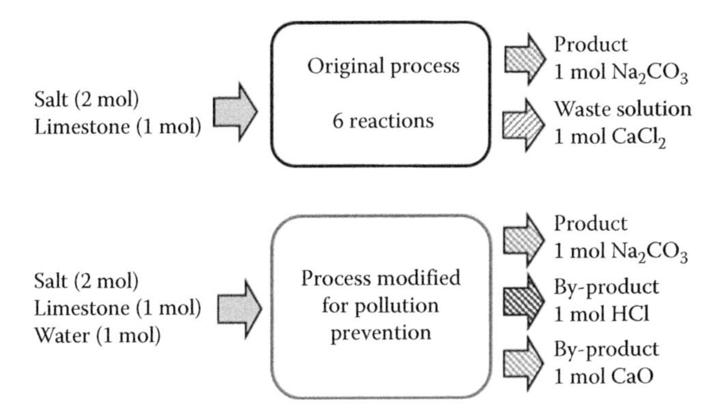

FIGURE 10.6 Comparison of the original Solvay process with a less-polluting process.

Reaction 5—Bicarbonate production

$$NH_4OH + CO_2 + NaCl \rightarrow NH_4Cl + NaHCO_3$$

Reaction 6—Bicarbonate decomposition

$$2NaHCO_3 \rightarrow Na_2CO_3 + CO_2 + H_2O$$

New net reaction

$$2NaCl + CaCO_3 + H_2O \rightarrow Na_2CO_3 + 2HCl + CaO$$

Fortunately, the CaO is formed directly as a dry solid rather than in dilute solution, as was the $CaCl_2$ in the original process, and the HCl is a marketable gas.

Figure 10.6 compares the process as modified for pollution prevention with the original process. The theoretical stoichiometric amounts of the raw materials and products are shown. Excess salt is needed in the actual process.

Example 10.11 AE of the Solvay Process

The overall reaction of the Solvay process is

$$2NaCl + CaCO_3 \rightarrow Na_2CO_3 + CaCl_2$$

	$2NaCl$	$CaCO_3$	Na_2CO_3	$CaCl_2$
Reacting mass (kg)	117	100	106	111

The desired product is Na_2CO_3. The $CaCl_2$ is useless waste.

$$\text{Atom economy} = \frac{\text{Molar mass of desired products}}{\text{Molar mass of all reactants}} \times 100\%$$

$$= \frac{106 \text{ kg } Na_2CO_3}{117 \text{ kg NaCl} + 100 \text{ kg } CaCO_3} \times 100\% = \frac{106}{217} \times 100\% = 48.8\%$$

The outputs of the revised soda ash process are Na_2CO_3, the desired product, plus HCl and CaO, both useful by-products.

The AE for Na_2CO_3 using the revised soda ash process is

$$2NaCl + CaCO_3 + H_2O \rightarrow Na_2CO_3 + 2HCl + CaO$$

Masses (kg) 117 100 18 106 73 56

$$\text{Atom economy} = \frac{106 \text{ kg } Na_2CO_3}{117 \text{ kg NaCl} + 100 \text{ kg } CaCO_3 + 18 \text{ kg} H_2O} \times 100\% = \frac{106}{235} \times 100\% = 45.1\%$$

While this is lower than for the Solvay process, the by-products of the reaction (HCl and CaO) are useful materials, not wastes. So, the effective AE for the revised process is 100%.

10.8 CASE STUDY: RED MUD

This hypothetical case study illustrates how an engineer can use inductive reasoning to develop an initial plan of attack on a pollution problem by exploiting property differences to separate materials. The presentation is in the form of a scenario in which an engineer and an industrial chemist gradually develop an understanding of the problem and its solution. The solution is to change the physical properties of the materials by chemical reactions to enable simple and direct separations (Berthouex and Rudd, 1977).

K. J. Bayer in 1888 proposed a process for leaching alumina (Al_2O_3) from crude bauxite using sodium hydroxide solution. Crude bauxite is about 50% alumina. Red mud is a waste that is produced in large amounts when processing bauxite to manufacture alumina, from which aluminum metal is manufactured. In 2012 over 90 million metric tons of 99.5% pure Al_2O_3 were produced by the Bayer leaching process shown in Figure 10.7. Red-mud wastes from the thickening and filtration operations are particularly troublesome, and must be contained to prevent local environmental damage. Let us see what might be done with these wastes.

Engineer: The chemical analysis of the dried red mud from the Bayer process (Table 10.5) shows it to be rich in iron and aluminum. Perhaps the red mud can be converted into a secondary raw material source. We generate enormous amounts of red mud. Can you suggest any useful chemical transformations?

Industrial chemist: On the surface, the problem does not seem to involve chemistry. Why not merely separate the alumina from the red mud for recycle to the Bayer process?

Engineer: That's the problem. The aluminum and iron oxides are difficult to separate. We can't recycle the red mud to the Bayer process because it contains too much iron oxide, and the red mud can't be used as iron ore because it contains too much alumina.

Are there chemical differences in Fe_2O_3 and Al_2O_3 that can be exploited? Could we vaporize one of the oxides while the others remain solid, or make one of them water soluble while the others remain insoluble, or make one of them agglomerate into nodules while the others remain a fine powder? If they are to be separated, they must differ in some way.

Industrial chemist: I recall that the manufacture of pure titanium dioxide (TiO_2) uses rutile ore in a chlorination process. The titanium dioxide in the ore is chlorinated to form $TiCl_4$. The boiling point of $TiCl_4$ is not much greater than that of water, so during the chlorination the $TiCl_4$ comes off as a gas that is later burned to form TiO_2 powder. Perhaps we could use a similar process to separate the iron and aluminum oxides. Table 10.6 shows the physical properties of the oxides and chlorides of the material in red mud.

Engineer: This is excellent. If the red mud is chlorinated, the insoluble solids can be converted into gases or material soluble in water. Now we can begin thinking about leaching and vaporization operations to separate the red mud into its components. How would we go about forming the chloride salts?

Industrial chemist: The chlorination of metal oxides is particularly interesting. The direct chlorination reaction generally is not very satisfactory. However, the reaction of $TiCl_4$ with Fe_2O_3 is rapid and complete.

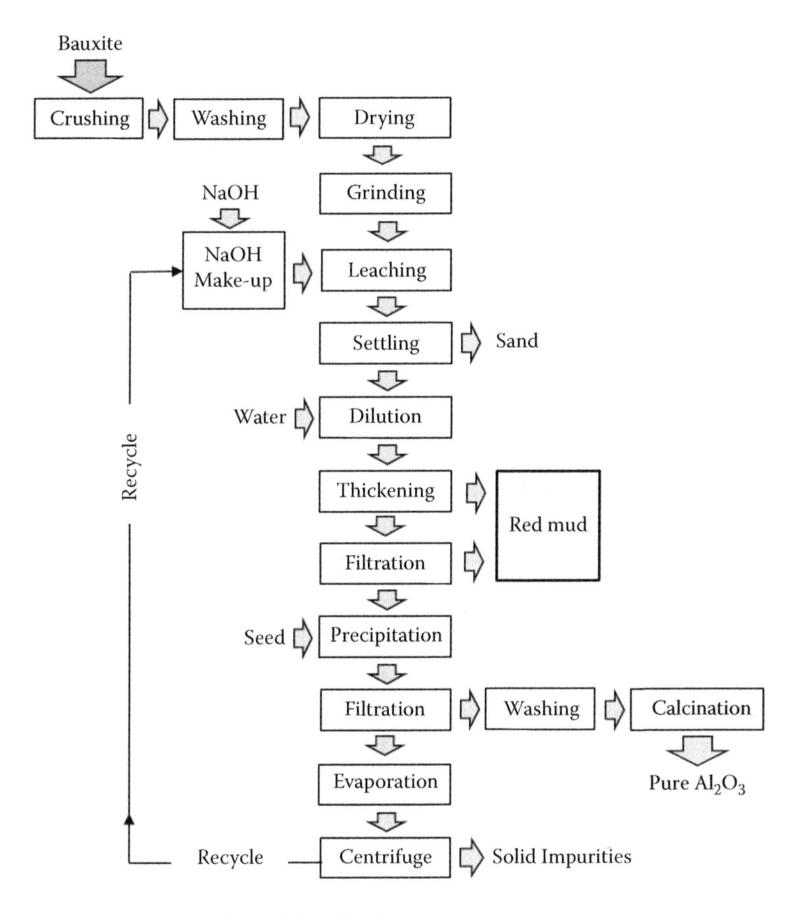

FIGURE 10.7 The Bayer process for making alumina.

TABLE 10.5
Typical Analysis of Red Mud

Species	Al_2O_3	Fe_2O_3	SiO_2	TiO_2	Na_2O	CaO
Weight percent	26	50	9	6	7	2

TABLE 10.6
Properties of the Oxides and Chlorides of Material in Red Mud

Oxide Species	Boiling Point (°C)	Solubility in Water	Chloride Species	Boiling Point (°C)	Solubility in Water
Al_2O_3 (s)	2200	Insoluble	$AlCl_3$ (g)	183	Soluble
Fe_2O_3 (s)	1565	Insoluble	$FeCl_3$ (g)	319	Soluble
SiO_2 (s)	2300	Insoluble	$SiCl_4$ (g)	58	Decomposes
TiO_2 (s)	High	Insoluble	$TiCl_4$ (g)	136	Soluble
Na_2O (s)	High	Insoluble	NaCl (g)	1413	Soluble
CaO (s)	2850	Insoluble	$CaCl_2$ (s)	1600	Soluble

$$3TiCl_4(g) + 2Fe_2O_3(s) \rightarrow 4FeCl_3(g) + 3TiO_2(s)$$

The chloride forming affinities from the oxides are as follows:

Na (highest) Ca Fe Ti Al Si (lowest)

The oxide of any metal on the left side of the list of chloride forming affinities will capture the chloride from any metal farther to the right on the list. We can exchange chlorine and oxygen atoms in this way. To form the chlorides directly, carbon monoxide must be used in addition to chlorine.

$$Fe_2O_3(s) + 3CO(g) + 3Cl_2(g) \rightarrow 2FeCl_3(g) + 3CO_2(g)$$

Engineer: We now have some useful tools. Let's see what suggests itself. A direct chlorination of the oxides would allow the $AlCl_3$, $FeCl_3$, $SiCl_4$, and $TiCl_4$ to vaporize leaving the NaCl and $CaCl_2$. However, this really doesn't help much since most of the material would appear as the chloride gases to be separated.

Industrial chemist: Perhaps the chloride–oxide replacement reaction could be used to advantage. If you think about it, what really happens is that a metal–chloride gas is changed to a metal–oxide solid at the same time a second metal–oxide solid is changed to a metal–chloride gas. The reaction seems to be a mechanism for reversing the volatility of the metals, changing gases to solids and solids to gases. Let's take 100 tons of the red-mud oxides and covert to ton-moles (see Table 10.7) to see just how these exchanges might work.

Engineer: The iron and aluminum dominate this problem so we should begin by seeing if we can synthesize a means of separating a mixture of Fe_2O_3 and Al_2O_3. The only oxide exchange reaction involving Fe and Al is

$$2AlCl_3(g) + Fe_2O_3(s) \rightarrow Al_2O_3(s) + 2FeCl_3(g)$$

The Fe_2O_3 and Al_2O_3 exist in a mole ratio of 0.31/0.26 = 6/5 in the red mud. If we processed a red-mud mixture of 6 moles Fe_2O_3 and 5 moles Al_2O_3 with 12 moles $AlCl_3$ we would form 12 moles of $FeCl_3$ and 11 moles of Al_2O_3. The separation is easy! If only the $AlCl_3$ can be generated. The flow sheet would look like Figure 10.8.

Industrial chemist: The $AlCl_3$ can be formed directly from the Al_2O_3 by chlorination.

$$Al_2O_3 + 3CO(g) + Cl_2(g) \rightarrow 2AlCl_3(g) + 3CO_2(g)$$

TABLE 10.7
Composition of 100 Tons of Red Mud

Oxide Species	Molar Mass	Mass %	Tons	Ton-Moles
Al_2O_3 (s)	102	26	26	0.26
Fe_2O_3 (s)	160	50	50	0.31
SiO_2 (s)	60	9	9	0.15
TiO_2 (s)	80	6	6	0.08
NaO_2 (s)	62	7	7	0.11
CaO (s)	56	2	2	0.04

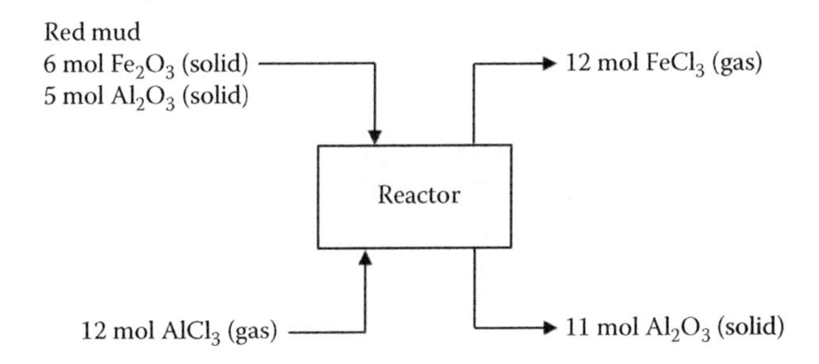

FIGURE 10.8 Fe_2O_3 and Al_2O_3 oxide exchange.

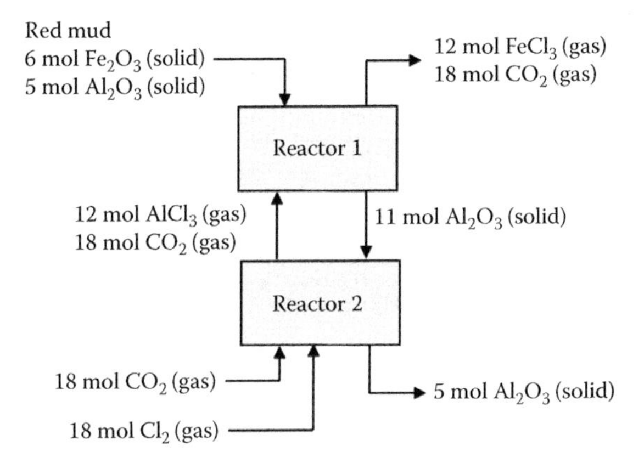

FIGURE 10.9 Formation of $AlCl_3$ gas in reactor 2 integrated with the oxide–chloride exchange reaction of reactor 1.

Engineer: This looks reasonable. Put it together and Figure 10.9 is the reaction scheme. What happens to the other materials? The sodium and calcium chlorides are not volatile and ought to appear as impurities in the solid Al_2O_3, leaving the second reactor. The titanium oxide has a greater affinity for chloride than the aluminum, so $TiCl_4$ ought to be formed in the first reactor and contaminate the $FeCl_3$ product. The silicon dioxide is the lowest on the chloride forming affinity list and ought to appear as an oxide with the alumina.

Now, the NaCl and $CaCl_2$ are water soluble and can be washed from the Al_2O_3–SiO_2 mixture. The $TiCl_4$ and $FeCl_3$ boil at 136°C and 319°C, so separation by condensation should be feasible.

Industrial chemist: We could burn the $FeCl_3$ in hydrogen to obtain pure iron. The Halomet process uses this reaction.

$$FeCl_3(g) + 1.5H_2(g) \rightarrow Fe(s) + 3HCl(g)$$

If it is economical, the HCl could be converted to the Cl_2 we need to power the chlorination reaction. The old Deacon process used this reaction

$$2HCl + 0.5O_2 \rightarrow Cl_2 + H_2O$$

The $TiCl_4$ could be burned to form TiO_2 pure enough for use as a pigment in paint manufacture.

$$TiCl_4 + O_2 \rightarrow TiO_2 + 2Cl_2$$

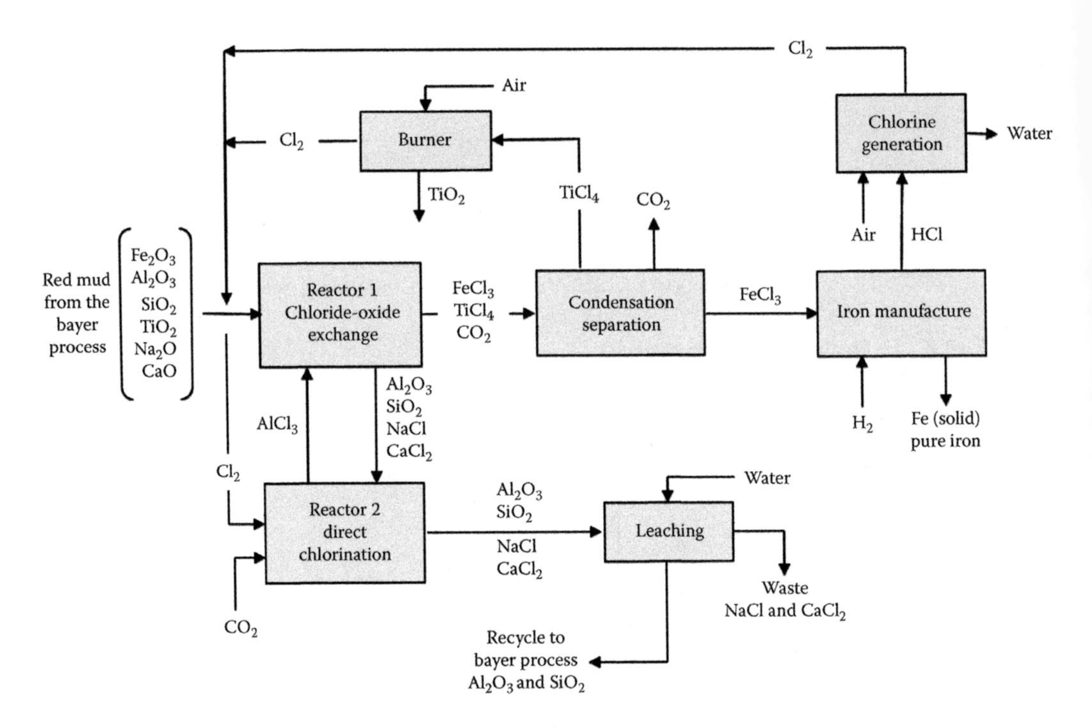

FIGURE 10.10 Generation of pure iron, titanium dioxide pigment, and Al_2O_3–SiO_2 for the Bayer process.

Engineer: Figure 10.10 is a roughed-out plan for the conversion of the red mud into iron, paint pigment, and a recycle stream of Al_2O_3–SiO_2 back to the Bayer process. Of course, we have a long way to go before we can determine the economic and technical feasibility of this plan. We have some more details to work out. How can we obtain H_2 for the iron making, and the Cl_2 for the direct chlorination? Nevertheless, the chemistry seems to be working in our favor. Let's explain our idea to the boss.

Company president: This is a good process from the viewpoint of chemistry and separations. It fits with decision to follow the ISO 14000 international standards on environmental management. We need to do this to compete internationally. And we do it to be good citizens.

It's great to get rid of the red-mud wastes, but be careful not to create some new environmental problem. Suppose the new process created an unacceptable risk for losses of chlorine or some hazardous chlorinated compound. This would stop the project. Please think about this while you work out the other remaining details.

This example has shown the kind of thinking necessary to synthesize a processing system from diverse concepts. We have changed the physical properties of the materials using chemical reactions to enable simple and direct separations. Chemical transformation, leaching, condensation, and other operations fit together to form a tailor-made process to handle red-mud waste. Environmental considerations and process design go hand in hand. Such processes are not found complete in textbooks, but rather must be synthesized by the engineer using the principles described in this text.

10.9 NYLON AND ADIPIC ACID SYNTHESIS

Adipic acid ($C_6H_{10}O_4$) is a white crystalline dicarboxylic acid that is used in the manufacture of nylon. Adipic acid is a \$6–7 billion per year business, with a production of around 2.2 million tons/year.

Nylon 66 is made from polymerization of adipic acid and hexamethylene diamine (HMD). The primary reaction is

$$\underset{\text{adipic acid}}{C_6H_{10}O_4} + \underset{\text{HMD}}{C_6H_{16}N_2} \rightarrow \underset{\text{Nylon 66}}{C_{12}H_{22}O_2N_2} + 2H_2O$$

One mole of adipic acid is needed for every mole of nylon 66 produced (the conversion of adipic acid to nylon 66 is 99.99%).

Three commercial chemical synthesis routes are shown in Figure 10.11.

Adipic acid has historically been manufactured using either benzene or phenol, but today cyclohexane is the dominant feedstock of choice. As shown in Figure 10.12, cyclohexane is oxidized to form a mixture of cyclohexanol and cyclohexanone (ketone-alcohol oil, or KA oil). The KA oil is oxidized to adipic acid in presence of nitric acid.

The process is problematic because about 0.264 kg nitrous oxide (N_2O) is produced per kg of adipic acid. N_2O is a major source of stratospheric nitric oxide and has a global warming potential many times more than carbon dioxide. N_2O abatement technology must be used with this manufacturing process and the required removal efficiency is 99%; the allowable discharge is only 0.00264 kg N_2O/kg adipic acid (Mainhardt, 2000).

Chemists have looked for synthesis routes that do not produce N_2O. One emerging alternative involves the carbonylation (addition of carbon monoxide) of butadiene. The reaction proceeds as follows:

$$\underset{\text{butadiene}}{CH_2{=}CHCH{=}CH_2} + \underset{\text{carbon monoxide}}{2CO} + \underset{\text{water}}{2H_2O} \rightarrow \underset{\text{adipic acid}}{HOOC(CH_2)_4COOH}$$

This process has the advantage of not producing nitrous oxide.

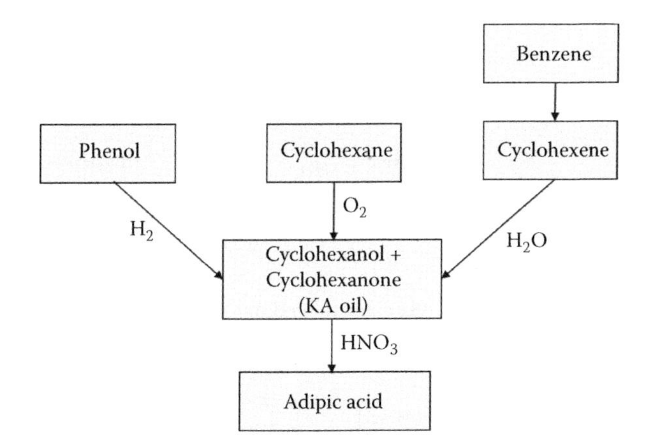

FIGURE 10.11 Chemical synthesis routes for the manufacture of adipic acid.

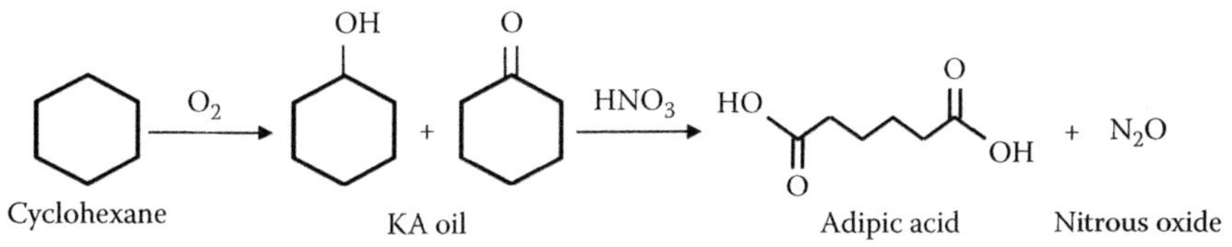

Cyclohexane $\xrightarrow{O_2}$ KA oil $\xrightarrow{HNO_3}$ Adipic acid + N_2O (Nitrous oxide)

FIGURE 10.12 A commercial method of making adipic acid from cyclohexane produces nitrous oxide.

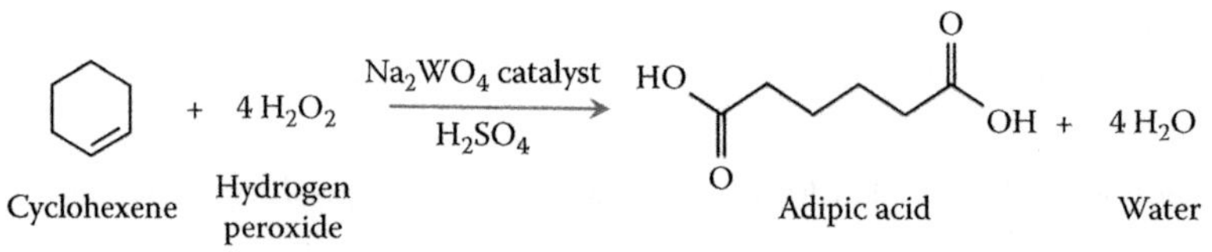

Cyclohexene + $4\,H_2O_2$ (Hydrogen peroxide) $\xrightarrow[H_2SO_4]{Na_2WO_4\ \text{catalyst}}$ Adipic acid + $4\,H_2O$ (Water)

FIGURE 10.13 Green adipic acid synthesis by cyclohexene oxidation with hydrogen peroxide in the presence of a catalyst. (From Usui, Y. and Sato, K., 2003.)

Another green route, shown in Figure 10.13, reacts cyclohexene with hydrogen peroxide in the presence of a tungsten catalyst, uses no solvents and has no undesirable by-products. The stoichiometry indicates that a reactant mass of 218 kg ($C_6H_{10} + H_2O_2$) yields a product mass of 146 kg adipic acid. The actual yield of adipic acid is 90% of the stoichiometric amount.

Adipic acid can also be manufactured from glucose using genetically modified *Escherichia coli*. Genes from other bacteria are inserted so the normal metabolic pathway of *E. coli* is altered to produce a high yield of muconic acid, which is enzyme catalyzed to produce adipic acid. An early route, via catechol, is shown in Figure 10.14. All reactions take place within the genetically modified *E. coli* cell.

Figure 10.15 shows a modification of this bio-based process with a hypothetical flowsheet and stoichiometry. The *E. coli*-produced muconic acid is harvested from the cells and then hydrogenated to adipic acid by catalysis.

Another non-petrochemical-based method of producing adipic acid is shown in Figure 10.16. First, biomass is converted to glucose. Glucose is converted to glucaric acid using selective oxidation, and glucaric acid is converted to adipic acid via selective hydrogenation.

The bulk of global production of adipic acid still uses the petrochemical-based methods shown in Figure 10.11; however, the trend is toward the bio-based processes, which carry a lighter environmental impact.

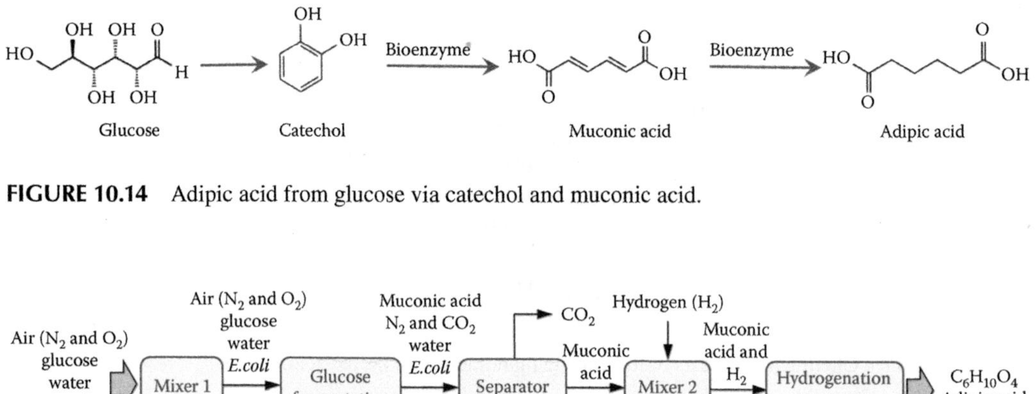

Glucose → Catechol $\xrightarrow{\text{Bioenzyme}}$ Muconic acid $\xrightarrow{\text{Bioenzyme}}$ Adipic acid

FIGURE 10.14 Adipic acid from glucose via catechol and muconic acid.

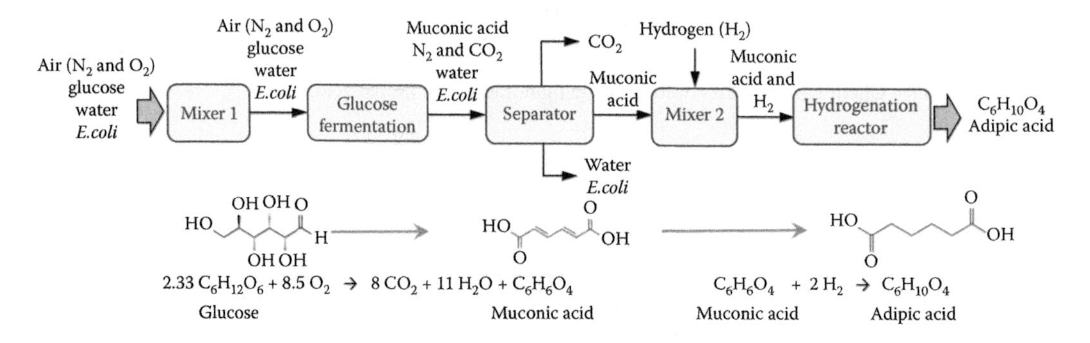

FIGURE 10.15 Bio-based synthesis of production of muconic acid from glucose using genetically modified *E. coli* with subsequent catalysis to adipic acid. (From Boussie, T. 2013.)

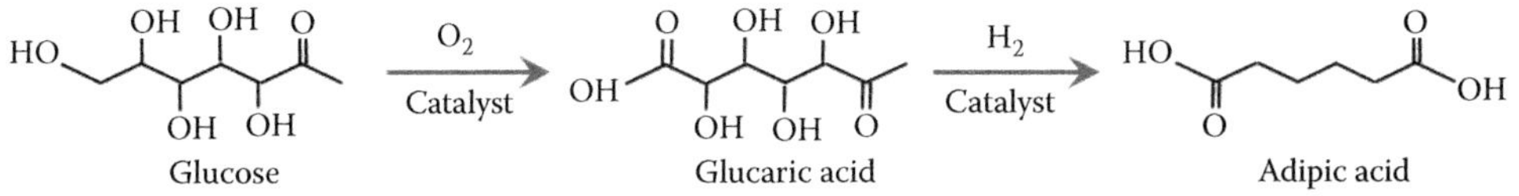

Glucose $\xrightarrow[\text{Catalyst}]{O_2}$ Glucaric acid $\xrightarrow[\text{Catalyst}]{H_2}$ Adipic acid

FIGURE 10.16 Adipic acid production from aerobic oxidation of glucose to glucaric acid.

10.10 CONCLUSION

This chapter discussed four related ideas about the chemistry of manufacturing processes.

- The redesign of process chemistry to reduce or eliminate pollution. The examples were the Solvay soda ash process, the processing of red mud from bauxite aluminum refining, and the production of adipic acid for the manufacture of nylon.
- The basic principles of green chemistry, in the form of the 12 rules promulgated by Anastas and Warner in 1998. In a few words, the rules collectively eliminate the use of and production of harmful chemicals and make the process chemistry yield more product from less raw material.
- Measures of efficiency of chemical, including atom efficiency, the E-Factor, and EMY. The E-Factor and the EMY incorporate information about the mass of waste produced by a reaction and the amount of non-benign chemicals that are used in the process.
- How catalysis and biochemistry are being used to promote greener chemistry.

This kind of chemistry is vastly different from chemistry in the laboratory, where reactions are carried out in small batches, solvent use can be extravagant without causing any major problems, where products and wastes can be easily separated, and waste disposal is easily managed. Usually the chemistry during product or process development is not highly efficient, but it must be made more so as production moves toward full scale. Doing this is aided by understanding that process chemistry can, at times, be modified to make the process more productive and less polluting.

Appendix A: Atomic Numbers and Atomic Masses

Actinium	Ac	89	227.0278
Aluminum	Al	13	26.98154
Americium	Am	95	(243)
Antimony	Sb	51	121.75
Argon	Ar	18	39.948
Arsenic	As	33	74.9216
Astatine	At	85	(210)
Barium	Ba	56	137.33
Berkelium	Bk	97	(247)
Beryllium	Be	4	9.01218
Bismuth	Bi	83	208.9804
Boron	B	5	10.81
Bromine	Br	35	79.904
Cadmium	Cd	48	112.41
Calcium	Ca	20	40.08
Californium	Cf	98	(251)
Carbon	C	6	12.011
Cerium	Ce	58	140.12
Cesium	Cs	55	132.9054
Chlorine	Cl	17	35.453
Chromium	Cr	24	51.966
Cobalt	Co	27	58.9332
Copper	Cu	29	63.546
Curium	Cm	96	(247)
Dysprosium	Dy	66	162.50
Einsteinium	Es	99	(254)
Erbium	Er	68	167.26
Europium	Eu	63	151.96
Fermium	Fm	100	(257)
Fluorine	F	9	18.99840
Francium	Fr	87	(223)
Gadolinium	Gd	64	157.25
Gallium	Ga	31	69.72
Germanium	Ge	32	72.59
Gold	Au	79	196.9665
Hafnium	Hf	72	178.49
Helium	He	2	4.0026
Holmium	Ho	67	164.9304
Hydrogen	H	1	1.0079
Indium	In	49	114.82
Iodine	I	53	126.9045
Iridium	Ir	77	192.22
Iron	Fe	26	55.847
Krypton	Kr	36	83.80
Lanthanum	La	57	138.9055
Lawrencium	Lr	103	(260)
Lead	Pb	82	207.2
Lithium	Li	3	6.941
Lutetium	Lu	71	174.967
Magnesium	Mg	12	24.305
Manganese	Mn	25	54.9380
Mendelevium	Md	101	(258)
Mercury	Hg	80	200.59
Molybdenum	Mo	42	95.94
Neodymium	Nd	60	144.24
Neon	Ne	10	20.179
Neptunium	Np	93	237.0482
Nickel	Ni	28	58.70
Niobium	Nb	41	92.9064
Nitrogen	N	7	14.0067
Nobelium	No	102	(259)
Osmium	Os	76	190.2
Oxygen	O	8	15.9994
Palladium	Pd	46	106.4
Phosphorus	P	15	30.97376
Platinum	Pt	78	195.09
Plutonium	Pu	94	(244)
Polonium	Po	84	(209)
Potassium	K	19	39.0983
Praseodymium	Pr	59	140.9077
Promethium	Pm	61	(145)
Protactinium	Pa	91	231.0389
Radium	Ra	88	226.0254
Radon	Rn	86	(222)
Rhenium	Re	75	186.207
Rhodium	Rh	45	102.9055
Rubidium	Rb	37	85.4678
Ruthenium	Ru	44	101.07
Samarium	Sm	62	150.4
Scandium	Sc	21	44.9559
Selenium	Se	34	78.96
Silicon	Si	14	28.0855
Silver	Ag	47	107.868
Sodium	Na	11	22.98977
Strontium	Sr	38	87.62
Sulfur	S	16	32.06
Tantalum	Ta	73	180.9479
Technetium	Tc	43	(97)
Tellurium	Te	52	127.60
Terbium	Tb	65	158.9254

(*Continued*)

(Continued)

Thallium	Tl	81	204.37
Thorium	Th	90	232.0381
Thulium	Tm	69	168.9342
Tin	Sn	50	118.69
Titanium	Ti	22	47.90
Tungsten	W	74	183.85
Uranium	U	92	238.029
Vanadium	V	23	50.9414
Xenon	Xo	54	131.30
Ytterbium	Yb	70	173.04
Yttrium	Y	39	88.9059
Zinc	Zn	30	65.38
Zirconium	Zr	40	91.22

Appendix B: Equivalent Weights

Table B1.1 gives the equivalent weights for cations, anions, and compounds commonly used in pollution prevention and control.

TABLE B1.1
Table of Equivalent Weights for Ions and Compounds Commonly Used in Pollution Prevention and Control

Ion or Compound	Formula	Number of ± Charges	Molar Mass (g/mol)	Equivalent Weight (g/eq)
		Cations		
Aluminum	Al^{3+}	3	27.0	9.0
Ammonium	NH_4^+	1	18.0	18.0
Calcium	Ca^{2+}	2	40.0	20.0
Copper	Cu^{2+}	2	63.6	31.8
Hydrogen	H^+	1	1.0	1.0
Ferrous iron	Fe^{2+}	2	55.8	27.9
Ferric iron	Fe^{3+}	3	55.8	18.6
Magnesium	Mg^{2+}	2	24.4	12.2
Manganese	Mn^{2+}	2	55.0	27.5
Potassium	K^+	1	39.1	39.1
Sodium	Na^+	1	23.0	23.0
		Anions		
Bicarbonate	HCO_3^-	1	61.0	61.0
Carbonate	CO_3^{2-}	2	60.0	30.0
Chloride	Cl^-	1	35.5	35.5
Fluoride	F^-	1	19.0	19.0
Iodide	I^-	1	126.9	126.9
Hydroxide	OH^-	1	17.0	17.0
Nitrate	NO_3^-	1	62.0	62.0
Phosphate (tribasic)	PO_4^{3-}	3	95.1	31.7
Phosphate (dibasic)	HPO_4^{2-}	2	96.0	48.0
Phosphate (monobasic)	$H_2PO_4^-$	1	97.0	97.0
Sulfate	SO_4^{2-}	2	96.0	48.0
Bisulfate	HSO_4^-	1	97.1	97.1
Sulfite	SO_3^{2-}	2	80.0	40.0
Bisulfite	HSO_3^-	1	81.1	81.1
Sulfide	S^{2-}	2	32.0	16.0
		Compounds		
Alum	$Al_2(SO_4)_3 \cdot 18H_2O$	6	666.0	111.0
Aluminum sulfate (anhydrous)	$Al_2(SO_4)_3$	6	342.0	57.0
Aluminum hydroxide	$Al(OH)_3$	3	78.0	26.0
Aluminum oxide	Al_2O_3	6	102.0	17.0

(Continued)

TABLE B1.1 ***(Continued)***

Table of Equivalent Weights for Ions and Compounds Commonly Used in Pollution Prevention and Control

Ion or Compound	Formula	Number of ± Charges	Molar Mass (g/mol)	Equivalent Weight (g/eq)
Ammonia	NH_3	1	17.0	17.0
Sodium aluminate	$Na_2Al_2O_4$	6	163.8	27.3
Calcium bicarbonate	$Ca(HCO_3)_2$	2	162.2	81.1
Calcium carbonate	$CaCO_3$	2	100.0	50.0
Calcium chloride	$CaCl_2$	2	111.0	55.5
Calcium hydroxide	$Ca(OH)_2$	2	74.0	37.0
Calcium oxide	CaO	2	56.0	28.0
Calcium sulfate (anhydrous)	$CaSO_4$	2	136.2	68.1
Calcium sulfate (gypsum)	$CaSO_4 \cdot 2H_2O$	2	172.2	86.1
Calcium phosphate	$Ca_3(PO_4)_2$	6	310.2	51.7
Carbon dioxide	CO_2	2	44.0	22.0
Chlorine	Cl_2	2	71.0	35.5
Ferrous sulfate (anhydrous)	$FeSO_4$	2	152.0	76.0
Ferric sulfate	$Fe_2(SO_4)_3$	6	399.6	66.6
Magnesium oxide	MgO	2	40.4	20.2
Magnesium bicarbonate	$Mg(HCO_3)_2$	2	146.4	73.2
Magnesium carbonate	$MgCO_3$	2	84.4	42.2
Magnesium chloride	$MgCl_2$	2	95.2	47.6
Magnesium hydroxide	$Mg(OH)_2$	2	58.4	29.2
Magnesium phosphate	$Mg_3(PO_4)_2$	6	262.8	43.8
Magnesium sulfate (anhydrous)	$MgSO_4$	2	120.4	60.2
Manganese hydroxide	$Mn(OH)_2$	2	89.0	44.5
Silica	SiO_2	2	60.0	30.0
Sodium bicarbonate	$NaHCO_3$	1	84.0	84.0
Sodium carbonate	Na_2CO_3	2	106.0	53.0
Sodium chloride	$NaCl$	1	58.4	58.4
Sodium hydroxide	$NaOH$	1	40.0	40.0
Sodium nitrate	$NaNO_3$	1	85.0	85.0
Trisodium phosphate	Na_3PO_4	3	164.1	54.7
Disodium phosphate	Na_2HPO_4	2	142.0	71.0
Monosodium phosphate	NaH_2PO_4	1	120.0	120.0
Sodium silicate	Na_2SiO_3	2	122.0	61.0
Sodium metaphosphate	$NaPO_3$	1	102.0	102.0
Sodium sulfate	Na_2SO_4	2	142.0	71.0
Sodium sulfite	Na_2SO_3	2	126.0	63.0
Sulfuric acid	H_2SO_4	2	98.0	49.0

Appendix C: Computer Programs for Chemical Equilibrium

A chemical equilibrium problem consists of a set of n mass-balance equations in n unknown free concentrations. These nonlinear equations must be solved by a method of successive approximations. The Newton–Raphson method is a popular choice.

The following software applications are useful for equilibrium calculations.

- Chemical Equilibrium Calculator (http://www.changbioscience.com/biochem/keq.html)
- Geochem EZ (freeware). A multipurpose chemical speciation program, used in plant nutrition and in soil and environmental chemistry research to perform equilibrium speciation computations, allowing the user to estimate solution ion activities and to consider simple complexes and solid phases.
- HySS. Titration simulation and speciation calculations.
- EQS4WIN. A powerful computer program originally developed for gas-phase equilibria but subsequently extended to general applications. Uses the Gibbs energy minimization approach.
- CHEMEQL. A comprehensive computer program for the calculation of thermodynamic equilibrium concentrations of species in homogeneous and heterogeneous systems. Many geochemical applications.
- WinSGW. A Windows version of the SOLGASWATER computer program, which is an algorithm for the computation of aqueous multicomponent, multiphase equilibria.
- Visual MINTEQ. A Windows version of MINTEQA2 (ver 4.0). MINTEQA2 is a chemical equilibrium model for the calculation of metal speciation, solubility equilibria, etc., for natural waters.
- MINEQL. A chemical equilibrium modeling system for aqueous systems. Handles a wide range of pH, redox, solubility, and sorption scenarios.

References

Ahlborg, U.G. et al. 1994. Toxic equivalency factors for dioxin-like PCBs, *Chemosphere*, 28, 1049–1067.
Ahluwalia, V.K. 2009. *Green Chemistry: Environmentally Benign Reaction*, CRC Press, Boca Raton, FL.
Akesrisakul, K. and Jiraprayuklert, A. 2007. The application of statistical techniques to reduce ethylene glycol in waste water produced by esterification reaction in polyester manufacturing process, *J. Sci. Tech.*, 29, 181–190.
Allen, D.T. and Rosselot, K.S. 1996. *Pollution Prevention for Chemical Processes*, Wiley-Interscience, New York.
Amato, I. 1993. The slow birth of green chemistry, *Science*, 259, 1538–1541.
Amirtharajah, A. and O'Melia, C.R. 1990. Coagulation process: destabilization, mixing, and flocculation. In *Water Quality and Treatment*, Pontius, F. W., ed. Denver, CO: McGraw-Hill, Inc., 269–367.
Anastas, P.T. and Farris, C.A. eds. 1994. *Benign By Design: Alternative Synthetic Design for Pollution Prevention*, American Chemical Society Press, Washington, DC.
Anastas, P.T. and Warner, J.C. 1998. *Green Chemistry: Theory and Practice*, Oxford University Press, Oxford, UK.
Anastas, P.T. and Zimmerman, J.B. 2003. Design through the 12 principles of green engineering, *Environ. Sci. Technol.*, 37(5), 94A–101A.
Ander, N.G. 2000. *Practical Process Research & Development*, Academic Press, New York.
Aqion. 2014. *Hydrochemistry and Water Analysis*, Free Hydrochemistry Software, www.aqion.de.
Aspen Technology. 2013. *Aspen Plus Technology Brochure*, Burlington, MA.
Aube, B. 2004. *The Science of Treating Acid Mine Drainage and Smelter Effluents*, EnvirAube, Quebec, Canada.
AWWA. 2010. *Water Quality and Treatment: A Handbook on Drinking Water 6/E*, McGraw-Hill, New York.
AWWA and ASCE. 2012. *Water Treatment Plant Design 5/E*, McGraw-Hill, New York.
Balis, J.F., et al. 2008. Metals precipitation from effluents: A review, *Pract. Period. Hazard., Toxic, Radioact. Waste Manage., ASCE*, 12, 135–149.
Benjamin, M.M. 2014. *Water Chemistry*, 2nd ed. Waveland Press, Long Grove, IL.
Berthouex, P.M. and Brown, L.C. 2002. *Statistics for Environmental Engineering*, CRC Press, Boca Raton, FL.
Berthouex, P.M. and Brown, L.C. 2013. *Pollution Prevention and Control: Human Health and Environmental Quality*, Bookboon.com, London.
Berthouex, P.M. and Brown, L.C. 2014. *Pollution Prevention and Control: Material and Energy Balances*, Bookboon.com, London.
Berthouex, P.M. and Rudd, D.F. 1977. *The Strategy of Pollution Control*, John Wiley & Sons, New York.
Betz. 1991. *Betz Handbook of Industrial Water Conditioning*, 9th ed., Betz Laboratories.
Bhattacharyya, D. et al. 1986. *Sulfide Precipitation of Nickel and Other Heavy Metals from Single and Multi-Metal Systems*, U.S. EPA, Washington, DC.
Bigda, R.J. 1995. Consider Fenton's chemistry for wastewater treatment, *Chem. Eng. Prog.*, 91, 62–66.
Birnbaum, L.S. 1994. The mechanism of dioxin toxicity: Relationship to risk assessment, *Environ. Health Perspect.*, 102, 157–167.
Birnbaum, L.S. and DeVito, M.J. 1995. TEF's: A practical approach to a real-world problem, *Toxicology*, 105, 391–401.
Bishop, P.L. 2004. *Pollution Prevention: Fundamentals and Practice*, Waveland Press, Long Grove, IL
Boussie, T. 2013. Scale-up and commercialization of bio-based adipic acid and HMD, *4th Biobased Chemicals Commercialization and Partnering Conference*, Sept. 16–17, San Francisco.
Box, G.E.P., Hunter, J.S., and Hunter, W.G. 2005. *Statistics for Experimenters: Design, Innovation, and Discovery*, 2nd ed., John Wiley & Sons, New York.
Breen, J.J. and Dellarco, M.J. 1992. *Pollution Prevention in Industrial Processes: The Role of Process Analytical Chemistry*, American Chemical Society Symposium Series 508, Washington, DC.
Brezonik, P. 1993. *Chemical Kinetics and Process Dynamics in Aquatic Systems*, CRC Press, Boca Raton, FL.
Brezonik, P. and Arnold, W. 2011. *Water Chemistry: An Introduction to the Chemistry of Natural and Engineered Aquatic Systems*, Oxford University Press, London.
Brown, T.E., LeMay, H.E., Jr. and Bursten, B.E. 2006. *Chemistry: The Central Science*, 10th ed., Pearson – Prentice Hall, Upper Saddle River, NJ.
Butler, J.N. and Cogley, D.R. 1998. *Ionic Equilibrium: Solubility and pH Calculations*, Wiley-Interscience, New York.

Cameron, D.C. et al. 1998. Metabolic engineering of propanediol pathways, *Biotechnol. Prog.*, 14, 116–125.

Constable, C.J. et al. 2002. Metrics to 'green' chemistry; which are the best?, *Green Chemistry*, 4, 521–527.

Cortright, R. and Crittenden, J.C. 1998. *Research in Clean Reaction Technologies*, Center for Clean Treatment and Technology, Michigan Technological University, Houghton, MI.

Crittenden, J.C. et al. 2012. *MWH's Water Treatment: Principles and Design,* 3rd ed., John Wiley & Sons, New York.

Cushnie, G.C., Jr. 1985. *Electroplating Wastewater Pollution Control Technology*, Noyes Publications, Park Ridge, NJ.

Davis, M.L. 2010. *Water and Wastewater Engineering: Design Principles and Practice*, McGraw Hill, Inc., New York.

Degremont. 1991. *Water Treatment Handbook*, Vols. 1 and 2, 7th ed., Lavoisier. Cachon Cedex, France.

Devi, A. et al. 2013. A review on spent pickling liquor, *Int. J. Environ. Sci.*, 4, 284–295.

Douglas, J.M. 1992. Process synthesis for waste minimization, *Ind. Eng. Chem. Res.*, 31, 238–243.

Drabkin, M. 1988. The waste minimization assessment: A useful tool for the reduction of industrial hazardous wastes, *J. Air Pollut. Control Assoc.*, 38, 1530–1541.

Dunn, P.J. et al. 2004. The development of an environmentally benign synthesis of sildenafil citrate (Viagra) and its assessment by green chemistry metrics, *Green Chem.*, 6, 43–48.

Eckenfelder, W.W. 2000. *Industrial Water Pollution Control*, McGraw-Hill, New York.

Emerson, S. and Hedges, J. 2008. Carbonate chemistry (Chapter IV), *Chemical Oceanography and the Marine Carbon Cycle*, Cambridge University Press, Cambridge, UK.

Faust, S.D. and Aly, O.M. 1998. *Chemistry of Water Treatment,* CRC Press, Boca Raton, FL.

Fendorf, S.E. and Li, G. 1996. Kinetics of chromate reduction by ferrous iron, *Environ. Sci. Technol.,* 30, 1614–1617.

Ferguson, J.F., Eastman, J., and Jenkins, D. 1973. Calcium phosphate precipitation at slightly alkaline pH valules, *J. Water Pollut. Control Fed.*, 45, 620.

Fiedler, H. 2009, *Polychlorinated Biphenyls (PCBs): Uses and Environmental Releases*, Bavarian Institute for Water Research, Augsburg, Germany.

Fiedler, H. et al. 1994. *Environmental Fate of Organochlorines in the Aquatic Environment,* 199. ECO-INFORMA Press, Bayreuth.

Florida Lakewatch. 2004. *A Beginner's Guide to Water Management—Color,* Info. Circular 108, Gainesville, FL.

Freeman, H. 1994. *Industrial Pollution Prevention Handbook*, McGraw-Hill, New York.

Fusi, L. et al. 2012. Determining calcium carbonate neutralization kinetics from experimental laboratory data, *J. Math. Chem.,* 50, 2492–2511.

Ghosh, P. et al. 2012. Determination of reaction rate constant for p-chlorophenol and nitrobenzene reacting with •OH during oxidation by $Fe(II)/H_2O_2$ system, *Int. J. Chem. Tech. Res.*, 4(1), 116–123.

Haas, C.N. and Vamos, R.J. 1995. *Hazardous and Industrial Waste Treatment*, Prentice Hall, Englewood Cliffs, NJ.

Haynes, W.M. 2014. *CRC Handbook of Chemistry and Physics*, 95th ed., CRC Press, Boca Raton, FL.

He, F. and Lei, L.C. 2004. Degradation kinetics and mechanisms of phenol in photo-Fenton process, *J. Zhejiang Univ. Sci.*, 5, 198–205.

Hendricks, D.W. 2006. *Water Treatment Processes: Physical and Chemical*, CRC Press, Boca Raton, FL.

Henze, M. et al. 2008. *Biological Wastewater Treatment: Principles, Modelling and Design*, IWA Publishing, London, UK.

Hermanowicz, S.W. 2006. Chemical fundamentals of phosphorus precipitation, *WERF Boundary Condition Workshop*, Washington, DC.

Hill, J.H. and Petrucci, R.H. 1999. *General Chemistry: An Integrated Approach*, 2nd ed., Prentice Hall, Upper Saddle River, NJ.

Himmelblau, D.M. and Riggs, J.B. 2004. *Basic Principles and Calculations in Chemical Engineering*, 7th ed., Prentice Hall, Englewood Cliffs, NJ.

Hollod, G.J. and McCartney, R.F. 1988. Waste reduction in the chemical industry, *J. Air Pollut. Control Assoc.*, 38, 174–179.

Hribernik, A. et al. 2009. Application of 2^k Factorial Design in Wastewater Decolorization Research, *XIX IMEKO World Congress Fundamental and Applied Metrology*, Sept. 6–11, 2009, Lisbon, Portugal.

Hydromantis. 2014. *Hydromantis Environmental Systems Software*, www.hydromantis.com, Hamilton, Ontario, Canada.

Jenkins, D. and Hermanowicz, S.W. 1991. Principles of chemical phosphate removal. In *Phosphorus and Nitrogen Removal from Municipal Wastewate*r, 2nd ed., Richard Sedlak, ed., Lewis Publisher, 91–110, Boca Raton, FL.

Jenkins, D. et al. 1971. Chemical processes for phosphate removal, *Water Res.*, 5, 369–389.
Jimenez-Gonzolez, C. and Constable, D.J.C. 2011. *Green Chemistry and Engineering: A Practical Design Approach*, John Wiley & Sons, New York.
Kang, L.S. and Cleasby, J.L. 1995. Temperature effects on flocculation kinetics using Fe(III) coagulant, *J. Environ. Eng. Div., ASCE*, 121, 893–901.
Kosusko, M. and Nunex, C.M. 1990. Destruction of volatile organic compounds using catalytic oxidation, *J. Air Waste Manage. Assoc.*, 40(2), 254–259.
Kvech, S. and Edwards, M. 2002. Solubility controls on aluminum in drinking water at relatively low and high pH, *Water Res.*, 36, 4356–4368.
LaGrega, M.D. 2010. *Hazardous Waste Management*, Waveland Press, Lopg Grove, IL.
Lancaster, M. 2002. *Green Chemistry; An Introductory Text*, Royal Society of Chemistry, Cambridge.
Leudecke, C., Hermanowicz, S.W. and Jenkins, D. 1988. Precipitation of ferric phosphate in activated sludge: A chemical model and its verification, *Water Sci. Technol.*, 21, 352–337.
Lewis, A.E. 2010. Review of metal sulphide precipitation, *Hydrometallurgy*, 104, 222–234.
Lide, D.R. ed. 2009. *CRC Handbook of Chemistry and Physics, 2009–2010*, 90th ed., CRC Press, Boca Raton, FL.
Mackay, D. et al. 2006. *Physical-Chemical Properties and Environmental Fate for Organic Chemicals*, 2nd ed., vol. 1–4, CRC Press, Boca Raton.
Manahan, S.E. 2004. *Environmental Chemistry*, 8th ed., CRC Press.
Manahan, S.E. 2010. *Green Chemistry and the Ten Commandments of Sustainability*, 3rd ed., ChemChar Research.
Marteel-Parrish, A.E. and Abraham, M.A. 2014. *Green Chemistry and Engineering: A Pathway to Sustainability*, John Wiley & Sons, Hoboken, NJ.
Masters, G.M. and Ela, W.P. 2007. *Introduction to Environmental Engineering and Science*, 3rd ed., Prentice Hall, Englewood Cliffs, NJ.
Metcalf and Eddy. et al. 2007. *Water Reuse: Issues, Technology, and Applications*, McGraw-Hill, New York.
Mohan, G.R. et al. 2011. Development of a process model for recovery of nutrients from wastewater by precipitation as struvite, *Florida Water Resour. J.*, 17–22.
Morel, F.M.M. and Hering, J.G. 1993. *Principles and Applications of Aquatic Chemistry*, Wiley-Interscience, New York.
Murphy, R.M. 2007. *Introduction to Chemical Processes: Principles, Analysis and Synthesis*, McGraw-Hill, Boston.
Myers, R.J. 1986. The new low value for the second dissociation constant for H_2S: Its history, its best value, and its impact on the teaching of sulfide equilibria, *J. Chem. Educ.*, 64, 687–690.
Nalco. 1992. *The Nalco Guide to Cooling-Water Systems Failure Analysis*, Flynn, D., ed. McGraw-Hill, New York.
Nalco. 2009. *The Nalco Water Handbook*, 3rd ed., Flynn, D., ed. McGraw-Hill, New York.
Neethling, J.B. 2013. Optimizing chemical phosphorus removal, *2013 Technical Conference and Exposition*, Ohio Water Environmental Association, Mason, OH.
Ohlinger, K.N. et al. 1998. Predicting struvite formation in digestion, *Water Research*, 32, 3607–3614.
Ohlinger, K.N. et al. 1999. Kinetics effects on preferential struvite accumulation in wastewater, *ASCE Jour. Enviro Engr.*, 125, 730–737.
Pernitsky, D.J. 2003. Coagulation 101, *Tech Transfer Conference*. Calgary, Alberta, Canada
Pernitsky, D.J. and Edzwald, J.K. 2003. Solubility of polyaluminum coagulants, *J. Water Supply Research: Res. Technol.,-AQUA*, 52(6), 395–406.
Pushkarev, V.V. et al. 1983. *Treatment of Oil-Containing Wastewater*, Allerton Press, New York.
Rich, L.G. 1963. *Unit Processes in Sanitary Engineering*, John Wiley & Sons, Inc., New York, NY.
Rittman, B.E. and McCarty, P.L. 2001. *Environmental Biotechnology, Principles and Applications*, McGraw-Hill, New York.
Safe, S. 1990. Polychlorinated biphenyls (PCBs), Dibenzo-p-dioxins (PCDDs), Dibenzofurans (PCDFs), and related compounds: Environmental and mechanistic considerations which support the development of toxic equivalency factors (TEFs), *CRC Crit. Rev. Toxicol.*, 21, 51–88.
Safe, S. 1994. Polychlorinated biphenyls (PCBs): Environmental impact, biochemical and toxic responses and implications for risk assessment, *CRC Crit. Rev. Toxicol.*, 24, 87149.
Sarai, D.S. 2005. *Basic Chemistry for Water and Wastewater Operators*, American Water Works Association, Denver, CO.
Sato, K., Aoki, M., and Noyori, R.A. 1998. Green route to adipic acid: Direct oxidation of cyclohexenes with 30 percent hydrogen peroxide. *Science*, 28, 1646–1647.

Sawyer, C.N., McCarty, P.L., and Parkin, G.F. 2003. *Chemistry for Environmental Engineering and Science*, 5th ed., McGraw-Hill, New York.

Schnoor, J.L. 1996. *Environmental Modeling: Fate and Transport of Pollutants in Water, Air, and Soil*, Wiley-Interscience, New York.

Schwarzenbach, R., Gschwend, P.M., and Imboden, D.M. 2002. *Environmental Organic Chemistry*, 2nd ed., Wiley-Interscience, New York.

Sedlack, R. 1991. Principles of chemical phosphate removal. In *Phosphorus and Nitrogen Removal from Municipal Wastewate*r, 2nd ed., Lewis Publisher, Boca Raton, FL.

Shah, A.R. and Ploeser, J.H. 1999. Reusing rinse wastewater at a semiconductor plant, *JAWWA*, 91, 58–65.

Shammas, N.K. and Wang, L.K. 2010. *Fair, Geyer, and Okun's. Water and Wastewater Engineering; Water Supply and Wastewater Removal*, John Wiley & Sons, New York.

Shei, A. and McDuff, A. 2014. *ICE Tables*. CHEMWiki, UC Davis, Davis, CA.

Smith, R. 1995. *Chemical Process Design*, McGraw-Hill, New York.

Snoeyink, V.L. 2006. *Water Chemistry*, John Wiley & Sons, New York.

Snoeyink, V.L. and Jenkins, D. 1980. *Water Chemistry*, John Wiley & Sons, New York.

Snurer, H. 2008. *Sludge Production from Chemical Precipitation*, Lund Tekniska Hogskola, Lund, Sweden.

Speight, J. 2017. *Lange's Handbook of Chemistry*, 17th ed., McGraw-Hill, New York.

Stix, G. 1993. Turning green: Can industrial chemistry trade benzene for sugar, *Sci. Am.*, 269, 104–106.

Stumm, W. and Morgan, J.J. 1996. *Aquatic Chemistry: Chemical Equilibria and Rates in Natural Waters*, 3rd ed., Wiley-Interscience, New York.

Takács, I. et al. 2004. pH in wastewater treatment plant modelling. *Proceedings 4th IWA World Water Conference*, Marrakech, Morocco, Sept. 19–24, 2004.

Takács, I. et al. 2006. Chemical phosphorus removal to extremely low levels: Experience of two plants in the Washington, DC area, *Water Sci. Technol.*, 53, 21–28.

Tchobanoglous, G.F. et al. 2003. *Wastewater Engineering Treatment, Disposal and Reuse*, McGraw-Hill, Boston, MA.

Temkar, P.M. 1990. Calcium carbonate scale dissolution in water stabilized by carbon dioxide, USACERL Technical Report N-90/01, U.S. Army Corps of Engineers.

Tunay, O. et al. 2010. *Chemical Oxidation Applications for Industrial Wastewaters*, IWA Publishing, London, UK.

U.S. EPA. 1993. Guides to pollution prevention: Municipal pretreatment programs, EPA/625/R-93/006, Washington, DC.

U.S. EPA. 1994. The product side of pollution prevention: Evaluating the potential for safe substitutes, EPA/600-R-94/178, Washington, DC.

U.S. EPA. 2008. Framework for application of the toxicity equivalence methodology for polychlorinated dioxins, furans, and biphenyls in ecological risk assessment, EPA/100/R-08/004, Risk Assessment Forum, Washington, DC.

U.S. EPA. 2010. Recommended toxicity equivalence factors (TEFs) for human health risk assessments of 2, 3, 7, 8-tetrachlorodibenzo-p-dioxin and dioxin-like compounds, EPA 100/R 10/005, Dec. 2010, Washington, DC.

Usui, Y. and Sato, K. 2003. A green method of adipic acid synthesis: Organic solvent- and halide-free oxidation of cycloalkanones with 30% hydrogen peroxide, *Green Chem.*, 5, 373–375.

Van Benschoten, J.E. and Edzwald, J.K. 1990. Chemical aspects of coagulation using aluminum salts–1. Hydrolytic reactions of alum and polyaluminum chloride, *Water Res.*, 24, 1519–1526.

Weber, W.J. 2000. *Environmental Systems and Processes: Principles, Modeling, and Design*, Wiley-Interscience, New York.

Whang, J.S. 1982. Soluble-sulfide precipitation for heavy metals removal from wastewaters, *Environ. Prog.*, 1, 110–113.

WHO. 1997. WHO toxic equivalency factors (TEFs) for dioxin-like compounds for humans and wildlife. Presented at *DIOXIN'97, 17th International Symposium on Chlorinated Dioxins and Related Compounds*, Indianapolis, IN.

Yonge, D. et al. 2011. Enhancing contaminant removal in storm water detention basins by coagulation, Research Report T9234–11, Washington State Transportation Commission.

Zhang, T. et al. 2010. Releasing phosphorus from calcium for struvite fertilizer production from anaerobically digested dairy effluent, *Water Environ. Res.*, 82, 34–43.

Index

Note: Page numbers followed by "*f*" indicate figures; those followed by "*t*" indicate tables.

D

M

N

O

R

S

X

Y

Z

9781138106321